Exploring Medicinal Bamboos

Therapeutic properties of bamboo have been known since ancient times as a part of several traditional systems of medicine and folk remedies in different parts of the world. Bamboo contains several active phytochemicals and minerals known for antioxidative and cardioprotective properties. This book is a comprehensive guide on botanical, ethnomedicinal, and pharmacological aspects of several medicinally active bamboo species.

Exploring Medicinal Bamboos focuses on phytochemically active ingredients of bamboos, imparting therapeutic potentials to each species. There are dedicated chapters on species, such as *Bambusa* spp., *Lophatherum gracile*, *Sasa senanensis*, *Melocanna baccifera*, and *Phyllostachys* spp., providing a deep understanding of the medicinal potential of these plants. The book also expands knowledge on bamboos as medicinally active plants, including their distribution, propagation, and utilization.

A volume in the *Exploring Medicinal Plants* series, this book is useful to academicians, researchers, and students involved with non-timber forest products, bamboo, and medicinal plants. Additionally, botanists, ecologists, phytochemists, and professionals in herbal drug and nutraceutical formulations will find the topics valuable.

Exploring Medicinal Plants

Series Editor: Azamal Husen, Wolaita Sodo University, Ethiopia

Medicinal plants render a rich source of bioactive compounds used in drug formulation and development; they play a key role in traditional or indigenous health systems. As the demand for herbal medicines increases worldwide, supply is declining as most of the harvest is derived from naturally growing vegetation. Considering global interests and covering several important aspects associated with medicinal plants, the Exploring Medicinal Plants series comprises volumes valuable to academia, practitioners, and researchers interested in medicinal plants. Topics provide information on a range of subjects including diversity, conservation, propagation, cultivation, physiology, molecular biology, growth response under extreme environment, handling, storage, bioactive compounds, secondary metabolites, extraction, therapeutics, mode of action, and healthcare practices.

Led by Azamal Husen, PhD, this series is directed to a broad range of researchers and professionals consisting of topical books exploring information related to medicinal plants. It includes edited volumes, references, and textbooks available for individual print and electronic purchases.

Medicinal Plants for the Management of Neurodegenerative Diseases
Edited by Jamal Akhtar, Maryam Sarwat and Fouzia Bashir

Nutraceuticals Inspiring the Contemporary Therapy for Lifestyle Diseases
Edited by Mala Trivedi, Sachidanand Singh, Parul Johri, and Pedro López-Sánchez

Plants as Medicine and Aromatics: Uses of Botanicals
Edited by Mohd Kafeel Ahmad Ansari, Mushtaq Ahmad, and Gary Owens

Exploring Traditional Wild Edible Plants
Edited by Vibhor Agarwal, Sachidanand Singh, and Rhaul Datta

Propagation to Pharmacopeia: Modern Approaches in Medicinal Plants
Edited by Abha Manohar K, Gopal Shukla, Saikat Gantait, A P Das, and Sumit Chakravarty

Exploring Medicinal Orchids
Edited by Azamal Husen and Manu Pant

Symbiotic Association of Microorganisms with Medicinal and Herbal Plants
Edited by Allah Ditta, Sajid Mehmood, and Azamal Husen

Tissue Culture Techniques and Medicinal Plants: Enhancing Propagation and Production
Edited by Azamal Husen and Manu Pant

Genetic Improvement and Conservation Practices of Medicinal Plants
Edited by Manu Pant, Sayed Sartaj Sohrab and Azamal Husen

Sustainable Use of Plants and their Products in Neurodegenerative Disease Management
Edited by Johnson Olaleye Oladele, Ebenezer I. O. Ajayi, María Liliana Flores-López, and Oluwaseyi E. Okoro

Wild Edibles of Deserts: Food Security and Sustainability
Salman Majeed, Muhammad Zafar, Mushtaq Ahmad, Mohd Kafeel Ahmad Ansari, Shazia Sultana, and Andrea Pieroni

Impact of Climate Change on Medicinal and Herbal Plant micro-RNA
Kanchanlata Tungare, Parul Johri, Sachidanand Singh, and Surojeet Das

Exploring Medicinal Bamboos
Edited by Manu Pant and Azamal Husen

For more information about this series, please visit: www.routledge.com/Exploring-Medicinal-Plants/book-series/CRCEMP

Exploring Medicinal Bamboos

Edited by
Manu Pant and Azamal Husen

CRC Press
Taylor & Francis Group
Boca Raton London New York

CRC Press is an imprint of the
Taylor & Francis Group, an **informa** business

Designed cover image: Shutterstock

First edition published 2026
by CRC Press
2385 NW Executive Center Drive, Suite 320, Boca Raton FL 33431

and by CRC Press
4 Park Square, Milton Park, Abingdon, Oxon, OX14 4RN

CRC Press is an imprint of Taylor & Francis Group, LLC

ISBN: 9781032771625 (hbk)
ISBN: 9781032780306 (pbk)
ISBN: 9781003485933 (ebk)

DOI: 10.1201/9781003485933

Typeset in Times
by KnowledgeWorks Global Ltd.

Dedicated to Niranjan Kumar Pant (elder brother of Manu Pant) for his constant encouragement to achieve academic milestones.

Contents

About the Editors

Manu Pant is serving as a full professor and is head of the Department of Biosciences at Graphic Era (Deemed to be University), Dehradun, Uttarakhand, India. She completed her master's degree from Banasthali Vidyapith in Rajasthan, India, and her doctorate from HNB Garhwal University (a central university) in Srinagar, India. She has extensive experience in the field of plant sciences, having worked as a research fellow at the Forest Research Institute and the Indian Council of Forestry Research and Education on nationally funded projects. Her research work on different plant species has been funded by the state governments, earning her recognition and awards at the national level. She has several peer-reviewed international papers to her credit, published patents, and has delivered numerous oral and poster presentations, keynote addresses, and invited talks at national and international scientific events. She has also edited two books in the series *Exploring Medicinal Plants* (Taylor & Francis Group, USA); and is near to completion of some more important book projects which will be published by the reputed publishing house. Professor Pant has conducted many national and international conferences, seminars, and trainings related to the thrust areas in plant sciences. She has been supervising Ph.D., B.Tech, MSc, and BSc students in their research work on different aspects of medicinal plants, ornamental crops, and indigenous crop species. For her work in the field of plant sciences, Dr. Pant has also received appreciation from state- and national-level bodies.

Azamal Husen has served as full professor and head of the Department of Biology, University of Gondar, Ethiopia; and is currently working as a foreign delegate at Wolaita Sodo University, Wolaita, Ethiopia. He is also working as an adjunct professor at Graphic Era (Deemed to be University), Dehradun, Uttarakhand, India. He also worked as a full professor at Sankalchand Patel University, Visnagar, India, and visiting faculty of the Forest Research Institute and the Doon College of Agriculture and Forest at Dehra Dun, India. His research and teaching experience of 25 years encompasses nanobiotechnology, plant physiology, environmental botany, medicinal plants, and clonal propagation. He has conducted research sponsored by the World Bank, the National Agricultural Technology Project, the Indian Council of Agriculture Research, the Indian Council of Forest Research Education, and the Japan Bank for International Cooperation. Husen has published extensively (over 350, which includes research papers, books, book chapters, etc.) and served on the editorial board and as reviewer of reputed journals published by Elsevier, Frontiers Media, Taylor & Francis, Springer Nature, RSC, Oxford University Press, Sciendo, the Royal Society, CSIRO, PLOS, MDPI, John Wiley & Sons, and UPM Journals. He is on the advisory board of Cambridge Scholars Publishing, United Kingdom. He is a fellow of the Plantae group of the American Society of Plant Biologists and a member of the International Society of Root Research, Asian Council of Science Editors, and International Natural Product Sciences. He has given numerous talks and presentations in many conferences and workshops. Husen has also given various orientation courses for academic staff; prepared different modules; and developed BSc, MSc, and Ph.D. curriculum. Moreover, he has received a recognition award from the University of Gondar, Ethiopia, for a decade of service in teaching, research, and community services. He is editor-in-chief of the *American Journal of Plant Physiology*, and a book series editor of *Exploring Medicinal Plants* (Taylor & Francis Group, USA); *Plant Biology, Sustainability, and Climate Change* (Elsevier, USA);

and *Smart Nanomaterials Technology* (Springer Nature, Singapore). He is also working as a handbook series editor, namely *Nanotechnology and Its Applications* (Springer Nature, Singapore). He has achieved the distinguished honor of being recognized as one of the "World's Top 2% Scientists" for the years 2022, 2023, and 2024 by Stanford University, USA. This recognition has also been prominently featured in the Elsevier Data Repository. Moreover, he has also worked as a special issue guest editor on various topics, such as "Advances in Plant Metabolomics: Insights, Applications, and Integration" (*Plant Physiology and Biochemistry*), "Exploring Nanotechnology in Medicinal Plants" (*Journal of Herbal Medicine*), and "Integrating Nanotechnology with Plant Compounds: Enhancing Abiotic Stress Resilience" and "Crop Improvement via Signaling Molecules" (*Frontiers in Plant Science*).

List of Contributors

Ankita Acharya
Department of Pharmaceutical Technology
JIS University
West Bengal, India

Oluwafunmilayo Dorcas Adegbaju
Food and Business Department
Reaseheath College
Cheshire, United Kingdom

Tomi Lois Adetunji
Unit for Environmental Sciences and Management
Faculty of Natural and Agricultural Sciences
North-West University
Potchefstroom, South Africa

Sabulal Baby
Phytochemistry and Phytopharmacology Division
Jawaharlal Nehru Tropical Botanic Garden and Research Institute
Kerala, India

Ashish Baldi
Department of Pharmaceutical Sciences & Technology
Maharaja Ranjit Singh Punjab Technical University
Punjab, India

Kiran Bisht
Department of Biotechnology
Graphic Era (Deemed to be University)
Uttarakhand, India

Rita Bungla
Department of Biotechnology
Graphic Era (Deemed to be University)
Uttarakhand, India

Anshika Chauhan
Department of Biotechnology
Graphic Era (Deemed to be University)
Uttarakhand, India

Nagendra Singh Chauhan
Drugs Testing Laboratory Avam Anusandhan Kendra
Raipur, Chhattisgarh, India

Nirmala Chongtham
Department of Botany
Panjab University
Chandigarh, India

Moulima Das
Department of Pharmaceutical Technology
JIS University
West Bengal, India

Hemant Dasila
Department of Microbiology
Akal College of Basic Sciences
Himachal Pradesh, India

Anitha Rani Elavarasan
AMET (Deemed to be University)
Tamil Nadu, India

Sumitha Elayaperumal
Department of Biotechnology and Bioinformatics
JSS Academy of Higher Education and Research
Karnataka, India

Aanika Gupta
Department of Biotechnology
Graphic Era (Deemed to be University)
Uttarakhand, India

Aribam Indira
Department of Botany
Panjab University
Chandigarh, India

Babita Joshi
Department of Botany
Panjab University
Chandigarh, India

Nikita Kalyan
Department of Botany
Panjab University
Chandigarh, India

Ravjot Kaur
Department of Pharmaceutical Sciences & Technology
Maharaja Ranjit Singh Punjab Technical University
Punjab, India

Konnath Chacko Koshy
Plant Genetic Resources Division
Jawaharlal Nehru Tropical Botanic Garden and Research Institute
Kerala, India

Ashwani Koul
Department of Biophysics
Panjab University
Chandigarh, India

Jitender Madan
Department of Pharmaceutics
National Institute of Pharmaceutical Education and Research
Telangana, India

Damini Maithani
College of Horticulture
Sardar Vallabhbhai Patel University of Agriculture and Technology
Meerut, India

Sneha Makwana
Department of Pharmaceutical Sciences
Sardar Patel University
Gujarat, India

Sadari Manusha
Department of Biotechnology and Bioinformatics
JSS Academy of Higher Education and Research
Karnataka, India

Varsha Nagabhushan
Department of Biotechnology and Bioinformatics
JSS Academy of Higher Education and Research
Karnataka, India

Olamide Omolafe Ogunremi
Institute for Sustainable Development
First Technical University
Ibadan, Oyo State, Nigeria

Chijioke Olisah
Institute for Coastal and Marine Research (CMR)
Nelson Mandela University
Gqeberha, South Africa
and
Research Centre for Toxic Compounds in the Environment (RECETOX)
Masaryk University
Brno, Czech Republic

Vaibhav Painuly
Department of Biotechnology
Graphic Era (Deemed to be University)
Uttarakhand, India

Hema Pandey
Department of Biotechnology
Graphic Era (Deemed to be University)
Uttarakhand, India

Shivam Pandey
Department of Chemistry
Uttaranchal University
Dehradun, India

Ravish Patel
Ramanbhai Patel College of Pharmacy
Charotar University of Science and Technology
Gujarat, India

Viral Patel
Ramanbhai Patel College of Pharmacy
Charotar University of Science and Technology
Gujarat, India

Joohee Pradhan
Department of Pharmaceutical Sciences
Mohanlal Sukhadia University
Rajasthan, India

Nikita Pundir
Department of Microbiology
Dev Bhoomi Uttarakhand University
Uttarakhand, India

Shefali Pundir
Department of Chemistry
Uttaranchal University
Dehradun, India

Mithun Sing Rajput
Department of Pharmacology
Nirma University
Gujarat, India

Varshini Ramesh
Department of Biotechnology and Bioinformatics
JSS Academy of Higher Education and Research
Karnataka, India

Purnita Raturi
Department of Horticulture
G.B. Pant University of Agriculture and Technology
Uttarakhand, India

Manan Raval
Ramanbhai Patel College of Pharmacy
Charotar University of Science and Technology
Gujarat, India

Anjali Rawat
Department of Biotechnology
Graphic Era (Deemed to be University)
Uttarakhand, India

Abhinav Sharma
Department of PDP
Graphic Era (Deemed to be University)
Uttarakhand, India

Aditi Sharma
Department of Biotechnology
Graphic Era (Deemed to be University)
Uttarakhand, India

Himani Sharma
Department of Chemistry
Uttaranchal University
Dehradun, India

Pallavi Singh
Department of Biotechnology
Graphic Era (Deemed to be University)
Uttarakhand, India

Yuvaraj Sivamani
Department of Pharmaceutical Chemistry
Cauvery College of Pharmacy
Karnataka, India

Uma Maheshwari Srinivasa
CSIR-Central Food Technological Research Institute
Karnataka, India

Anjali Tiwari
Department of Environmental Science
G.B. Pant University of Agriculture and Technology
Uttarakhand, India

Viabhav Kumar Upadhyay
Department of Microbiology
College of Basic Sciences and Humanities
Dr. Rajendra Prasad Central Agricultural University
Bihar, India

Preface

Bamboo is an important source of non-wood forest products which offers a variety of social, environmental, and economic benefits. The high-fiber content of bamboo leads to its extensive utilization in paper and pulp industries while the physical properties of the culms make it suitable for construction purposes, crafting instruments, and artifacts. Bamboo shoots have a long history of being used as a cuisine in many parts of the world, where they are consumed raw, cooked, or in fermented and dry forms. Bamboo shoots have nutritional advantages over other vegetarian foods as they contain high fiber, low fat, high protein, and bioactive content. Contrastingly, the therapeutic potential of this "green gold" has been explored marginally. Bamboos have been an important component of traditional systems of medicine and folk remedies in different parts of the world. While the medicinal properties of bamboo have been known since ancient times, its application in herbal treatments has been limited. The phytochemical potential of bamboo leaves has been evaluated, bolstering its traditional usage in treating a plethora of diseases and infections. Bamboo buds and flowers have also been shown to exhibit therapeutic properties. Researchers have identified a range of active phytochemicals and minerals in bamboo that impart exceptional antioxidant properties to the plant, making it a suitable raw material for medications and therapies to treat inflammation and ailments related to the heart, liver, and nervous system. More recently, bamboo products such as charcoal, vinegar, clothes, and cosmetic products have been prepared and found to exhibit antimicrobial and antioxidant properties.

Despite these developments in bamboo research, the documented literature on the medicinal significance of bamboo species is scarce and scattered. The present book is an attempt to provide a compendium of the latest developments on the medicinal properties of bamboo and its utilization. The traditional practices using various parts of the bamboo plant for therapeutic purposes and the latest scientific evidence on the phytochemical potential of bamboos, medicinal properties of parts of the plant, medicinally active bamboo species, their active ingredients, and pharmaceutical potential are some of the crucial topics that are being elaborated in this book.

We express our sincere thanks to all the authors who have contributed to this book in the form of well-articulated and comprehensive chapters and appreciate them for incorporating the suggestions to make their contributions more impactful. We also thank our publishers for bringing out this book on the important aspect of the medicinal value of bamboo. We are hopeful that the book will open new vistas in the field of medicinal bamboos.

Manu Pant
Azamal Husen

1 The Medicinal Significance of Bamboos

Manu Pant, Aditi Sharma, Anjali Rawat, Kiran Bisht, Abhinav Sharma, and Azamal Husen

1.1 INTRODUCTION

Bamboo is a perennial, giant woody evergreen plant that belongs to the family Poacea and subfamily Bambusoideae (Hodkinson et al., 2010; Panda, 2011; Scurlock et al., 2000; Sungkaew et al., 2009; Yeasmin et al., 2015) (Figure 1.1). The bamboo subfamily is divided into three tribes: (i) Olyreae – includes herbaceous bamboo, generally found in the tropical forests of South and Central America. Globally, about 123 bamboo species belong to this tribe (Clark et al., 2015; Triplett et al., 2014; Yormann et al., 2020). (ii) Arundinarieae – includes temperate woody bamboo growing at higher

FIGURE 1.1 Bamboos growing in their natural habitat.

DOI: 10.1201/9781003485933-1

altitudes in tropical areas. It consists of 581 species, mostly exclusive to Eurasia, and only 3 species can be found in North America (Nayak et al., 2016; Triplett et al., 2014; Yormann et al., 2020). (iii) Bambuseae – includes tropical woody bamboos that occur in tropical areas but can also grow outside these regions. Bamboo species under this tribe are distributed in major parts of the globe and consist of 976 species (Triplett et al., 2014; Yormann et al., 2020).

The human race has been utilizing bamboo for its multidisciplinary benefits since time immemorial (Chongtham et al., 2011). Bamboo species are a significant part of traditional Asian medicine, especially in China and Japan, and their medicinal properties and uses are described in various traditional systems of medicine like in Ayurveda, Siddha, and Unani. (Panee, 2015; Roy and Pramanick, 2021). Bamboo has been used to make "Lemang bamboo" which is the traditional food of Sumatera, Indonesia, made from glutinous rice that is cooked with *Schizostachyum brachycladum,* commonly known as "Sacred Bali bamboo" (Jufrida et al., 2021). In Atharva Veda, *Bambusa* bamboos are described to treat snakebite, scorpion poisoning, disturbed sleep, headache, flanks and abdominal pain. Besides, Kani tribes of the Kanyakumari district in the south-western ghats of India continue to use seeds of *Bambusa arundinacea* to enhance fertility and trade them to pharmaceutical companies to manufacture fertility-enhancing drugs (Kiruba et al., 2007). The Kotwali tribes of Gujrat (in western India) use dry or green leaves of bamboo, when boiled with water, to treat urinary system–related diseases and its ashes along with jaggery or honey to treat cough. To give quick relief to a person with deafness, a powder-like substance found in bamboo joints is dropped in the ears. Bamboo leaves are also given to animals during delivery to expel their placenta (Patel, 2005). People of the Shiwalik region of Nepal use various bamboo species like *Bambusa bambos, Bambusa tulda, Bambusa nutans*, etc., to cure diseases like jaundice, typhoid, gastritis, hypertension and diabetes. The powder of culm sheath hair is popularly used to apply on fresh cuts and wounds, while the sap of internodal regions of *B. nutans* and *B. tulda* is used to cure nocturnal enuresis and jaundice. The inhabitants of this region also use young culms of *B. nepalensis* and *B. tulda* to treat blood sugar levels, snake and scorpion bites, hypertension and weight loss and as toothbrushes to reduce toothache during an oral infection (KC et al., 2024).

Since bamboo species are a prominent part of several ethnomedicinal systems, it becomes important to understand their therapeutic value, apart from their well-established usage for other purposes, for sustainable utilization for medicinal purposes.

1.2 HABIT AND HABITAT

Geographically, bamboos are found in the tropical, subtropical and mild temperate zones of the world (Hossain et al., 2015). Mostly they are grown in the tropics with a latitude of 46°N in Japan to 47°S in Argentina at an altitude of 4,000 m above sea level (Panda, 2011). They are most commonly found in Asia, where about half of the species of bamboos are found, in America and fewer species in Africa (Bystriakova and Kapos, 2005; Goyal et al., 2013; Hodkinson et al., 2010; Panda, 2011). Naturally, bamboos are not found in cold continents like Europe and Antarctica but a few species have been introduced in the mild-temperate zone of Europe (Hakeem et al., 2015; Panda, 2011; Shukla et al., 2012). They can adapt to harsh environmental conditions and hence can be found variably in dry areas, areas with plains to high-altitude mountain regions and in almost all types of soils, except alkaline soil, deserts and marshes. Bamboos can grow in a temperature range of 8.8°C to 36°C but some species can also be grown at a temperature as low as –20°C. India has the maximum area under bamboo cultivation around the world, covering approximately 11,361 km^2, followed by China with 5,444 km^2 (Goyal et al., 2012).

Bamboo is considered the fastest-growing and longest grass in the world (Shukla et al., 2012). Some bamboos can grow at a rate of 1 m per day while some species like *Bambusa oldhamii munro* can even grow at a high rate of 4 ft per day (Wu et al., 2011). The height of the bamboo varies among species, with some as short as 30 cm or as tall as the giant timber bamboo *Dendrocalamus giganteus* reaching up to 40 m (Shukla et al., 2012).

FIGURE 1.2 (a) Bamboo shoots; (b) bamboo culms.

1.3 BOTANICAL FEATURES

Bamboo is devoid of a central trunk, unlike other woody plants, and the plant body is composed of flowers, leaves, branches, culms, roots, and rhizomes (Banik, 2015; Liese and Tang, 2015) (Figure 1.2a,b). The rhizome is a well-developed, underground portion of bamboo-bearing roots which spread to form an associated network giving rise to a new rhizome (Banik, 2015; Das et al., 2008; Hardin et al., 2009; Liese and Tang, 2015). There are two types of rhizome systems which can be observed: one is the pachymorph type with clumping bamboo and the other is the leptomorph type with non-clumping bamboo. These two systems act as defining features in growth, branching pattern and distance between bamboo culms. Some bamboos may involve both of these systems; hence, called the amphimorph type (Banik, 2015; Das et al., 2008; Waite, 2009). Culms are the upper-soil portion of bamboo that is hollow and cylindrical, consisting of nodes and internodes (Chaowana, 2013; Liese and Tang, 2015). Its diameter and height range from 0.25 inches to 12 inches and from 1 foot to 120 feet, respectively. Bamboo majorly exhibits gregarious, irregular and annual types of flowering patterns (Bhattacharya et al., 2006; Bhattacharya et al., 2009; Das et al., 2008). It is observed that all clumps in a specific location flower at the same time and after 2 to 3 years of flowering, the bamboo dies (Das et al., 2008; Scurlock et al., 2000). As far as the woody bamboos are concerned, they hardly flower (Mukherjee et al., 2010; Ramanayake et al., 2007).

1.4 MEDICINAL PROPERTIES OF DIFFERENT BAMBOO SPECIES

Phytochemical analysis of various species of bamboo has confirmed the presence of amino acids, proteins, carbohydrates, flavonoids, phenols, alkaloids, tannins, etc., and these phytochemicals are associated with various medicinal properties like anti-cancerous, anti-microbial, anti-diabetic and anti-inflammatory, some of which are listed in Table 1.1. These phytochemicals account for a plethora of medicinal and therapeutic properties of bamboo, some of which are discussed below.

TABLE 1.1
Medicinal Properties of Some Bamboo Species

S. No.	Medicinal Bamboo	Common name	Plant Part	Nutrients and Phytochemicals Present	Medicinal Property	Key References
1.	*Bambusa arundinacea*	Katang	Seeds, shoots, leaves, extract	Apigenin, luteolin, quercetin, ferulic acid, tricin, tannin, saponin, terpenoids, phytosterol, reducing sugar, rutin, gallic acid, stigmasterol, orientin, iso-orientin, iso-vitexin, vitexin, caffeic acid, coumaric acid, gallic acid, chlorogenic acid, glucopyranoside, oleic acid, linolenic acid, stearic acid, tricin, vanillic acid	Anti-hypertensive, anti-hypolipidemic, anti-anemic, anti-obesity, estrogenic effect, anti-diabetic, prebiotic activity, anti-microbial, anti-oxidant, anti-inflammatory	Badwaik et al. (2015); Goyal et al., 2017; Nirmala et al., 2014; Pande et al., 2018; Rathod et al., 2011; Silva et al., 2012; Sriraman et al., 2015; Tripathi et al., 2015; Wang et al., 2020; Zubair et al., 2013
2.	*Bambusa balcooa*	Bhima				
3.	*Bambusa bamboos*	Indian Thorny bamboo				
4.	*Bambusa nutans*	Jatia Baans				
5.	*Bambusa tulda*	Barak				
6.	*Bambusa vulgaris*	Common bamboo				
7.	*Bambusa textilis*	Chinese Temperate bamboo				
8.	*Dendrocalamus calostachyus*	Hukri Baans	Shoots, leaves	Chlorogenic acid, luteolin, caffeic acid, vitexin, rutinoside apigenin, orientin, iso-orientin, iso-vitexin, aglycone, rutinoside tricin, tricin, scopoline, scopolitin	Insulin regulatory, Anti-allergic, Anti-hypertensive, Anti-diabetic, prebiotic activity, cardioprotective, vasodilatory, Anti-microbial, Anti-oxidant, Anti-inflammatory	Bisht et al., 2015; Li et al., 2018; Nirmala et al., 2014; Van Hoyweghen et al., 2012; Wang et al., 2013; Wang et al., 2020
9.	*Dendrocalamus giganteus Munro*	Giant bamboo				
10.	*Dendrocalamus hamiltonii*	Kalo Baans				
11.	*Dendrocalamus hookeri Munro*	Hooker's bamboo				
12.	*Dendrocalamus oldamic McClure*	Taiwan Giant bamboo				
13.	*Dendrocalamus sikkimensis*	Sikkim bamboo				
14.	*Fargesia hsuehiana*	Hsueh bamboo	Leaves	Chlorogenic acid, farobin-A, farobin-B, iso-orientin, feruloylquinic acid, cinnamic acid, luteolin, tricin, glucoside	Anti-oxidant	Keski-Saari et al., 2008; Van Hoyweghen et al., 2012; Wang et al., 2012
15.	*Fargesia rufa*	Green Panda bamboo				
16.	*Fargesia robusta*	Red sheath bamboo				
17.	*Gigantochloa amplexifolia*	–	Shoots	Apigenin, sinapoyl hexose	Anti-oxidant, Anti-allergic, Anti-inflammatory, Anti-microbial, cardioprotective, vasodilatory activity	Bisht et al., 2015; Van Hoyweghen et al., 2012
18.	*Gigantochloa albociliata Munro*	Clumping bamboo				
19.	*Gigantochloa apus*	Tabashir bamboo				
20.	*Indocalamus latifolius*	Broad leaf bamboo	Leaves	Chlorogenic acid, orientin, apigenin, latifoluisine-A, iso-orientin, L-phenylalanine, iso-vitexin, vitexin, caffeic acid, coumaric acid	Anti-oxidant	Ni et al., 2013; Sun et al., 2015; Sun et al., 2016

21.	*Merostachys pluriflora*	-		Gallic acid, vitexin, ferulin, vanillic acid, benzoic acid, caffeic acid, ferulic acid, luteolin, phenylacetic acid, apigenin, coumaric acid, syringic acid	Anti-oxidant, allelopathic activity	Faria and Grombone-Guaratini, 2011; Gagliano et al., 2018; Jose et al., 2016
22.	*Merostachys riedeliana*	-				
23.	*Phyllostachys edulis*	Moso bamboo	Shoots, leaves, culms, shaving	Stigmasterol, orientin, furfural, iso-orientin, arbutin, galacturonic acid, coumaric acid, arabinose, iso-vitexin, ferulic acid, tachioside, vitexin, glucoside, benzoic acid	hypoglycemic, Anti-microbial, Anti-tumor, free radical scavenging, Anti-inflammatory	Bisht et al., 2015; Jin et al., 2011; Liu et al., 2018; Nirmala et al., 2018; Park et al., 2007; Park and Jhon, 2010; Patel et al., 2016; Tanaka et al., 2013; Yoshimura et al., 2017; Zhao et al., 2017
24.	*Phyllostachys manii*	Manii bamboo				
25.	*Phyllostachys bambusoides*	Madake bamboo				
26.	*Phyllostachys parvifolia*	Golden bamboo				
27.	*Phyllostachys heterocycla*	Tortoise-shell bamboo				
28.	*Phyllostachys nigra*	Black bamboo				
29.	*Sasa borealis*	Northern bamboo	Leaves	Glucoside, arabinoside, tricin, luteolin, rutin, chlorogenic, cinnamic acid, taxifolin, naringenin, ferulin, sinapinic acid, iso-orientin, iso-vitexin, vitexin, caffeic acid, tangeretin, coumaric acid	Anti-oxidant, Anti-UV, Anti-diabetic, Anti-adipogenicity, Anti-HIV, phagocytic, Anti-inflammatory, hepato-protective, Anti-tumor	Herath et al., 2018; Matsuta et al., 2011; Nirmala et al., 2018; Park et al., 2007; Seki et al., 2010; Yang et al., 2017

1.4.1 Anti-Cancer Property

Anti-cancerous properties of bamboo are primarily due to the phytochemicals like phenols, flavonoids, alkaloids, terpenoids, etc. (Nirmala et al., 2018). Zhao et al. (2010; 2013a, b) reported the anti-cancer properties of bamboo salt, a traditional functional food of Korean origin, used for medical treatments in many parts of the world, and purple bamboo salt (a nine-times-baked bamboo salt). The studies confirmed that the salts could lead to apoptosis of cancerous cells *in vitro* and *in vivo* (TCA8113, human tongue squamous carcinoma; HCT-116 and HT-29 human colon cancer) by upregulation of caspase-3, caspase-9 and Bax genes and downregulating Bcl-2. Further, the upregulation of IκB-α mRNA expression and downregulation of NF-κB-p65, CoX-2 and iNOS mRNA expression was found to enhance anti-inflammatory activities; hence, showing anti-cancerous activity. The *Phyllostachys* spp. has been shown to have anti-cancerous compounds like limonene and 1-epi-Bicyclosesquiphellandrene which exhibit anti-angiogenic, proapoptotic and anti-cancerous effects against the PC-3 (prostatic small cell carcinoma) cell line (Kim et al., 2014).

Bamboo shaving, a Chinese traditional medicine, is rich in compounds like vincristine (alkaloid) and friedelin (triterpenoids), where vincristine exhibited an anti-cancerous effect against A549 (hypotriploid alveolar basal epithelial cells) with an inhibition rate of 24.8% at 0.063 mg/mL dosage; friedelin showed a 50% inhibition rate at 60 μmol/L dosage of extract in 24 hours against THP-1 (human leukemia monocytic cell line) (Lu et al., 2010).

Another set of studies reported the synthesis of silica nanoparticles from the ashes of bamboo leaves, showing selective cytotoxicity towards a lung cancerous cell line (A549) without harming the normal cells. It was effective at a dosage of 750 μg/mL, leading to cell membrane damage ultimately causing oxidative stress to cancer cells (Ahamed, 2013; Irfan et al., 2014). Besides, metallic silver nanoparticles like BaAgNPs and BnAgNPs were biosynthesized using *in vitro* grown leaves of *Bambusa arundinacea* and *Bambusa nutans*, respectively. Both bamboo species are known to have flavonoid and phenolic content that are potential anti-cancer agents. The synthesized AgNPs condensed cytoplasm, shrunken cells and blotted plasma membrane of PC-3 cells and the use of these nanoparticles are strongly suggested as a nanomedicine against prostate cancer (Kalaiarasi et al., 2015).

1.4.2 Anti-Oxidant Activity

The anti-oxidant property of different species of bamboo is due to the presence of phenols and flavonoids (iso-orientin, apigenin, c-glucosides, homo-orientin, vitexin, iso-vitexin), phenolic acids (chlorogenic acid, caffeic acid, ferulic acid, *p*-coumaric acid), polyphenols (3-*O*-3′ methyl caffeoyl, 5-*O*-caffeoyl-4-methyl, 3-*O*-caffeoyl-1-methyl), essential oils (has a small anti-oxidant effect), tannins, amino acids, minerals (potassium, phosphorus, copper, selenium, manganese, zinc, iron, magnesium) and vitamins (thiamine, riboflavin, vitamin C) and β-carotene which are present in various parts of the bamboo that contain these phytochemicals, like the shoot, culm, rhizome, branch and leaves (Gagliano et al., 2018; Gong et al., 2015; Hu et al., 2011; Jin et al., 2011; Tanaka et al., 2014; Wu et al., 2012).

Phenols contain hydroxyl groups that scavenge reactive oxygen and nitrogen species. It chelates metal like iron and forms a complex with ferrous ions that restricts metal and lipid interaction and keeps cell membranes intact. Similarly, flavonoids prevent lipid peroxidation, chelate metal ions and scavenge free radicals. Additionally, bamboo extract converts unstable free radicals by donating electrons or hydrogen ions and preventing radical chain reactions (Macwan et al., 2010).

1.4.3 Cardiovascular Activity

The ability of bamboo to fight against cardiovascular diseases is associated with phenols like ferulic acid and *p*-coumaric acid, phytosterols, terpenes, bamboo oil, silica and vitamins C and E. Major flavonoids like orientin are responsible for such cardiovascular properties which lower the blood triglyceride and low-density lipoprotein (LDL) cholesterol while increasing the dosage-dependent HDL cholesterol (Ying et al., 2017). Bamboo leaves contain flavonoids that lower serum triglyceride, increase high-density lipoprotein content and regulate blood lipid concentration; thus, decreasing the risk of atherosclerosis (Palombini and Nogueira, 2023). In a study, flavonoids decrease TNF-α levels in *Rattus norvegicus,* reducing thrombocyte aggregation. Flavonoids decrease the blood monocyte count and prevent foam cell formation; thus, decreasing vascular cell adhesion molecules-1 and monocyte chemo-attachment protein-1, slowing down the progression of atherosclerosis (Soesanto et al., 2019; Surati et al., 2023). Phenols like ferulic acid and *p*-coumaric acid, along with flavonoids, have anti-oxidant properties with a vasodilatory effect due to the increased production of nitric oxide (NO) in serum which donates hydrogen ions and regulates blood pressure in kidney blood vessels (Devi and Pamba, 2015). Besides, bamboo culm was found to enhance hepatic anti-oxidant enzyme activity and hence decrease hepatic lipid peroxidation and protein carbonylation. Bamboo contains vitamin C which has an anti-oxidant property that neutralizes reactive oxygen and restricts lipid peroxidation; also, vitamin E is a fat-soluble anti-oxidant inhibiting LDL oxidation (Nirmala et al., 2014; Palombini and Nogueira, 2023). Other than this, bamboo shoots contain an inhibitory peptide, angiotensin-converting enzyme, which has the Asp-Tyr key active compound and a hypotensive effect. It also contains triterpenes like friedelin which have a vasodilatory effect. Bamboo shoots contain silica that lowers blood lipid and cholesterol and enhances the ability of potassium and calcium to bring down hypertension and help regulate the heartbeat, respectively. The effect of increasing the concentration of bamboo oil was observed in a study on the level of total cholesterol and triglyceride concentration. It was observed that an increase in bamboo oil concentration by up to 80% decreased the level of total cholesterol from 139.6 to 80.4 mg/mL, which showed the medicinal effect of bamboo oil on fatty livers and atherosclerosis induced by a high cholesterol diet (Devi and Pamba, 2015; Palombini and Nogueira, 2023). Bamboo contains phytosterols that can lower LDL cholesterol by restricting cholesterol intake in the intestine and recirculated cholesterol secreted by the bile of the liver for cholesterol elimination; hence, influencing sterol biosynthesis in the body (Devi and Pamba, 2015; Gong et al., 2015).

1.4.4 Anti-Inflammatory Activity

The anti-inflammatory effect of bamboo is due to the presence of flavonoids, coumaric acid and thaumatin-like protein. The methanolic extract of *Bambusa vulgaris* can inhibit the permeability of small vessels during acute inflammation and also has anti-lipoperoxidation activity that reduces oxidative stress and hence shows an anti-inflammatory effect. Additionally, the extract contains flavonoids that restrict enzymes involved in the production of chemical mediators for inflammation. It also inhibits chemical mediators: serotonin, histamine, formalin and prostaglandin-like compound. Thus, it showed an anti-inflammatory effect against acute inflammation like vascular permeability caused by acetic acid, and formaldehyde-induced paw edema (Carey et al., 2009).

In another study, the methanolic extract of *Bambusa arundinacea,* when attributed with a non-steroidal anti-inflammatory agent (NSAIA), modern medicine had produced the best anti-inflammatory drug with the least toxicity and significant potential against rheumatoid arthritis with peptic ulcers (Muniappan and Sundararaj, 2003). In a study, *Sasa albomargimata* inhibited LPS-induced NO and interleukin (IL-6 and IL-1β) production and iNOs secretion at non-toxic concentrations. Besides, it contains coumaric acid which has radical scavenging activity, inhibiting migration and tube formation in HUVEC (human umbilical vein endothelial cell) that inhibits the VEGF-VEGFR-2 pathway and also inhibits LPS downstream signaling (Kojima et al., 2022).

1.4.5 Anti-Microbial Activity

Bamboo has been found to inhibit the growth of micro-organisms like *Staphylococcus aureus, Escherichia coli, Bacillus subtilis, Aspergillus niger, Penicillium citrinum, Pseudomonas aeruginosa* and *Saccharomyces cerevisiae.* The shoot skin of *Phyllostachys pubescens* has shown an anti-bacterial effect against *Escherichia coli* and *Staphylococcus aureus* associated with the presence of stigmasterol, dihydrobrassicasterol and β-sitosterol compound in the shoot skin. The double bond in the sterol compound present in the shoot skin interacts with protein and peptide bonds and kills the microbes (Tanaka et al., 2014).

Another species of bamboo, *Dendrocalamus asper* was found effective against diarrheagenic *E. coli*. The leaf extract of *Dendrocalamus asper* contains fatty acid, ester and alcoholic and aldehyde groups that either hydrolyze the lipase enzyme or oxidize itself to form acid and thus destroy the bacteria (Mulyono et al., 2012).

The aqueous extract of a bamboo shaving contains anthraquinone-like emodin, rhein and aloe-emodin which have an anti-microbial effect against bacteria like *Staphylococcus aureus* and *Bacillus subtilis* and fungi like *Saccharomyces cerevisiae, Aspergillus niger* and *Penicillium citrinum* (Zhang et al., 2010).

Silver nanoparticles (AgNPs) derived from *Bambusa aurendinacea* leaves show anti-bacterial activity against gram-positive bacteria like *Staphylococcus epidermis, Bacillus subtilis* and gram-negative bacteria like *Pseudomonas aeruginosa, Escherichia coli and Shigella bigdii.* AgNPs increase the permeability of the cell membrane of the bacteria and enter into the bacterial cell which then solidifies protein, thus altering the enzyme activity and inhibiting respiration in bacteria and ultimately killing it (Jayarambabu et al., 2023).

Bamboo also shows anti-viral activity; sodium copper chlorophyllin made from bamboo leaf extract inhibits the adsorption of HIV and influenza virus adsorption on the surface of the cells. Pyroligneous acid extracted from bamboo is used as a disinfectant against encephalomyocarditis virus (Ito et al., 2016).

1.4.6 Anti-Diabetic Activity

One of the leading health problems across the world is diabetes mellitus and as per estimation, there will be 79.4 million patients who will suffer from diabetes in the year 2030. In this search line, the seed and root extract of *Bambusa arundinaceae* were administered to alloxan-induced diabetes in Wistar rats. The compound alloxan decreased the production of endogenous insulin by damaging insulin-secreting β-cells. The bamboo extract prohibited the formation of free radicals generated by alloxan and increased the secretion of insulin from pancreatic β-cells that diffused out from the pancreas to blood and hence showed a hypoglycemic effect (Macharla et al., 2012).

The leaf of *Bambusa arundinaceae* contains flavonoids, tannins, steroids, phenols and glycosides. Oxidative stress releases reactive oxygen species that cause lipid peroxidation, which is a characteristic of chronic diabetes. The leaves of *Bambusa arundinaceae* increase glutathione which acts on reactive oxygen species and prevents lipid oxidation. It also recovered the activity of catalase enzyme and superoxide dismutase which is reduced during diabetes. It contains β-sitosterol glucoside and stigmasterol that has a synergetic effect and hence lowers blood sugar levels (Nazreen et al., 2011).

β-pyran, a highly branched polysaccharide, was isolated from the bamboo species *Leleba oldhami* and has the caliber to reduce blood sugar, serum triglyceride and total cholesterol which recover serum insulin and improve the body weight of streptozotocin-induced diabetic mice (Zheng et al., 2016).

Soluble dietary fiber extracted from bamboo shoot shells has been found to increase the insulin level in blood and has high glucose-adsorbing capacity; thus, it traps glucose and delays the diffusion of glucose into the blood for a longer period. α-amylase and α-glucosidase are essential components that are involved in the breakdown and adsorption of starch, increasing blood glucose levels. Soluble dietary fiber also prohibited the activity of α-amylase and α-glucosidase; hence, decreasing blood sugar levels (Zheng et al., 2019).

Diabetes causes nephropathy in the affected person which can be improved by administrating bamboo leaf extract. Diabetes nephropathy is characterized by weight loss, increased levels of blood urea nitrogen (BUN), serum creatinine (Scr) and 24 h UP. The administration of bamboo leaf extract reduced the effect of oxidative stress by enhancing the activity of T-superoxide dismutase (T-SOD) and diminishing the level of malondialdehyde (MDA). Diabetic nephropathy is induced mainly by oxidative stress causing the inhibition of the AKT signaling pathway which in turn causes apoptosis of renal cells, activating PGSK-3β and increasing other factors associated with apoptosis like BAX and cleaved-capase-3. The leaf extract enhanced the activity of the AKT signaling pathway, reducing cleaved-caspase-3, PGSK-3β and BAX (Kato et al., 2006; Mansuroğlu et al., 2015; Ying et al., 2017; Zheng et al., 2016; Zhou et al., 2015).

1.5 FUTURE PROSPECTS

While bamboo has been widely used for its medicinal properties, there are unexplored valuable species, non-analyzed pharmaceutically important phytochemicals, under-studied mechanisms, research at microscopic levels and many more. While a majority of studies have been done on Asian bamboos, those growing in region of America and Africa are yet to be explored. This highly potent, multi-utility grass rich in chemical composition diversity at the species level requires analysis for other phytochemicals and chemical compounds as most studies primarily focus on flavonoids and alkaloids. To add to it, the current studies lack specific explanations for mechanisms associated with the therapeutic effects of bamboo in the human body. Some studies have been conducted to assess the therapeutic potential of bamboo-based products; however, there is a major scope of studies related to molecular and genetic levels analysis of the long-term effects of using bamboo and its products for medicinal purposes (Table 1.2). There is also a need for increasing awareness about the

TABLE 1.2
Therapeutic Potential of Some Bamboo-Based Products

S. No.	Bamboo Product	Medicinal Property	Model Used	Therapeutic Value	References
1.	Bamboo salt	Gastro-protective	Male Sprague-Dawley (SD) rats.	Contains ions Mn^{2+} which react with free radical that damages gastric mucosa. It contains many hydroxy groups that neutralize acid in the stomach. Salt also decreases the volume of gastric secretion and elevates the reduced pH.	Nassini et al., 2010; Zhao et al., 2012; Zhao et al., 2013a
2.	Non-steroidal anti-inflammatory agent (NSAIA) isolated from *Bambusa arundinacea*	Anti-ulcer	Carrgeenin-induced edema in rats	It contains steroid component are Anti-ulcer agents that has sparging effect. It inhibits edema like *Borduella pertussis* and paw edema.	Muniappan and Sundararaj, 2003
3.	Bamboo shaving polysaccharide (BSP)	Anti-obesity	High fat diet (HFD) induced obese mice.	It improves and regulates the composition of gut microbiota and hence prevents HFD-induced obesity. Obesity is characterized by an increase in adipose tissue, lipid fat accumulation and insulin resistance. BSP decreases the weight and size of adipose tissue. Decreases lipid content in the liver and plays a crucial role in lipid metabolism. It improves the homeostasis of glucose by improving the sensitivity of insulin.	Chen et al., 2018

(Continued)

TABLE 1.2 *(Continued)*
Therapeutic Potential of Some Bamboo-Based Products

S. No.	Bamboo Product	Medicinal Property	Model Used	Therapeutic Value	References
4.	PN3 from stem of *Phyllostachys nigra*	Neuro-protective	PC12 cells	Neurodegenerative diseases are caused due to oxidative stress. PN-3 avoid generation of reactive oxygen species and ameliorates injury caused by oxidative stress by activating Nrf2 factor (nuclear factor erythroid 2–related factor 2) which produces anti-oxidant enzymes like glutamate-cysteine ligase, NADHP quinone oxidoreductase and heme oxygenase that protect PC12 cells.	Yang et al., 2022
5.	Black bamboo rhizome extract		Cerebral ischemia injury model rats.	Extract increases expression of CREB and BDNF that promotes survival, generation, proliferation and growth of neurons. It increases expression of Bcl that prevents cell apoptosis and reduces the expression of Caspase-3 enzymes that initiate apoptosis.	Kirkland et. al., 2007; Ma et al., 2019; Yi et al., 2021

therapeutic effects of bamboo, the establishment of proper marketing links and the promotion of herbal medications to generate significant economic benefits from the bamboo industry.

1.6 CONCLUSION

Apart from providing shelter, food and bioprospection to humankind, bamboo has been exploited for its medicinal use in different forms. Bamboo has been traditionally used to cure diabetes, snake-bite, jaundice, fertility, cough, fever and many other incurable diseases. Various studies have shown that bamboo has inherent medicinal properties like anti-cancerous, anti-diabetic, anti-oxidant, anti-inflammatory, cardiovascular, anti-microbial, anti-ulcer, anti-obesity, gastroprotective and neuro-protective activity. Such ingrained medicinal properties are the result of phytochemical constituents like alkaloids, tannins, flavonoids, steroids and phenols. These phytochemicals work at the molecular level and modulate the expression of various pathways and proteins; thus, regulating the activity of various enzymes involved in metabolic functions and contributing to the inherent therapeutic property of bamboo. These properties make bamboo a worthy raw material for the food and pharmaceutical industries. However, there is still a huge scope in exploring medicinal bamboos, calling for concerted efforts in this direction.

REFERENCES

Ahamed, M. 2013. Silica nanoparticles-induced cytotoxicity, oxidative stress and apoptosis in cultured A431 and A549 cells. *Human & Experimental Toxicology 32*(2): 186–195.

Badwaik, L.S., Gautam, G. & Deka, S.C. 2015. Influence of blanching on antioxidant, nutritional and physical properties of bamboo shoot. *The Journal of Agricultural Sciences 10*(3): 140–150.

Banik, R.L. 2015. Morphology and growth. In: *Bamboo: The Plant and Its Uses*. Springer International Publishing, Cham, 10, 43–89.

Bhattacharya, S., Das, M., Bar, R. & Pal, A. 2006. Morphological and molecular characterization of *Bambusa tulda* with a note on flowering. *Annals of Botany 98*(3): 529–535.

Bhattacharya, S., Ghosh, J.S., Das, M. & Pal, A. 2009. Morphological and molecular characterization of *Thamnocalamus spathiflorus* subsp. *spathiflorus* at population level. *Plant Systematics and Evolution 282*: 13–20.

Bisht, M.S., Nirmala, C. & Meetei, O.S. 2015. Bamboo shoots for food in North-East India: Conventional and contemporary. In: Proceedings of 10th World Bamboo Congress, Theme: Food and Pharmaceuticals. Korea.

Bystriakova, N. & Kapos, V. 2005. Bamboo diversity: The need for a red list review. *Biodiversity 6*(4): 12–16.

Carey, W.M., Dasi, J.M.B., Rao, N.V. & Gottumukkala, K.M. 2009. Anti-inflammatory activity of methanolic extract of *Bambusa vulgaris* leaves. *International Journal of Green Pharmacy 3*(3): 234–238.

Chaowana, P. 2013. Bamboo: An alternative raw material for wood and wood-based composites. *Journal of Materials Science Research 2*(2): 90–102.

Chen, Y., Jin, L., Li, Y., Xia, G., Chen, C. & Zhang, Y. 2018. Bamboo-shaving polysaccharide protects against high-diet induced obesity and modulates the gut microbiota of mice. *Journal of Functional Foods 49*: 20–31.

Chongtham, N., Bisht, M.S. & Haorongbam, S. 2011. Nutritional properties of bamboo shoots: Potential and prospects for utilization as a health food. *Comprehensive Reviews in Food Science and Food Safety 10*(3): 153–168.

Clark, L.G., Londono, X. & Ruiz-Sanchez, E. 2015. Bamboo Taxonomy and habitat. In: Liese W. & Kohl M. (eds) *Tropical Forestry, Bamboo: The Plant and Its Uses*. Springer International Publishing, Switzerland, 10, 1–30.

Das, M., Bhattacharya, S., Singh, P., Filgueiras, T.S. & Pal, A. 2008. Bamboo taxonomy and diversity in the era of molecular markers. *Advances in Botanical Research 47*: 225–268.

Devi, O.J. & Pamba, P. 2015. Anti-hypertensive activity of bamboo shoot: A review. *Asian Journal of Pharmaceutical and Clinical Research 8*(1): 46–47.

Faria, D.A. & Grombone-Guaratini, M.T. 2011. Avaliação do potencial alelopático de Merostachys pluriflo-raMunroex EG Camus um bambu nativo da Mata Atlântica. *Congresso De Ecologia Do Brasil 10*: 1–2.

Gagliano, J., Grombone-Guaratini, M.T. & Furlan, C.M. 2018. Antioxidant potential and HPLC-DAD profile of phenolic compounds from leaves and culms of Merostachys plurifora. *South African Journal of Botany 115*: 24–30.

Gong, J., Xia, D., Huang, J., Ge, Q., Mao, J., Liu, S. & Zhang, Y. 2015. Functional components of bamboo shavings and bamboo leaf extracts and their antioxidant activities in vitro. *Journal of Medicinal Food 18*(4): 453–459.

Goyal, A.K., Dubey, A.K. & Sen, A. 2012. Inventorying bamboo biodiversity of North Bengal: A case study. *International Journal of Fundamental and Applied Sciences 1*(1): 2–5.

Goyal, A.K., Kar, P. & Sen, A. 2013. Advancement of bamboo taxonomy in the era of molecular biology: A review. *Biology of Useful Plants and Microbes*: 197–208.

Goyal, A.K., Middha, S.K., Usha, T. & Sen, A. 2017. Analysis of toxic, anti-diabetic and antioxidant potential of *Bambusa balcooa* Roxb. leaf extracts in alloxan-induced diabetic rats. *3 Biotech 7*: 1–11.

Hakeem, K.R., Ibrahim, S., Ibrahim, F.H. & Tombuloglu, H. 2015. Bamboo biomass: Various studies and potential applications for value-added products. *Agricultural Biomass Based Potential Materials 2015*: 231–243.

Hardin, I.R., Wilson, S.S., Dhandapani, R. & Dhende, V. 2009. An assessment of the validity of claims for "Bamboo" fibers. *AATCC Review 9*(10): 33–36.

Herath, K.H.I.N.M., Cho, J., Kim, A., Eom, T.K., Kim, J.S., Kim, J.B., Doh, Y.H. & Jee, Y. 2018. Phenolic acid and favonoid-rich fraction of *Sasa quelpaertensis* Nakai leaves prevent alcohol induced fatty liver through AMPK activation. *Journal of Ethnopharmacology 224*: 335–348.

Hodkinson, T.R., Ní Chonghaile, G., Sungkaew, S., Chase, M.W., Salamin, N. & Stapleton, C.M. 2010. Phylogenetic analyses of plastid and nuclear DNA sequences indicate a rapid late Miocene radiation of the temperate bamboo tribe Arundinarieae (*Poaceae, Bambusoideae*). *Plant Ecology & Diversity 3*(2): 109–120.

Hossain, M.F., Islam, M.A. & Numan, S.M. 2015. Multipurpose uses of bamboo plants: A review. *International Research Journal of Biological Sciences 4*(12): 57–60.

Hu, C.J., Liao, H.B., Yuan, K., Hu, J.H. & Lin, X.C. 2011. Chemical composition and antioxidant activity of essential oil of bamboo leaves from four species in China. *Asian Journal of Chemistry 23*(6): 2543.

Irfan, A., Cauchi, M., Edmands, W., Gooderham, N.J., Njuguna, J. & Zhu, H. 2014. Assessment of temporal dose-toxicity relationship of fumed silica nanoparticle in human lung A549 cells by conventional cyto-toxicity and 1H-NMR-based extracellular metabonomic assays. *Toxicological Sciences 138*(2): 354–364.

Ito, A., Tsuneki, A., Yoshida, Y., Ryoke, K., Kaidoh, T. & Kageyama, S. 2016. In-vitro inhibition of cytopathic effect of influenza virus and human immunodeficiency virus by bamboo leaf extract solution and sodium copper chlorophyllin. *Yonago Acta Medica 59*(1): 61.

Jayarambabu, N., Velupla, S., Akshaykranth, A., Anitha, N. & Rao, T.V. 2023. *Bambusa arundinacea* leaves extract-derived Ag NPs: Evaluation of the photocatalytic, antioxidant, antibacterial, and anticancer activities. *Applied Physics A 129*(1): 13.

Jin, Y.C., Yuan, K. & Zhang, J. 2011. Chemical composition, and antioxidant and antimicrobial activities of essential oil of *Phyllostachys heterocycla* cv. *Pubescens* varieties from China. *Molecules 16*(5): 4318–4327.

Jose, C.M., Brandão Torres, L.M., Torres, M.A.M.G., Shirasuna, R.T., Farias, D.A., dos Santos, N.A. & Grombone-Guaratini, M.T. 2016. Phytotoxic effects of phenolic acids from *Merostachys riedeliana*, a native and overabundant Brazilian bamboo. *Chemoecology 26*: 235–246.

Jufrida, J., Basuki, F.R., Oksaputra, M.F. & Fitaloka, O. 2021. Ethnoscience analysis of "lemang bamboo" Sumatera traditional food. *Journal of Physics: Conference Series 1731*(1): 1–7.

Kalaiarasi, K., Prasannaraj, G., Sahi, S.V. & Venkatachalam, P. 2015. Phytofabrication of biomolecule-coated metallic silver nanoparticles using leaf extracts of in vitro-raised bamboo species and its anticancer activity against human PC3 cell lines. *Turkish Journal of Biology 39*(2): 223–232.

Kato, M., Yuan, H., Xu, Z.G., Lanting, L., Li, S.L., Wang, M., Hu, M.C.T., Reddy, M.A. & Natarajan, R. 2006. Role of the Akt/FoxO3a pathway in TGF-β1–mediated mesangial cell dysfunction: A novel mechanism related to diabetic kidney disease. *Journal of the American Society of Nephrology 17*(12): 3325–3335.

KC, B.M., Lamichhane, J., Khanal, S.N. & Gauchan, D.P. 2024. Traditional utilization of bamboo in the Central Siwalik region, Nepal. *PLoS One 19*(1): 1–33.

Keski-Saari, S., Ossipov, V., Julkunen-Tiitto, R., Jia, J., Danell, K., Veteli, T., Guiquan, Z., Yaowu, X. & Niemelä, P. 2008. Phenolics from the culms of five bamboo species in the Tangjiahe and Wolong Giant Panda Reserves, Sichuan, China. *Biochemical Systematics and Ecology 36*(10): 758–765.

Kim, J.S., Lee, H.C., Jo, J.S., Jung, J.Y., Ha, Y.L. & Yang, J.K. 2014. Evaluation of antioxidant and anticancer activity of steam extract from the bamboo species. *Journal of the Korean Wood Science and Technology 42*(5): 543–554.

Kirkland, R.A., Saavedra, G.M. & Franklin, J.L. 2007. Rapid activation of antioxidant defenses by nerve growth factor suppresses reactive oxygen species during neuronal apoptosis: Evidence for a role in cytochrome c redistribution. *Journal of Neuroscience 27*(42): 11315–11326.

Kiruba, S., Jeeva, S., Das, S. & Kannan, D. 2007. Bamboo seeds as a means to sustenance of the indigenous community *4*:86–93.

Kojima, S., Hakamata, M., Asanuma, T., Suzuki, R., Tsuruda, J.I., Nonoyama, T., Lin, Y., Fukatsu, H., Koide, N. & Umezawa, K. 2022. Cellular anti-inflammatory and antioxidant activities of Bamboo *Sasa albomarginata* leaf extract and its constituent coumaric acid methyl ester. *The Scientific World Journal 2022*(1): 1–10.

Li, X., Fu, B., Guo, J., Ji, K., Xu, Y., Dahab, M.M. & Zhang, P. 2018. Bamboo shoot fiber improves insulin sensitivity in high-fat diet-fed mice. *Journal of Functional Foods 49*: 510–517.

Liese, W. & Tang, T.K.H. 2015. Properties of the bamboo culm. *Bamboo: The Plant and Its Uses 10*: 227–256.

Liu, H., He, P., He, L., Li, Q., Cheng, J., Wang, Y., Yang, G. & Yang, B. 2018. Structure characterization and hypoglycemic activity of an arabinogalactan from *Phyllostachys heterocycla* bamboo shoot shell. *Carbohydrate Polymers 201*: 189–200.

Lu, B., Liu, L., Zhen, X., Wu, X. & Zhang, Y. 2010. Anti-tumor activity of triterpenoid-rich extract from bamboo shavings (*Caulis bamfusae* in Taeniam) *African Journal of Biotechnology 9*(38): 6430–6436.

Ma, J., Bao, L., Xia, X., Feng, Q., Zhou, Y., Wang, Y. & Cao, Z. 2019. MiR-128b promotes cerebral infarction by regulating the expressions of BCL-2 and CAPASE 3. *World Neurosurgery 123*: e245–e251.

Macharla, S.P., Goli, V. & Nath, A.R. 2012. Antidiabetic activity of *Bambusa arundinaceae* root extracts on alloxan induced diabetic rats. *Asian Journal of Research in Pharmaceutical Science 2*(2): 73–75.

Macwan, C., Patel, H.V. & Kalia, K. 2010. A comparative evaluation of in vitro antioxidant properties of Bamboo *Bambusa arundinacea* leaves extracts. *Journal of Cell and Tissue Research 10*(3): 2413–2417.

Mansuroğlu, B., Derman, S., Yaba, A. & Kızılbey, K. 2015. Protective effect of chemically modified SOD on lipid peroxidation and antioxidant status in diabetic rats. *International Journal of Biological Macromolecules 72*: 79–87.

Matsuta, T., Sakagami, H., Satoh, K., Kanamoto, T., Terakubo, S., Nakashima, H., Kitajima, M., Oizumi, H. & Oizumi, T. 2011. Biological activity of luteolin glycosides and tricin from *Sasa senanensis* Rehder. *In Vivo 25*(5): 757–762.

Mukherjee, A.K., Ratha, S., Dhar, S., Debata, A.K., Acharya, P.K., Mandal, S., Panda, P.C. & Mahapatra, A.K. 2010. Genetic relationships among 22 taxa of bamboo revealed by ISSR and EST-based random primers. *Biochemical Genetics 48*: 1015–1025.

Mulyono, N., Lay, B.W., Rahayu, S. & Yaprianti, I. 2012. Antibacterial activity of petung bamboo (*Dendrocalamus asper*) leaf extract against pathogenic *Escherichia coli* and their chemical identification. *International Journal of Pharmaceutical & Biological Archive 3*(4): 770–778.

Muniappan, M. & Sundararaj, T. 2003. Antiinflammatory and antiulcer activities of *Bambusa arundinacea. Journal of Ethnopharmacology 88*(2–3): 161–167.

Nassini, R., Andrè, E., Gazzieri, D., De Siena, G., Zanasi, A., Geppetti, P. & Materazzi, S. 2010. A bicarbonate-alkaline mineral water protects from ethanol-induced hemorrhagic gastric lesions in mice. *Biological and Pharmaceutical Bulletin 33*(8): 1319–1323.

Nayak, L. & Mishra, S.P. 2016. Prospect of bamboo as a renewable textile fiber, historical overview, labeling, controversies and regulation. *Fashion and Textiles 3*(1): 2.

Nazreen, S., Kaur, G., Alam, M.M., Haider, S., Hamid, H. & Alam, M.S. 2011. Hypoglycemic activity of *Bambusa arundinacea* leaf ethanolic extract in streptozotocin induced diabetic rats. *Pharmacologyonline 1*: 964–972.

Ni, Q., Wang, Z., Xu, G., Gao, Q., Yang, D., Morimatsu, F. & Zhang, Y. 2013. Altitudinal variation of antioxidant components and capability in *Indocalamus latifolius* (Keng) McClure leaf. *Journal of Nutritional Science and Vitaminology 59*(4): 336–342.

Nirmala, C., Bisht, M.S., Bajwa, H.K. & Santosh, O. 2018. Bamboo: A rich source of natural antioxidants and its applications in the food and pharmaceutical industry. *Trends in Food Science & Technology 77*: 91–99.

Nirmala, C., Bisht, M.S. & Laishram, M. 2014. Bioactive compounds in bamboo shoots: Health benefits and prospects for developing functional foods. *International Journal of Food Science & Technology 49*(6): 1425–1431.

Palombini, F. L., & Nogueira, F. M. (eds). 2023. *Bamboo Science and Technology*. Springer Nature, Singapore, 10, 1–397.

Panda, H. 2011. *Bamboo Plantation and Utilization Handbook*. Asia Pacific Business Press Inc, Delhi.

Pande, H., Kumar, B. & Varshney, V.K. 2018. HPLC-ESI-QTOF-MS analysis of phenolic compounds, antioxidant capacity and α-glucosidase inhibitory effect of Bambusa nutans leaves. *Indian Journal of Chemistry 57B*: 988–996.

Panee, J. 2015. Potential medicinal application and toxicity evaluation of extracts from bamboo plants. *Journal of Medicinal Plant Research 9*(23): 681.

Park, E.J. & Jhon, D.Y. 2010. The antioxidant, angiotensin converting enzyme inhibition activity, and phenolic compounds of bamboo shoot extracts. *LWT-Food Science and Technology 43*(4): 655–659.

Park, H.S., Lim, J.H., Kim, H.J., Choi, H.J. & Lee, I.S. 2007. Antioxidant flavone glycosides from the leaves of *Sasa borealis. Archives of Pharmacal Research 30*: 161–166.

Patel, A.B. 2005. Traditional bamboo uses by the tribes of Gujarat. *Indian Journal of Traditional Knowledge 4*(2): 179–184.

Patel, M., Mehta, P., Bakshi, S. & Tewari, S. 2016. Effect of *Phyllostachys parvifolia* leaf extract on ionizing radiation-induced genetic damage: A preliminary in vitro cytogenetic study. *Journal of Ayurveda and Integrative Medicine 7*(3): 138–140.

Ramanayake, S.M.S.D., Meemaduma, V.N. & Weerawardene, T.E. 2007. Genetic diversity and relationships between nine species of bamboo in Sri Lanka, using random amplified polymorphic DNA. *Plant Systematics and Evolution 269*: 55–61.

Rathod J.D., Pathak Nimish, L., Patel Ritesh, G., Jivani, N.P. & Bhatt Nayna, M. 2011. Phytopharmacological properties of *Bambusa arundinacea* as a potential medicinal tree: An overview. *Journal of Applied Pharmaceutical Science 1*(10): 27–31.

Roy, P. & Pramanick, K. 2021. A review on pharmacological activity of bamboo. *World Journal of Pharmaceutical and Medical Research 7*(8): 89–92.

Scurlock, J.M., Dayton, D.C. & Hames, B. 2000. Bamboo: An overlooked biomass resource? *Biomass and Bioenergy 19*(4): 229–244.

Seki, T., Kida, K. & Maeda, H. 2010. Immunostimulation-mediated anti-tumor activity of Bamboo (*Sasa senanensis*) leaf extracts obtained under 'Vigorous' condition. *Evidence-Based Complementary and Alternative Medicine 7*(4): 447–457.

Shukla, R., Sumit, G., Sajal, S., Dwivedi, P.K. & Mishra, A. 2012. Medicinal importance of bamboo. *International Journal of Biopharm & Phytochemical Research 1*(1): 9–15.

Silva, M.A.D., Roder, M.A., Pereira, M.A.R., Neves, F.T.A. & Constantino, D. 2012. Avaliação farmacognóstica qualitativa das folhas do vegetal *Bambusa textilis*. *Revista Salusvita (Online) 31*(3): 237–246.

Soesanto, E., Dharmana, E., Hadisaputro, S. & Muis, S.F. 2019. Effect of apus bamboo shoot extract on total interleukin 17 and leukocyte at atherosclerosis process. *Open Access Macedonian Journal of Medical Sciences 9*(2): 25–29.

Sriraman, S., Ramanujam, G.M., Ramasamy, M. & Dubey, G.P. 2015. Identification of beta-sitosterol and stigmasterol in *Bambusa bambos* (L.) Voss leaf extract using HPLC and its estrogenic effect in vitro. *Journal of Pharmaceutical and Biomedical Analysis 115*: 55–61.

Sun, J., Xun, H., Yu, J., Tang, F., Yue, Y.D. & Guo, X.F. 2015. Chemical constituents and antibacterial properties of *Indocalamus latifolius* McClure leaves, the packaging material for "Zongzi. *Molecules 20*(9): 15686–15700.

Sun, J., Yu, J., Xun, H., Yue, Y.D., Feng, T. & Guo, X.F. 2016. Two new compounds from the leaves of *Indocalamus latifolius*. *Journal of Asian Natural Products Research 18*(4): 360–365.

Sungkaew, S., Stapleton, C.M., Salamin, N. & Hodkinson, T.R. 2009. Non-monophyly of the woody bamboos (*Bambuseae; Poaceae*): A multi-gene region phylogenetic analysis of Bambusoideae s.s. *Journal of Plant Research 122*: 95–108.

Surati, S., Setyowatiningsih, L., Aulia, Q.A.Y. & Sutrisna, E.M. 2023. Effect of bamboo shoot (*Gigantochloa Nigrociliata*) extract on decreased blood pressure and Tnf-A in male Wistar rats (*Rattus Norvegicus*). *Indonesian Journal of Medicine 8*(1): 121–127.

Tanaka, A., Shimizu, K. & Kondo, R. 2013. Antibacterial compounds from shoot skins of Moso bamboo (*Phyllostachys pubescens*). *Journal of Wood Science 59*: 155–159.

Tanaka, A., Zhu, Q., Tan, H., Horiba, H., Ohnuki, K., Mori, Y., Yamauchi, R., Ishikawa, H., Iwamoto, A., Kawahara, H. & Shimizu, K. 2014. Biological activities and phytochemical profiles of extracts from different parts of bamboo (*Phyllostachys pubescens*). *Molecules 19*(6): 8238–8260.

Tripathi, Y.C., Jhumka, Z. & Anjum, N. 2015. Evaluation of total polyphenol and antioxidant activity of leaves of *Bambusa nutans* and *Bambusa vulgaris*. *Journal of Pharmacy Research 9*(4): 271–277.

Triplett, J.K., Clark, L.G., Fisher, A.E. & Wen, J. 2014. Independent allopolyploidization events preceded speciation in the temperate and tropical woody bamboos. *New Phytologist 204*(1): 66–73.

Van Hoyweghen, L., Beer, T.D., Deforce, D. & Heyerick, A. 2012. Phenolic compounds and anti-oxidant capacity of twelve morphologically heterogeneous bamboo species. *Phytochemical Analysis 23*(5): 433–443.

Waite, M. 2009. Sustainable textiles: The role of bamboo and a comparison of bamboo textile properties-part 1. *Journal of Textile and Apparel, Technology and Management 6*(2): 1–17.

Wang, J., Yue, Y.D., Tang, F. & Sun, J. 2012. TLC screening for antioxidant activity of extracts from fifteen bamboo species and identification of antioxidant flavone glycosides from leaves of Bambusa. textilis McClure. *Molecules 17*(10): 12297–12311.

Wang, S., Tang, F., Yue, Y., Yao, X., Wei, Q. & Yu, J. 2013. Simultaneous determination of 12 coumarins in bamboo leaves by HPLC. *Journal of AOAC International 96*(5): 942–946.

Wang, Y., Chen, J., Wang, D., Ye, F., He, Y., Hu, Z. & Zhao, G. 2020. A systematic review on the composition, storage, processing of bamboo shoots: Focusing the nutritional and functional benefits. *Journal of Functional Foods 71*: 1–13.

Wu, C.S., Chen, Y.J., Chen, J.J., Shieh, J.J., Huang, C.H., Lin, P.S., Chang, G.C., Chang, J.T. & Lin, C.C. 2012. Terpinen-4-ol induces apoptosis in human nonsmall cell lung cancer in vitro and in vivo. *Evidence-Based Complementary and Alternative Medicine 2012*(1): 1–13.

Wu, T.H., Liao, M.H., Kuo, W.Y., Huang, C.H., Hsieh, H.L. & Jinn, T.L. 2011. Characterization of copper/zinc and manganese superoxide dismutase in green bamboo (*Bambusa oldhamii*): Cloning, expression and regulation. *Plant Physiology and Biochemistry 49*(2): 195–200.

Yang, J.H., Choi, M.H., Shin, H.J., Na, C.S. & Ki, S.H. 2022. *Phyllostachys nigra* variety Henosis, a domestic Bamboo species, protects PC12 cells from oxidative stress-mediated cell injury through Nrf2 activation. *Biotechnology and Bioprocess Engineering 27*(3): 398–406.

Yang, J.H., Choi, M.H., Yang, S.H., Cho, S.S., Park, S.J., Shin, H.J. & Ki, S.H. 2017. Potent anti-inflammatory and antiadipogenic properties of bamboo (*Sasa coreana* Nakai) leaves extract and its major constituent flavonoids. *Journal of Agricultural and Food Chemistry 65*(31): 6665–6673.

Yeasmin, L., Ali, M.N., Gantait, S. & Chakraborty, S. 2015. Bamboo: An overview on its genetic diversity and characterization. *3 Biotech 5*: 1–11.

Yi, C.A., Jiang, Y.H., Wang, Y., Li, Y.X., Cai, S.C., Wu, X.Y., Hu, X.S. & Wan, X.G. 2021. Black bamboo rhizome extract improves cognitive dysfunction by upregulating the expression of hippocampal BDNF and CREB in rats with cerebral ischaemia-reperfusion injury. *Neuropsychiatric Disease and Treatment 17*: 2257–2267.

Ying, C., Mao, Y., Chen, L., Wang, S., Ling, H., Li, W. & Zhou, X. 2017. Bamboo leaf extract ameliorates diabetic nephropathy through activating the AKT signaling pathway in rats. *International Journal of Biological Macromolecules 105*: 1587–1594.

Yormann, G.E., Rúgolo, Z.E. & Apóstolo, N.M. 2020. Culms of *Bambusa vulgaris* cv. *vittata, Arundinaria simonii* and *Phyllostachys aurea* (*Poaceae, Bambusoideae*): Characterization of vascular bundles and fibers. *Flora 263*: 1–10.

Yoshimura, M., Ochi, K., Sekiya, H., Tamai, E., Maki, J., Tada, A., Sugimoto, N., Akiyama, H. & Amakura, Y. 2017. Identification of characteristic phenolic constituents in Mousouchiku extract used as food additives. *Chemical and Pharmaceutical Bulletin 65*(9): 878–882.

Zhang, J., Gong, J., Ding, Y., Lu, B., Wu, X. & Zhang, Y. 2010. Antibacterial activity of water-phase extracts from bamboo shavings against food spoilage microorganisms. *African Journal of Biotechnology 9*(45): 7710–7717.

Zhao, X., Deng, X., Park, K.Y., Qiu, L. & Pang, L. 2013b. Purple bamboo salt has anticancer activity in TCA8113 cells in vitro and preventive effects on buccal mucosa cancer in mice in vivo. *Experimental and Therapeutic Medicine 5*(2): 549–554.

Zhao, X., Jung, O.S. & Park, K.Y. 2012. Alkaline and antioxidant effects of bamboo salt. *Journal of the Korean Society of Food Science and Nutrition 41*(9): 1301–1304.

Zhao, X., Kim, S.Y. & Park, K.Y. 2013a. Bamboo salt has in vitro anticancer activity in HCT-116 cells and exerts anti-metastatic effects in vivo. *Journal of Medicinal Food 16*(1): 9–19.

Zhao, X., Song, J., Lee, J.H. & Kim, S.Y. 2010. Antioxidation and cancer cell (HT-29) antiproliferation effects of *Rubus coreanus* Miquel bamboo salt. *Journal of the Korean Cancer Prevention Society 15*(4): 306–312.

Zhao, Y., Li, H.X., Kim, Y.H., Cho, C.W., Hwang, S.Y., Oh, H.A., Kim, K.T. & Kang, J.S. 2017. Standardization of extract mixture of *Chaenomeles sinensis* and *Phyllostachys bambusoides* for anti-obesity by HPLC–UV. *Archives of Pharmacal Research 40*: 1156–1165.

Zheng, Y., Wang, Q., Huang, J., Fang, D., Zhuang, W., Luo, X., Zou, X., Zheng, B. & Cao, H. 2019. Hypoglycemic effect of dietary fibers from bamboo shoot shell: An in vitro and in vivo study. *Food and Chemical Toxicology 127*: 120–126.

Zheng, Y., Zhang, S., Wang, Q., Lu, X., Lin, L., Tian, Y., Xiao, J. & Zheng, B. 2016. Characterization and hypoglycemic activity of a β-pyran polysaccharides from bamboo shoot (*Leleba oldhami* Nakal) shells. *Carbohydrate Polymers 144*: 438–446.

Zhou, J., Sun, W., Yoshitomi, H., Li, L., Qin, L., Guo, X., Wu, L., Zhang, Y., Wu, X., Xu, T. & Gao, M. 2015. Qiwei granules alleviates podocyte lesion in kidney of diabetic KK-Ay mice. *BMC Complementary and Alternative Medicine 15*(1): 11.

Zubair, M., Bibi, Z., Rizwan, K., Rasool, N., Zahoor, A.F. & Riaz, M. 2013. In vitro antimicrobial and haemolytic studies of *Bambusa arundinaceae* leaves. *Journal of Applied Pharmaceutical Science 3*(4): 111–115.

2 Bamboo in Traditional and Folk Medicine

Tomi Lois Adetunji, Oluwafunmilayo Dorcas Adegbaju, Olamide Omolafe Ogunremi, and Chijioke Olisah

ABBREVIATIONS

CAM Complementary and Alternative Medicine
RDA Recommended Dietary Allowance
TCM Traditional Chinese Medicine
USDA United States Department of Agriculture

2.1 INTRODUCTION

Bamboo, with over 1,600 species distributed within over 120 genera, belongs to the family Poaceae (grass family) and subfamily Bambusoideae (Banerjee et al., 2021; Canavan et al., 2017). Within the subfamily Bambusoideae, bamboo is further divided into three tribes: Arundinarieae (temperate woody bamboos), Bambuseae (tropical woody bamboos), and Olyreae (herbaceous bamboos) genera, and species based on morphological, ecological, and molecular characteristics (Ahmad et al., 2021). Notable genera include *Bambusa*, *Phyllostachys*, *Dendrocalamus*, and *Arundinaria* (Figure 2.1). Bamboo is distributed across various regions of the world, spanning from tropical to temperate climates (Banerjee et al., 2021). They are estimated to cover more than 39 million hectares globally, making up between 4 to 10 percent of the total forest area in certain countries. They are most commonly found in Asia, particularly in countries like China, India, Indonesia, and Thailand, where they form extensive bamboo forests, playing significant roles in local ecosystems and cultures. In addition to Asia, bamboo also occurs naturally in other continents, including Africa, the Americas, and Oceania, except Antarctica and Europe (Banerjee et al., 2021). Bamboo is considered one of the world's most valuable natural plants known as "gold for the poor". Due to its significant economic value, it serves as an important economic component of forest resources (Gagliano et al., 2022; Tripathi, 2011).

Bamboo species have a long history of multiple uses rooted deeply in traditional and folk medicine practices across diverse cultures in the world. They are well revered for their versatility, abundance, and resilience. Consequently, they have become integral to human societies for ages, providing a rich repository of medicinal compounds that continue to shape traditional health care in different cultures around the world (Gagliano et al., 2022; Tripathi, 2011). Several ethnobotanical surveys done in different parts of the world (Benjamin et al., 2021; Ritonga et al., 2023), records reported in herbal books referring to ethnomedicinal systems, and scientific literature reports (Noremylia et al., 2023) all revealed that bamboo species have been extensively utilized by humans for therapeutic purposes, consumed as food, and also utilized as non-medical products. Different parts of bamboo plants, including shoots, roots, leaves, stems, seeds, and rhizomes, have medicinal properties that are utilized in traditional and folklore medicine (Noremylia et al., 2023).

Bamboo holds significant importance in traditional medicine across Asia, notably in countries like India, China, and Japan. For example, in Ayurvedic, Tibetan, traditional Chinese and Unani

 DOI: 10.1201/9781003485933-2

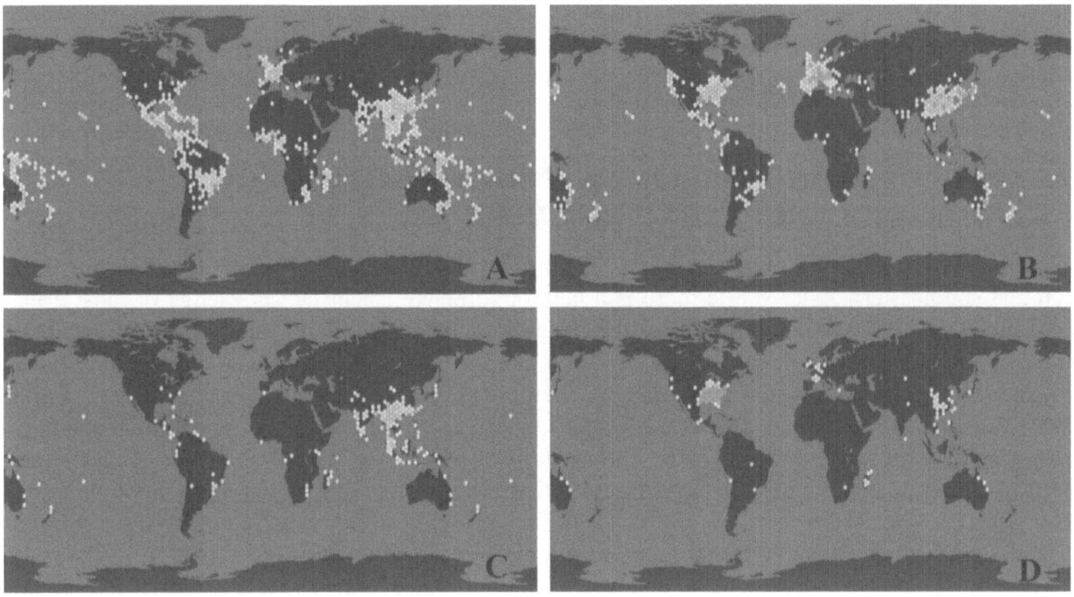

FIGURE 2.1 Global map of distribution of notable bamboo genera (a) *Bambusa*; © OpenStreetMap contributors, © OpenMapTiles, GBIF: Available at https://www.gbif.org/species/2705751; (b) *Phyllostachys* © OpenStreetMap contributors, © OpenMapTiles, GBIF: Available at https://www.gbif.org/species/2705978; (c) *Dendrocalamus* © OpenStreetMap contributors, © OpenMapTiles, GBIF: Available at https://www.gbif.org/species/2703313; (d) *Arundinaria* © OpenStreetMap contributors, © OpenMapTiles, GBIF: Available at https://www.gbif.org/species/2705960

medicine, "tabasheer", also spelt as "tabachir" or "tabashir", known as bamboo silica, is one of the main substances from bamboo that is utilized in traditional medicine for its purported health benefits (Tripathi, 2011). Tabasheer is a white, translucent substance found in the hollow stems of certain bamboo species such as *Bambusa bambos*, *Melocanna baccifera,* etc. Traditionally, tabasheer is believed to have cooling, diuretic, and anti-inflammatory properties. Tabasheer has been utilized to treat illnesses including digestive issues, fever, and respiratory disorders (Anjum et al., 2019).

A foot massager, called "ao-dake-humi" in traditional Japanese medicine, is made from a curved piece of green bamboo. Ao-dake-humi is utilized to massage and stimulate the bottoms (soles) of the feet among older citizens in Japan. This process of stimulating the sole of the feet is believed to promote well-being and overall health in elderly people (Minagawa et al., 2016). In Traditional Chinese Medicine (TCM), an expectorant ("Zhu Li") is made from bamboo sap and it is used for treating asthma and coughs as well as clearing phlegm. The leaves and stems of bamboo are also eaten to treat digestive disorders, lack of sleep, and restlessness and are used as diuretics (Tripathi, 2011). To improve digestion, bamboo shoots are eaten by people in Southeast Asian traditional medicine. Furthermore, to reduce inflammation and to treat pain in joints, bamboo stems and leaf extracts are consumed. The leaves are also consumed as herbal teas to improve well-being and overall health in Southeast Asian traditional medicine. Different bamboo species are also utilized in African traditional medicine. For instance, sap and leaf extracts from *Bambusa vulgaris* are used for several health purposes, including cleaning wounds and hastening wound healing; leaf decoctions or infusions are utilized to treat bronchitis, coughs, colds, and other respiratory conditions; and leaf decoctions are also used to reduce body temperature and to treat fevers (Honfo et al., 2015; Panee, 2015).

With the vast traditional ethnobotanical usage of bamboo in various cultures around the globe, as well as past and ongoing studies showing the prospects of bamboo as a valuable and sustainable plant in improving human health, we aim to provide in this chapter a comprehensive worldwide status of bamboo in folk and traditional medicine. This chapter also discusses the importance of conservation practices, sustainable harvesting, and the preservation of the ethnobotanical knowledge of bamboo for upcoming generations.

2.2 HISTORICAL AND CULTURAL SIGNIFICANCE OF BAMBOO IN TRADITIONAL HEALING PRACTICES

Various folk remedies continue to serve as cultural and traditional alternatives for people in eastern and southeastern Asia. While there is a lack of sufficient scientific knowledge of most drugs (Astutik et al., 2019), a comprehensive investigation of their traditional knowledge can facilitate the creation of novel or alternative treatments for a range of illnesses (Pan et al., 2014). Traditional medicine has been widely used by people of different cultures, with bamboo regarded as one of the most valuable due to its intensive usage (Astutik et al., 2019), further aided by its abundance and sustainable character (Bahru and Ding, 2021). Bamboo is extensively utilized in the management of many physical conditions, such as gastrointestinal issues, wound recovery, high blood pressure, and gynaecological disorders. The plant's medicinal properties have been recognized for over six decades, with a resurgence of research interest in recent years (Benjamin et al., 2021).

Bamboo holds significant cultural and historical importance, and is deeply intertwined with the beliefs and traditions of various societies. Currently, bamboo serves many purposes, such as construction, medicinal applications, regular consumption, renewable biosource, and participation in certain national rituals among numerous ethnic groups residing in Southeast Asian nations (Bishnu Maya et al., 2024). Traditional medicine uses a variety of herbal elements, including roots, seeds, flowers, leaves, and many more, in combination with some medicinal goods to create particular recipes that can treat specific ailments (Bahru and Ding, 2021). Overall, bamboo shoots and the silicate components of bamboo are regarded as good therapeutic foods. They are rich in vitamins and possess medicinal properties that can have beneficial effects on patients experiencing fever and cough, amongst other ailments, providing both nutritional and healing benefits (Gabriel-Ajobiewe et al., 2021; Santosh et al., 2018).

In the past, people in Asia, Africa, and Latin America utilized different bamboo species to treat a wide range of illnesses. Traditional practitioners all over the world have used different parts of the bamboo plant to treat infections, inflammations, and even long-term illnesses for thousands of years (Benjamin et al., 2021).

The utilization of bamboo for medicinal purposes holds significant value in the traditional knowledge, culture, and economy of numerous Asian nations. According to tradition, it is widely accepted that bamboo plays a significant role in a person's life, symbolizing their birth and death (Nongdam and Tikendra, 2014). Asian bamboos belong to four primary genera: *Phyllostachys*, *Schizostachyum*, *Dendrocalamus*, and *Bambusa*. Bamboo's physical and mechanical qualities, quick growth, large yield, and capacity to sequester carbon have made it an essential component of Asian daily life (Bahru and Ding, 2021; Mustafa et al., 2021). Throughout its lifetime, bamboo can be utilized in various ways, reflecting its versatility and potential. This belief has been passed down through generations and is recognized globally. In certain regions of Asia, bamboo is commonly referred to as "kalpavriksha" or "mambu," which holds a significant place in their mythology.

The Asian traditional values have made bamboo a symbol that has fostered and maintained human existence through the close and mutually beneficial ties between humans and plants. These relationships encompass spiritual demands, traditional knowledge, and therapeutic practices, as well as technological and economic assistance. Mambu was transformed into *Bambusa*, serving as the foundation for Linnaeus' naming *Bambusa* and the term "bamboo" in English (Ahmad et al., 2021). Due to its substantial cultural and medicinal importance in diverse traditional healing methods, bamboo has been utilized in numerous ancient medical systems. Each system incorporates the distinctive characteristics of bamboo to enhance overall health and wellness.

2.3 MEDICINAL USES OF BAMBOO IN MAJOR TRADITIONAL HEALING PRACTICES

The utilization of bamboo for medicinal purposes is a significant aspect of traditional knowledge that has been passed down through centuries worldwide. Whether explicable or not, traditional medicine is the culmination of all knowledge, abilities, and procedures derived from indigenous theories, beliefs, and experiences from various cultures that are applied to the preservation of health as well as the avoidance, diagnosis, enhancement, or treatment of physical and mental illnesses (Benjamin et al., 2021).

The versatility and sustainability of bamboo influence its great potential for ethnomedicinal recognition in different cultural communities all over the world. This has led to the exploration of several applications in traditional medicine, especially in East Asia, India, and Southeast Asia (Benjamin et al., 2021; Kong et al., 2020; Nirmala, 2018). Virtually all parts (leaves, stems, shoots, sap, nodes, internodes, and roots) of bamboo plants are being explored (Benjamin et al., 2021) for diverse diseases and health benefits (Wróblewska et al., 2019). The preparation and application techniques of bamboo parts (regardless of the species type) utilized are dependent on the local knowledge and traditions of each region (Sujarwo, 2018).

The traditional knowledge on the therapeutic applications of bamboo, however, is declining due to a reduction in the elderly population, who possess a significant amount of this expertise, while the younger generation is increasingly learning modern knowledge. Native Americans and other Asian communities have long utilized bamboo plants as a means of treating a variety of illnesses and conditions in complementary and alternative medicine (CAM) systems, including TCM, Ayurveda medicine, homeopathy, chiropractic, and Unani medicine (Benjamin et al., 2021).

2.4 BAMBOO IN TRADITIONAL CHINESE MEDICINE

Traditional Chinese medicine is an ancient medicinal system that has been extensively utilized in China, Korea, Japan, Vietnam, Europe, the Middle East, and America for several millennia (Park et al., 2007). With 626 species, China has the most bamboo species, followed by India with 115 species. This explains why various portions of bamboo have traditionally been utilized for the treatment and healing of many maladies in ancient Chinese medicine, Ayurveda, Tibetan medicine, and several other traditional medical and healthcare systems (Benjamin et al., 2021). Because the therapeutic properties of medicinal plants used in TCM are generally ascribed to their unique chemical constituents, different parts of the same bamboo species may have different pharmacological activities. To date, over 50 different types of compounds have been isolated from bamboo, including lignans, flavones, phenolic acids, amino acids, and steroids. Many of these compounds possess interesting biological activities, such as antibacterial, antifungal, antioxidant, anti-inflammatory, neuroprotective, and hepatoprotection (Cruz et al., 2021; Tang et al., 2021).

The traditional use of bamboo and its derivatives, described in some ancient texts, like the Charaka Samhita and Sushruta Samhita, recommended bamboo manna ("vansmá") and tabasheer ("tugaksirsa") to treat a variety of diseases (Loukas et al., 2010). Despite the description of several traditional medicinal uses of bamboo, very few of them have been validated scientifically. In TCM, bamboo shavings ("Zhú S") are used to remove heat from the heart and stomach and to resolve phlegm (Ma et al., 2021; Xiong et al., 2013). Over the years, several bamboo species have been used in TCM and they are considered to have similar properties.

Bamboo has significantly influenced TCM, particularly in the area of acupuncture therapies. "Chenji Wan" is a Chinese herbal medicine made from bamboo stem shavings, tabasheer, and a combination of other herbs. It is used to treat insomnia, restlessness, and impaired vision resulting from phlegm mist clogging the orifices (Benjamin et al., 2021). "Gualou Zhishi Tang" is another TCM formula made from bamboo stem shavings and sap in combination with other herbs. It is used to help lessen difficult-to-expel thick phlegm. Bamboo leaves are used to treat fever with light-headedness, blurry vision, headaches, cough, phlegm, wheezing, pyrexia accompanied by headaches, and dizziness. In TCM, bamboo leaves are prepared and consumed in the form of tea. The bamboo leaves are categorized into two types: "Kuzhuye" which means bitter bamboo leaves and "Danzhuye" which means bland bamboo leaves (Panee, 2015). The dried leaves harvested from *Pleioblastus amarus* (Keng) Keng f. has a pungent smell and is slightly bitter. Because of its coldness, it cures febrile disease (Yang, 2002). The dried leaves of *Lophatherum gracile* Brongn are less effective in treating heat; however, it has been explored for urinary dysfunction and eczema (Yang, 2002). The Chinese also use bamboo poultices for the treatment of wound infections and ulcers (Panee, 2015). Bamboo shoots mixed with palm jaggery are also used to treat gynecological conditions (Wang et al., 2020). Furthermore, bamboo shoots are used to induce labor in the latest stage of pregnancy (Wang et al., 2020).

2.5 BAMBOO IN AYURVEDA

Ayurvedic medicine is an ancient Indian science that has been in practice since 2500 B.C. (Jaiswal and Williams, 2017). Meaning "science of life", it offers a complete system to live a long, healthy life, with programs to rejuvenate the body through diet and nutrition. Ayurveda describes instrumentalization (shastra karma), treatment methods such as employing monocot plants like bamboo (vanaspati), and how to cure intestinal parasites in humans and disorders like glycosuria/diabetes (madhumeha) (Pandey et al., 2013), to its recent application in the treatment of COVID-19 symptoms (Girija and Sivan, 2022). Since ancient times, many traditional drugs associated with bamboo have been used for treating fever and detoxification in Indian Ayurveda medicine and Chinese herbal medicine. Ayurveda recommends bamboo and its products such as "banslochan", "tabasheer", and "Sitopaladi Churna" for the treatment of various ailments (Nirmala, 2018). Tabasheer, or "banslochan", is used as a cooling tonic and aphrodisiac, to treat asthma, coughs, and other debilitating diseases (Nirmala, 2018). Bamboo also acts as a therapeutic agent in Ayurveda, with the ability to inhibit inflammation and enhance natural immunity for a long time (Patwardhan and Kumar, 2021). A "Bangshil" tablet is an Ayurveda medicine made from the decoction of bamboo leaves (40–50 mL) and used to treat genitourinary tract infections. "Dadimashtaka choorna" from powdered bamboo leaves is used to treat gastrointestinal complaints like diarrhea and indigestion. "Pugakanda", another Ayurveda medicine from blanched raw bamboo shoots (3–5 g), mixed with hot water or buttermilk, is used for treating abdominal pain, nausea, dyspepsia, gastritis, bleeding diseases, anemia, and constipation. "Sitopaladi Churna" is a commonly prescribed remedy for respiratory conditions such as tuberculosis, sore throats, coughs, colds, and congestion (Benjamin et al., 2021; Soni et al., 2013). In the Kanyakumari district located in the southern part of India, the Kani people utilize *Bambusa arundinaceae* seeds for boosting fertility (Azeem et al., 2020). The oil generated from the seeds can be used to cure polycystic ovary syndrome (Soumya et al., 2016). In India, bamboo shoots are also used to cure dyspepsia and diarrhea (Singhal et al., 2013).

2.6 BAMBOO IN UNANI MEDICINE

Hippocrates laid the groundwork for the Unani medical system, making it one of the oldest medical systems in the world between 460 and 477 B.C. (Masic, 2024). The four fundamental tenets of Greco-Arab medicine form the foundation of the Unani medical system. These are the fundamental ideas behind the four components that make up the human body: blood, phlegm, yellow, and black bile. These elements are essentially composed of environmental characteristics like heat, cold, moisture, and dryness, and the balance and up-equation of these factors govern how the human body functions (Priyanka et al., 2024). Tabasheer is a white, liquid substance that is said to be deposited in the bamboo internodes by Unani medicine. After dividing the bamboo, it later solidifies and takes on a circular, dirham-like shape. The superior quality, white-colored, lightweight product, "tabasheer e sadaf", also known as "tabasheer e kabood", is well made. If it possesses a faint blue hue resembling that of sadaf, it is known as "tabasheer sadafi". Friction between the bamboo stem and the wind leads to the formation of tabasheer inside the bamboo stem by burning (ehterak). Tabasheer is insoluble in water and occasionally tainted with charred bone ash (Mobeen and Ahmad, 2021). Tabasheer possesses astringent, antispasmodic, and aphrodisiac properties. It is also used to soothe coughing bouts, act as a coolant and stimulant, and relieve asthma (Benjamin et al., 2021; Soni et al., 2013).

2.7 BAMBOO IN HOMEOPATHY

Bamboo and its dilutions are commonly used in homeopathy. After studying the materia medica and other repositories, it was discovered that bamboo has a wide variety of medicinal uses due to its inherent healing properties (Jain et al., 2023). Symptoms related to the mind, spine, extremities, and menstrual are known to be cured by bamboo decoction in homeopathy (Guo et al., 2021). Numerous experienced homeopaths have made significant contributions to the advancement of knowledge in this field and have shared their expertise in using bamboo for various health purposes (Iqbal et al., 2022; Jain et al., 2023).

Ankylosing spondylitis, also referred to as bamboo spine, is an inflammatory, autoimmune disorder of the spine characterized by severe stiffness and calcification. One of the main treatments for this ailment is bamboo. Bamboo is similarly effective in all rheumatic disorders where stiffness is the main complaint (Jain et al., 2023). *B. arundinacea* is one of the homoeopathic medicines used to cure various types of arthritis, migraine headaches caused by stress and neck tension, and victims of traumatic brain injury. Others in need of these gentle remedies are those who suffer from Alzheimer's disease. Bamboo is an important homoeopathic remedy for disorders of the menstrual period, including endometriosis (Bartram, 2015).

2.8 OTHER MEDICINAL TRADITIONAL USES OF BAMBOO IN DIFFERENT CULTURES ACROSS THE GLOBE

In the present-day ethnomedicine, bamboo plants have been identified for several health benefits and explored differently in various communities across the globe as antioxidant, anticancer (Wang et al., 2019), anti-inflammatory, antimicrobial, antidiabetic, and antiulcer agents for health care and healing purposes (Cheng et al., 2023; Gagliano et al., 2022; Huang et al., 2018; Panee, 2015; Sangeetha et al., 2015). Several studies have revealed that bamboo extracts can cure high blood pressure (Azeem et al., 2020; Wang et al., 2021) and skin diseases (Zhang et al., 2016). Considering the antimicrobial and antioxidant activities evaluated in some *Phyllostachys* species like *Phyllostachys kwangsiensis*, *P. heterocycla* f. *gracilis*, and *P. heterocycla* cv. *Pubescens,* it was affirmed that their essential oils and other bamboo-derived products have a high antimicrobial capacity, portraying bamboo as a great potential for the production of antibiotics (Wang et al., 2021; Wróblewska et al., 2019; Xu et al., 2016), anticancer drugs (Wang et al., 2019), and antiviral agents (Li et al., 2020). Furthermore, bamboo dietary fibers are sources of cellulose (Kumar et al., 2017; Rasheed et al., 2020), explored for the production of pharmaceutical products like capsules and tablets (Isukuru et al., 2023).

The Tibetan and Indo-Persian medical systems have explored bamboo as an essential tonic for the treatment of respiratory diseases (Emamverdian et al., 2020; Kong et al., 2020). Gonorrhea is being treated with bamboo shoots in Indonesia (Benjamin et al., 2021). Furthermore, the Japanese and Koreans have explored bamboo for the treatment of anti-inflammatory diseases (Santosh et al., 2021). The leaves of *Phyllostachys nigra* (Lodd. Munro) are utilized similarly as "Danzhuye" in TCM by the Japanese and can be utilized to treat coughs, hypertension (Basumatary et al., 2017), cardiovascular diseases, arteriosclerosis, osteoarthritis, osteoporosis, and leprosy (Sangeetha et al., 2015; Wróblewska et al., 2019). The global use of bamboo for ethnomedicinal purposes in different parts of the world is summarized in Table 2.1.

TABLE 2.1
Global Uses of Bamboo in Ethnomedicine

Country	Areas	Botanical Name	Common/ Local Names	Parts Used/Preparation	Ailments	Mode of Administration	Reference
Bangladesh	Chittagong Hill Tracts	*Bambusa multiplex* (Lour.) Raeusch. ex Schult.f	Thirwa	Leaves: Decoction with *Terminalia belerica* sliced fruits, *Terminalia chebula* Roots: as paste	Fever Skin diseases such as itches	Orally Topically	Benjamin et al., 2021
	Chittagong Hill Tracts	*Bambusa oldhamii* Munro	O-anna	Leaves: Decoction Roots: Paste	Fever Skin diseases such as itches	Orally Topically	Benjamin et al., 2021; Hanif et al., 2009
	Muktipara, Chuadanga Sadar, Chuandanga District, Assam	*Bambusa tulda* Roxb.	Mitinga/Bans	Succulent part of stems and leaves: Decoction with fruits, *Phyllanthus emblica* fruits, *Solanum violaceum* sliced leaves, and *Drynaia quercifolia* roots. Roots: Decoction	Cardiovascular disorders, weakness of heart, arthritis, rheumatism Piles, constipation	Orally	Hanif et al., 2009; Rahmatullah and Biswas, 2012
	Kushtia District	*Bambusa arundinacea* (Retz.) Willd	Bans	Leaves: Decoction Seeds: Supplement	Coughs, colds, and joint pains Sperm potency	Orally	Azeem et al., 2020; Sharkar et al., 2013
	Rangamati Hill District, Chittagong Hill Tracts	*Bambusa bambos* (L.) Voss	Kanta bans	Stems, leaves roots, sprouts, barks: Decoction	Inflammatory conditions, leukoedema, colds, and coughs	Orally	Rudra et al., 2021
Cambodia	Phnom Penh; Kandal; Takeo; Kampong Speu	*Bambusa* spp.	Russey Srok, Russey Prey	Woods: Decoction	Liver disorder	Orally	Chassagne et al., 2017
China	Chuxiong, Xishuangbanna, Yunnan Province	*Phyllostachys nigra* (Lodd ex Lindl.) Munro	Black bamboo	Leaves and shoots: Decoction	Malaria, coughs, colds, cancers	Orally	Panee 2015; Jaya et al., 2019
	NA	*Bambusa vulgaris* Schrad. ex J.C. Wendl	NA	Different bamboo parts boiled with *Nicolsonia renifolia, Dillenia indica and Mahonia spp.*	Hepatitis	Orally taken with sugar and rice	Benjamin et al., 2021

	Sansui County	*Bambusa pervariabilis* McClure	Zhuru	Shavings: Decoction	Clearing of heat, stops bleeding	Orally	Luo et al., 2020
	NA	*Dendrocalamus hamiltonii* Nees & Arn. ex Munro	Aqyul	Culms: Boiled with ginger and garlic Water extracted from burnt culm	Food poisoning and edema Fever treatment	Orally	Benjamin et al., 2021
	NA	*Phyllostachys edulis* (Carrière) J. Houz	Maozhu Moso	Leaves: Prepared as extracts and oil	Cancer	NA	Del Rosario et al., 2012; Jin et al., 2011
	NA	*Gigantochloa albociliata* Munro	NA	Roots and buds: Decoction	Dysmenorrhea, irregular menstruation, postpartum blood clots	Orally	Benjamin et al., 2021
	NA	*Schizostachyum funghomii* McClure	NA	Nodes: Decoction with *Areca catechu* petioles	Headaches, body aches, common colds, increases appetite	Orally	Benjamin et al., 2021
	NA	*Dendrocalamus* spp.	NA	Roots: Decoction with leaves of *Areca catechu* fruit or *Alpinia galanga*, *Gendarussa vulgaris* roots, and *Adhatoda vasica* roots	Induces urination	Orally	Benjamin et al., 2021
	NA	*Lophatherum gracile* Brongn	Bamboo leaf	Leaves and Stems: Extract	Stomach heat, urinary dysfunction, and eczema	Orally	Yang, 2002; Panee, 2015
	NA	*Pleioblastus amarus* (Keng) Keng f	NA	Leaves: Extract	Inflamed heart and lungs, fevers	NA	Kiruba et al., 2007; Panee, 2015; Ren et al., 2019
	NA	*Bambusa textilis* Mc Clure	NA	Leaves: Juice	Infectious diseases, cancers, obesity	Orally	Liu et al., 2016; Nirmala et al., 2018; Wang et al., 2012
India	Himalayas, Tiruvallur District, Tamil Nadu	*Bambusa vulgaris Schrad* ex J.C. Wendl.	Golden Bamboo/ buloh gading	Leaves: Decoction	Fevers and sore throats, diabetes	Orally with or without additives	Esakkimuthu et al., 2016; Panee, 2015
	Kanyakumari district, Himalayas, Palamalai Region of Eastern Ghats	*Bambusa bambos* (L.) Voss	Choya bans/ Moongil	Leaves and stems: Extract Seeds: Prepared as paste	Leukodema, urinary discharges, osteoporosis, skin diseases, rheumatism	Orally Topically	Kumar et al., 2023; Sawarkar et al., 2023; Silambarasan and Ayyanar, 2015

(Continued)

TABLE 2.1 *(Continued)*
Global Uses of Bamboo in Ethnomedicine

Country	Areas	Botanical Name	Common/ Local Names	Parts Used/Preparation	Ailments	Mode of Administration	Reference
	Una, Kangra, Hamirpur, Solan, Bilaspur and the sub-Himalayan tract of river Yamuana eastwards and Jammu region	*Bambusa nutans Wall. ex Munro*	Naal, Mokal	Leaves: Paste Young shoots and roots Sap	Skin diseases such as chicken pox and smallpox, cardiovascular diseases Applied as a paste to dog bites and cure for ringworms Remedy for tuberculosis	Topical	Kumar et al., 2023; Nirmala et al., 2014; Richa et al., 2012; Saini et al., 2015
	Savandurga Forest, Bangalore Rural District, Karnataka	*Bambusa bambos* (L.) Voss	Bidiru	Bamboo shavings or sap prepared with rice	Digestive issues, nausea and vomiting, diarrhea, abdominal pain, fevers, and respiratory problems	Orally	Udayan et al., 2005
	Pudukottai District, Tamil Nadu	*Bambusa arundinacea* (Retz.) Willd	Moongil	Young leaves, terminal buds: Paste prepared by grinding with turmeric leaves and *Aloe vera*	Fractured bone	Topically	Anbarashan et al., 2011
	Tirunelveli Hills of Western Ghats, Pachamalai Hills of Tamil Nadu	*Bambusa arundinacea* (Retz.) Willd	Moongil	Seeds and seed oil: Prepared as a paste	Rheumatism	Topically	Ayyanar and Ignacimuthu, 2011
	Udhampur District of Jammu and Kashmir Andhra Pradesh	*Bambusa arundinacea* (Retz.) Willd	Baans/Venduru	Roots: Decoction Leaves: Decoction	Kidney stones Asthma	Orally	Bhatia et al., 2014; Savithramma et al., 2007
	Amravati District, Maharashtra, Himalayas	*Dendrocalamus strictus* (Roxb) Nees	Math/Kaban	Gum: Soaked in water Gum as powder Shoots: Decoction Tabasheer: Tonic	Earaches Tuberculosis Fatigue	Topically as ear drop Orally taken with water	Jagtap et al., 2006; Kumar et al., 2023
	Chattisgarh	*Bambusa* spp.	Bans	Leaf juice, bark, and seeds: Decoction	Cancer	Orally taken with honey	Jain and Jain, 2010

Country	Location	Species	Local name	Parts used and preparation	Ailments	Mode of administration	References
	Chittorgarh District, Rajasthan, Manipur, Arunachal Pradesh	*Dendrocalamus strictus* (Roxb.) Nees	Bans/Saneibi	Leaf juice, silica content in the leaves: Decoction Young shoots: Prepared as poultice and paste	Uterus cleaning Injuries Dislodgment of worms, ulcers	Orally Topically	Jain et al., 2005; Sharma and Borthakur, 2008; Singh and Pandey, 1998; Singh et al., 2003
	Atoor Village, Kanyakumari District, Tamil Nadu	*Bambusa arundinacea* (Retz.) Willd	Mulmunkil	Leaves: Juice Roots: Decoction Juvenile shoots: Cooked as curry Whole plants prepared as paste (with turmeric and *Areca catechu*)	Coughing in children Skin diseases Indigestion Contusion and swellings	Orally taken with honey Orally Orally Topically	Jeeva and Femila, 2012
	North Cachar Hills of Assam	*Dendrocalamus giganteus* Munro	Wa Mishel, Kepaii/Mpin	Fresh culm: Made into wool Shoots and culm: Juice	Fresh cuts and wounds Diabetes and obesity	Topically Orally	Medhi et al., 2010; Wang et al., 2020
	Kathiyavadi Village, Vellore District, Tamil Nadu	*Bambusa vulgaris* Schrad. ex J.C. Wendl.	Mungil	Leaves: Extract	Reduced sperm count	Orally	Sathiyaraj et al., 2012
	Apatani Plateau, Arunachal Pradesh	*Cephallostachium capitatum* Munro	Yabing	Bamboo water	Dysentery and diarrhea	Orally	Sundriyal et al., 2002
	Subhartipuram, Meerut District, Uttar Pradesh	*Bambusa vulgaris* Schrad ex J.C. Wendl.	NA	Young shoots: Cooked	Rheumatism and malaria	Orally taken with rice	Singh et al., 2020
	Northeast	*D. giganteus,* Munro	Giant bamboo	Culm: Prepared as steroid drugs	Aphrodisiac	Orally	Bhatt et al., 2005
	Northeast	*Melocanna baccifera* (Roxb.) Kurz	NA	Tabashir: Prepared as tonic	Respiratory diseases	Orally	Azeem et al., 2020; Singh, 2019
	NA	*Schizostachyum pergracile* (Munro) R.B. Majumdar	NA	Leaves: Infusion	Stomach pain	Orally	Singh, 2019
Indonesia	Weh island, Pulang Pisau Regency, Central Kalimantan	*Bambusa vulgaris* Schrad. ex J.C. Wendl.	Trieng kuneng/ Awi Haur/ bambu kuning	All parts of yellow varieties Leaves: Decoction	Gout and gastric ulcers, fevers, and liver disease Malaria	Orally	Ritonga et al., 2023; Novaryatiin and Indah, 2019
	Weh Island	*Bambusa spinosa Roxb.*	Trieng duroe	Culms	Gout and liver disease	Orally	Ritonga et al., 2023

(Continued)

TABLE 2.1 (Continued)
Global Uses of Bamboo in Ethnomedicine

Country	Areas	Botanical Name	Common/ Local Names	Parts Used/Preparation	Ailments	Mode of Administration	Reference
	Weh Island	*Bambusa tuldoides* Munro	Trieng biasa	All parts Stems: Extract from shavings	Throat discomfort Fatigue	Orally	Ritonga et al., 2023; Zhang et al., 2006; Panee 2015
	Weh Island, Ngadisari Village, Sukapura District, Probolinggo	*Dendrocalamus asper* (Schult. & Schult.f.) Baker ex K.Heyne	Trieng betung/ bamboo betung	Young shoots Stems: Pounded	Hypertension and fever Muscle soreness	Orally	Ritonga et al., 2023; Jadid et al., 2020
	Weh Island	*Schizostachyum brachycladum* (Kurz ex Munro) Kurz	Buloh leumang	Culms	Lower back pain	NA	Ritonga et al., 2023
	Weh Island	*Schizostachyum silicatum* Widjaja	Igeuh	Leaves	Asthma	NA	Ritonga et al., 2023
Japan	Hokkaido	*Sasa senanensis* (Franch. & Sav.) Rehder	Japanese bamboo (*Kumaizasa)*	Leaves: Extract (Sasa health)	Tumors	Orally	Nirmala et al., 2018; Patel et al., 2016; Das, 2019; Santosh et al., 2021; Panee, 2015
	NA	*Phyllostachys parvifolia* C.D. Chu and H.Y. Chou	NA	Leaves	Tumors	NA	Seki and Maeda, 2010
Korea	Eastern Mountainous Region of North, Gayasan National Park	*Phyllostachys bambusoides* Siebold & Zucc	Wangdae/ Bamboo timber	Sprouts: Decoction Root: Prepared as sweet juice with fermented rice Culm: Exract Leaves and stems: Prepared as salt	Pollakiuria Lumbago Cardiovascular diseases Cancer	Orally	Lee et al., 2008; Singhal et al., 2013; Kim and Song, 2011; Song et al., 2014
	NA	*Phyllostachys henonis*	Tabasheer	Tabasheer: Prepared as powder	Excess salivation, respiratory disease, hemoptysis, and acute infections	Orally	Das et al., 2021

	Gayasan National Park	*Sasa coreana* Nakai	Sinidae	Leaves: Decoction and infusion	Common cold	Orally	Benjamin et al., 2021; Song et al., 2014
	Eastern Mountainous Region of North	*Sasa borealis* (Hack.) Makino	Joritdae	Root stems and leaves: Decoction	Hyperthermia, hypoglycemia, common cold	Orally	Benjamin et al., 2021
Laos	Nakai District, Khammouane Province	*Gigantochloa parvifolia* (Brandis ex Gamble) T.Q. Nguyen	Mai Sod	Leaves: Steam sauna, bath	Postpartum recovery, anemia, puerperal fever, lactagogue, perineal healing, uterus withdrawal	Topically	de Boer et al., 2011
	Paksan District, Bolikhamsai Province	*Bambusa vulgaris* Schrad. ex J.C. Wendl.	Mai Sang Kham	Roots: Boiled (with the root of Mak Feuang Phu)	Kidney stones	Orally	Libman et al., 2006
Malaysia	Kampung Menunggui, Kota Belud District	*Bambusa* sp.	Buluh	Young shoots: Boiled	Postpartum diet	Orally	Awang-Kanak et al., 2018
Nigeria	NA	*Bambusa vulgaris Schrad* ex J.C. Wendl.	NA	Leaves: Extract Leaves: Boiled or raw	Induce abortion, relieves labor pain Postpartum cleansing in livestock	Orally	Lans et al., 2007; Panee, 2015
Philippines	Central Philippines, Guimaras Island	*Schizostachyum lumampao* (Blanco) Merr	Bagakay	Stems: Burnt to ash	Infant's freshly cut umbilical cord	Topically	Ong and Kim, 2017
	Central Philippines, Guimaras Island	*Bambusa vulgaris* Schrad. ex J.C. Wendl.	Kawayan	Leaves: Boiled	Postpartum wash	Topically	Ong and Kim, 2017
	Batan island of Batanes	*Bambusa vulgaris* Schrad. ex J.C. Wendl.	Kawayan	Stems: Decoction Stems: Burnt with coconut oil to form ash	Allergies Skin diseases	Orally Topically	Abe and Ohtani, 2013
Thailand	Chawang District, Nakhon Si Thammarat Province Khao Luang Region	*Bambusa bambos* (L.) Voss	Phai Pa	Leaves: Prepared as powder and as bolus Young stems: Decoction as potion	Constipation Bilharzia leprosy and lipoma	Orally	Neamsuvan et al., 2016; Srisawat et al., 2016

NA* Not available

2.9 TRADITIONAL MEDICINAL PRODUCTS OF BAMBOO

2.9.1 Bamboo Salt

Bamboo salt is made by heating salt in bamboo nine times at high temperatures (850°C–1500°C) using pine firewood. During this process, impurities are removed and inorganic nutrients are retained at reasonable levels for the treatment of diseases (Sawarkar et al., 2023). Bamboo salt is produced as pills in Korea for the treatment of inflammatory diseases as well as bacterial and fungi infections (Das et al., 2021; Wang et al., 2020).

2.9.2 Bamboo Juice

Among the various bamboo species, *Phyllostachys edulis*, known as Moso bamboo, is mostly used for making juice. The procedure used for juice production involves the collection of fully developed bamboo between 2 and 3 years of age, followed by trimming of the stalks with sterilized devices, after which the internodes are drilled for sap collection into a sterilized jar. The sap is then filtered with a sterilized fine sieve to remove contaminants in order to obtain the pure and nutritious bamboo juice. Bamboo juice is used for treating bronchial asthma, cholecyst dysfunctions (Das et al., 2021), parasitic worms (Shukla et al., 2012), icterus (Kumar et al., 2017), ulcers, and osteoporosis. A dosage of 60 mL daily is prescribed for adults in the treatment of ulcers (Gao et al., 2020).

2.9.3 Bamboo Tea

This is produced through decoction. Bamboo material (leaves, shoots) is boiled for up to 2 hours for extraction of the useful content. Decoction from leaves of *B. vulgaris* is used for the treatment of fevers, sore throats, stomach disorders, and diarrhea; regulates menstrual cycles; and kills intestinal worms (Shukla et al., 2012). A cup can be consumed two to three times daily with (honey or lemon) (Panee, 2015) or without additives. Bamboo tea can also be used externally to treat smallpox, chicken pox, and skin diseases (Kumar et al., 2023). Decoction from bamboo shoots can be used with honey to cure respiratory diseases (Panee, 2015).

2.9.4 Bamboo Leaf Extract

Bamboo leaf extract has been recognized by the Chinese Ministry of Health and explored as a novel food fortification with a maximum dose of 0.5 g/kg in food products (Wang et al., 2020). It is used to treat osteoarthritis, osteoporosis (Banik, 2016; Sangeetha et al., 2015), respiratory dysfunction, and epilepsy (Gupta and Ranjan, 2016; Panee, 2015). The phenolic compounds present in the extract from bamboo shoots and leaves of some genera like *Phyllostachys, Dendrocalamus, Bambusa,* and *Gigantochloa* are responsible for their antioxidant and cardioprotective properties (Chongtham and Bisht, 2021; Nirmala et al., 2018; Wróblewska et al., 2019) for the treatment of inflammatory diseases, cardiovascular conditions, cancer, hypertension, and arteriosclerosis (Ma et al., 2022).

2.9.5 Bamboo Paste

Bamboo is consumed as a paste and can be used for food fortification due to the presence of phytosterol (Santosh et al., 2019). It serves as a preservative (Akinlabi et al., 2017). The shoot paste is applied to dog bites and the root paste is used to treat ringworm (Kumar et al., 2023), abscesses, and itches (Hanif et al., 2009). Bamboo paste is also used to cure infections from fungi (Singhal et al., 2013).

2.9.6 Bamboo Sap

This is a liquid product with various health uses. It is named Zhuli and is produced from fresh bamboo under a low temperature. Bamboo sap has been explored for medicinal purposes, pharmaceutical applications, and for food fortification (Chongtham and Bisht, 2021; Wróblewska et al., 2019). It is used to reduce phlegm-heat (Panee, 2015; Sangeetha et al., 2015) and treat neurological disorders such as epilepsy and paralysis (Panee, 2015). Bamboo sap is also used to cure phthisis (tuberculosis), jaundice in Indonesia, and skin diseases (Choudhury et al., 2012; Kumar et al., 2023). Bamboo sap is used to treat injuries and to hasten the healing of wounds because of its antibacterial property (Sawarkar et al., 2023).

2.9.7 Bamboo Poultice

Bamboo poultice is made from juvenile shoots. It is used for cleaning wounds and treating infections (Devi et al., 2017; Kumar et al., 2023; Panee, 2015). Bamboo poultice effectively dislodges worms and ulcers (Benjamin et al., 2021; Shukla et al., 2012).

2.9.8 Bamboo Char

Bamboo charcoal is manufactured by introducing dried bamboo into a kiln with a flame tunnel, heating the bamboo slowly to 600°C to 900°C, cooling to room temperature, and withdrawing the charcoal (Wang et al., 2012). Several bamboo plants have been used for bamboo charcoal, including *B. bambos, Dendrocalamus brandisii, D. stocksii, D. strictus, Phyllostachys pubescens, P. nigra*, and *P. bambusoides* (Sawarkar et al., 2023; Wang et al., 2015). Bamboo char has been explored for health purposes (Sawarkar et al., 2023) and for the production of medicine (Chaturvedi et al., 2023). Also, bamboo charcoal derived from the culm of some bamboo species can be used for the removal of pollutants from drinking water (Baskar et al., 2022; Isukuru et al., 2023; Kumar et al., 2023), and as a blood purifier (Chaturvedi et al., 2023).

2.9.9 Bamboo Vinegar

This is a brown-red transparent liquid produced during the heating of bamboo biomass in little or no presence of oxygen (Wang et al., 2015; Wróblewska et al., 2019). It is known as "Chikusakueki" in Japan, "Zhook-Ryeok" in Korean, and "Zhucu" in Chinese (Panee, 2015). Bamboo vinegar constitutes over 200 kinds of chemical compounds, such as phenols, alkanes, alcohols, and aldehydes, among others (Wang et al., 2012), and has been used for pharmaceutical purposes (Santosh et al., 2021; Sawarkar et al., 2023). Bamboo vinegar acts as a detoxifier and relief for poison ivy, aids digestion, and regulates cholesterol levels in the body. It is also used to treat snake and insect bites (Wang et al., 2015; Wróblewska et al., 2019), as well as treating skin diseases (Panee, 2015; Sawarkar et al., 2023). Bamboo vinegar is used as a food additive in food production (Wang et al., 2015). Ho et al. (2013) reported that bamboo vinegar could also ameliorate inflammatory diseases.

2.9.10 Bamboo Oil

This is extracted from the shoot of bamboo. Bamboo oil can be used to treat prostatitis (Lu et al., 2011). Reports suggest a cholesterol-reducing property in bamboo oil because of the presence of phytosterol in bamboo shoots (Lu et al., 2011; Sawarkar et al., 2023; Singhal et al., 2013)

2.9.11 Bamboo Shoot Soup

This is prepared as a soup to speed up measles appearance for fast recovery (Kumar et al., 2023; Singhal et al., 2021). It is consumed twice daily by the Chinese and Indians for the treatment of gastric ulcers and asthma (Kumar et al., 2017; Wang et al., 2020).

2.9.12 Bamboo Shavings

This is a sweet product with a cooling and medicinal effect (Chongtham and Bisht, 2021; Wróblewska et al., 2019). It is known as "Zhuru" in TCM (Sawarkar et al., 2023). It is prepared by shaving the dried bamboo stalk and cutting it into pieces (Panee, 2015; Wróblewska et al., 2019). This is used to cure nausea, vomiting, and hiccups resulting from stomach heat conditions (Panee, 2015), colds, and fevers (Das et al., 2021).

2.9.13 Finely Ground Powder (*Sitopaladi Churna*)

Sitopaladi churna is prepared by grinding together eight parts of bamboo tabasheer (powdery form) of *B. arundinacea*, four parts of *Piper longum* L. (black pepper), two parts of *Elettaria cardamomum* (L.) Maton (cardamom), and one part of *Cinnamomum zeylanicum* Blume (cinnamon) (Balkrishna et al., 2023; Das et al., 2021). This powdery mix is effective for coughs and colds associated with respiratory disorders (Nirmala, 2018). It can be mixed with honey or ghee for dry cough treatment (Panee, 2015) and tuberculosis. The dosage is 3–12 g (Shukla et al., 2012).

2.10 NUTRITIONAL COMPOSITION OF BAMBOO

Over the years, edible bamboo has been recognized as a plant-based diet (Santosh et al., 2021) in the food sector due to its high nutritional value found mostly in the stems, shoots, and leaves (Atanda, 2015). Several reports reveal a high amount of protein, amino acids, carbohydrates, vitamins, and minerals (Chongtham and Bisht, 2021; Isukuru et al., 2023; Santosh et al., 2021) in bamboo shoots. However, the fat content is low. The nutrients in bamboo shoots are highly dependent on the bamboo species and maturity (Wang et al., 2020). Most of the essential amino acids present in bamboo shoots for human health include lysine, serine, methionine, histidine, isoleucine, leucine, phenylalanine, and tyrosine, ranging between 3% and 7% of the recommended dietary allowance (RDA) (Sawarkar et al., 2023; Wang et al., 2020). Reports confirm that the freshly collected shoots of some bamboo species, such as *Dendrocalamus giganteus*, *P. pubescens*, and *Bambusa tulda*, are richer in amino acid content compared to their older shoots (Das, 2019).

Macronutrients such as potassium, phosphorus, sodium, calcium, and magnesium, as well as micronutrients like cobalt, copper, nickel, manganese, selenium, iron, and zinc, are abundantly found in bamboo shoots (Chongtham et al., 2021; Christian et al., 2015; Kumar et al., 2023; Wang et al., 2020). Bamboo salt has been reported to cure gastric ulcers due to its richness in several minerals, as highlighted above (Chongtham et al., 2021; Sawarkar et al., 2023). Most studies reveal potassium as the most abundant element, followed by phosphorus and magnesium. However, the variation of these elements depends on the bamboo species (Wang et al., 2020). Furthermore, bamboo possesses high amounts of thiamine, niacin, vitamin A, vitamin B1, vitamin B2, vitamin B3, vitamin B6, vitamin C, and vitamin E (Das, 2019; Kumar et al., 2023). Some studies report that fresh bamboo shoots have a higher amount of vitamin C than vitamin E (Wang et al., 2020).

Studies have revealed high phytosterol and fiber contents but low cholesterol in their leaves, shoots, and roots. Phytosterols are plant sterols similar in function and structure to cholesterol. Phytosterols reduce the risk of cardiovascular disease by lowering cholesterol levels in the body, as cardiac arrest results from high cholesterol (Nirmala et al., 2014; Sawarkar et al., 2023). Edible bamboo shoots are rich in lignin, which is associated with increased dietary fiber content (Xu et al., 2019). High dietary fiber was confirmed in *Melocanna baccifera, B. bambos, B. tulda, Dendrocalamus asper, D. giganteus*, and *Dendrocalamus hamiltonii*. It was further reported that exposure of these bamboo shoots to fermentation increased the fiber content (Das, 2019). Fermentation also increases other nutrients in bamboo (Benjamin et al., 2021).

According to the U.S. Department of Agriculture (USDA), a single cup of bamboo shoots is enough to meet the body's daily fiber requirement (USDA, 2022). This can prevent heart diseases, address gastrointestinal disorders, regulate blood pressure, and protect against cancer (Das, 2019; Santosh et al., 2021) due to the rich lignin content known to be a radical scavenger (Bhujbal et al., 2022). This establishes the fact that bamboo species possess therapeutic properties associated with antioxidant, antimicrobial, anti-inflammatory, antihelminthic, antidiabetic, and antiulcer activities (Singhal et al., 2013). These nutritional and therapeutic properties position bamboo as a great potential for healthy living (Chongtham et al., 2021).

Despite the presence of high nutritional compounds in bamboo, fresh bamboo shoots contain antinutritional compounds such as cyanogenic glycosides, thiocyanate, glucosinolate, phytate, and tannins that can be harmful to the body (Wang et al., 2020). Reports reveal high cyanogen content in the older shoots of some bamboo species compared to their fresh juvenile shoots (Pandey and Ojha, 2013), and it was confirmed that cyanogen levels are dependent on the type and age of the bamboo species (Wang et al., 2020). The persistent consumption of unprocessed bamboo shoots can lead to impairment of the pancreas, thyroid dysfunction, and cerebrospinal nervous system disorders (Sarkar et al., 2020).

It is important for young shoots of bamboo to undergo a detoxification process before consumption (Chongtham et al., 2021; Das, 2019; Kong et al., 2020; Rawat et al., 2016; Singhal et al., 2021). This is to expedite the removal of hydrogen cyanide, produced from taxiphyllin (Chongtham et al., 2021; Rawat et al., 2016). The process can either be by soaking or pickling (Chongtham and Bisht, 2021; Wang et al., 2020; Watanabe et al., 2021), drying, fermentation, or canning (Kong et al., 2020), which increases the shelf life of bamboo shoots (Chongtham et al., 2021; Das, 2019). Fermentation is the most commonly considered method (Singhal et al., 2021) for effectively removing cyanogenic compounds and improving the therapeutic properties of bamboo shoots. Additionally, boiling bamboo shoots for 20 minutes has been reported to efficiently remove 97% of cyanogenic content (Chongtham et al., 2021).

Recently, the global food market, popular Chinese cuisine, and the integration of cultures among various regions have contributed to the high rate of bamboo consumption. The annual bamboo consumption worldwide is over 2 million tons (Wang et al., 2020), where China recorded the largest variety of bamboo species (Das, 2019), followed by India, with a great potential to commercialize bamboo shoot products, especially fermented shoots (Das, 2019; Kikon, 2021). Fermented bamboo shoots are regularly consumed by many tribal communities in India (Manipur, Meghalaya, Sikkim, and Mizoram) with various preparation techniques. They are consumed as curry, pickles, or soup, and have been explored as food additives and for the fortification of food for quality purposes in the food system (Chongtham et al., 2021; Singhal et al., 2021).

Nepalese cuisine features bamboo shoots in various forms and recipes. For instance, fermented bamboo shoots are served with vegetables. “Alu Tama” consists of fermented bamboo shoots with turmeric and oil, cooked with potatoes and served with rice, while “Aloo Bodi Tama” consists of pickled bamboo shoots served with black-eyed beans and Tusa made from cooked juvenile bamboo shoots (Sawarkar et al., 2023). In India, tribal communities, including those in Assam, Amils, and Sambalpur, also ferment their bamboo shoots for consumption. Some parts of India use bamboo to make pancakes with rice flour to avoid crumbliness (Sawarkar et al., 2023). In Indonesia, a common dish known as Gulai Rebung is prepared as a curry from bamboo shoots by boiling the sliced bamboo shoots in spiced, thickened coconut milk (Santan). Sayur Lodeh is prepared by cooking bamboo shoots with vegetables in coconut milk. Lun Pia is a dish that involves frying bamboo shoots served with cooked vegetables (Sawarkar et al., 2023).

2.11 CONSERVATION OF BAMBOO FOR SUSTAINABLE USE

The increasing demand for bamboo products has led to overharvesting and habitat destruction, threatening the long-term sustainability of bamboo forests. Sustainable harvesting and conservation practices are essential to ensure the continued availability of bamboo resources for future generations. Moreover, since bamboo has been a cornerstone of traditional medicine for centuries, with various species possessing unique ethnomedicinal properties, the preservation of this knowledge is crucial for the development of new medicines and the conservation of cultural heritage. The morphological, ecological, and geographical diversity of bamboo species is remarkable, and in many places, plantations of a few species are displacing the local diversity of bamboo. Bamboo species introduction, management, and conservation may offer ways to preserve species diversity and encourage the possibility of using the resources for the long-term provision of various goods and services (Canavan et al., 2017).

Ex situ collection of bamboo species is possible all year round. This might offer techniques for removing propagules for seed propagation, tissue culture, and population building (Mustafa et al., 2021; Singh et al., 2013). The collection can be combined with other collections of single-site species to create a network that is complementary and influenced by each other. To advance sustainable development, especially in developing economies where there is a potential for collecting and analyzing samples for various purposes, societies responsible for collecting and providing services could exchange taxon protection records through a global biodiversity information database (Singh et al., 2013).

The development of bamboo-based enterprises is a result of the introduction of harvesting to convert bamboo into a variety of products. Excessive exploitation of bamboo resulted in the depletion of bamboo resources in certain nations. Collaboration between ecologists, botanists, and biologists is essential for the sustainable use and management of bamboo forests. To preserve the functions of bamboo plantations and prevent soil erosion, stabilize slopes, and deplete water supplies, many ecologists have advised managing them properly (Abebe et al., 2021; Akwada and Akinlabi, 2016). Documenting and preserving traditional knowledge about bamboo's ethnomedicinal uses are vital. This can be achieved through community-based initiatives, such as oral history projects and knowledge-sharing workshops.

2.12 HARVESTING TECHNIQUES FOR SUSTAINABLE BAMBOO MANAGEMENT

Sustainable harvesting practices, such as selective cutting and reforestation programs, must be implemented. Bamboo can be harvested either by cutting the entire culms or by first removing the branches and then dividing them, depending on the intended use and the specific species of bamboo (Sertse et al., 2011). Regeneration in both circumstances usually necessitates several months or years for the species to fully restore its previous size. This has an impact on the ecological stability of the environment and the long-term viability of bamboo populations. Forest industries and local users can harvest sustainable yields by utilizing bamboo management that does not involve any cutting regime or even non-extractive hazards to bamboos over time (Mizuki et al., 2014). Furthermore, cutting just appropriate ripe culm may be a necessary precondition for using bamboo in a non-destructive, sustainable manner (Maulana et al., 2019). Eliminating these fully grown culms will enhance the efficiency of water and nutrient usage within the nearby bamboo population, both in the present year and in the long-term process of ecological succession. Underdeveloped culms will not have reached their full size, and cutting will hinder the subsequent development of the culm (Anokye et al., 2014). The beneficial end products are primarily prepared and consumed during the years after the natural senescence of the culm after the full-sized culms have been harvested (Anokye et al., 2014). Additionally, conservation efforts, like protected areas and community-led management initiatives, should be established.

2.12.1 Non-Destructive Harvesting Methods

The optimal period for harvesting involves gathering fully developed culms that have recently concluded their vegetative phase, during which they have amassed the maximum in reserves. Culms that have emerged from the soil surface but have not yet fully developed their branches (tops fully crown out) are considered to be in a state suitable for harvesting (Darabant et al., 2016). The time interval from when bamboo penetrates the soil surface to when they start flowering is crucial in determining when they tend to senesce. Utilizing this phenophase as a reliable warning of approaching senescence ensures minimal depletion of valuable bamboo biomass during harvesting operations, thereby maximizing the carbon yield during the vegetative inter-culm phase (Darabant et al., 2016). During this phase, a substantial quantity of organic content in leaves is transported to the culm, which is then followed by stages of flowering and reproduction. In rhizomes and roots, most carbon and nutrients from the parent plant's reserves are transported to the seeds (Mizuki et al., 2014). It is imperative for growers to understand this unique growth-reproductive cycle to maximize the use of the matured culms (Sertse et al., 2011). With a high frequency of occurrence, and at irregular intervals, bamboo blooming causes a huge die-off and draws a variety of pests. With this in mind, most nations harvest it early – mostly 5 to 10 years (Virtanen et al., 2022).

2.13 CONCLUSION

Evidence from the literature review on the global status of bamboo in folk and traditional medicine highlights the vital role of bamboo species in promoting well-being and the overall health of humans across the world. Because of their widespread distribution and numerous species, bamboo has been used in different traditional medicinal systems, including Chinese, Ayurveda, African, Japanese, and in other parts of the world. Bamboo species are used in the ethnomedicinal treatment of illnesses ranging from digestive disorders, respiratory issues, inflammation, and promoting wound healing, among others. As researchers continue to study the bioactive components and pharmacological properties of bamboo, its therapeutic significance in traditional and modern medicine is likely to grow. However, there is a need to prioritize sustainable harvesting practices of bamboo species for their continued use in traditional and modern medicine.

REFERENCES

Abe, R. and Ohtani, K. 2013. An ethnobotanical study of medicinal plants and traditional therapies on Batan Island, the Philippines. *Journal of Ethnopharmacology 145*: 554–565.

Abebe, S., Minale, A.S. and Teketay, D. 2021. Socio-economic importance of the bamboo resources in the Lower Beles River Basin, North-Western Ethiopia. *Environment, Development and Sustainability 24*: 12162–12181. https://doi.org/10.1007/s10668-021-01938-4

Ahmad, Z., Upadhyay, A., Ding, Y., Emamverdian, A. and Shahzad, A. 2021. Bamboo: Origin, habitat, distributions and global prospective. In: Ahmad, Z., Ding, Y. & Shahzad, A. (eds) *Biotechnological Advances in Bamboo*. Springer, Singapore. https://doi.org/10.1007/978-981-16-1310-4_1

Akinlabi, E.T., Anane-Fenin, K. and Akwada, D.R. 2017. *Bamboo: The Multipurpose Plant* (1st ed). Springer, Switzerland. https://doi.org/10.1007/978-3-319-56808-9

Akwada, D.R. and Akinlabi, E.T. 2016, August 1–15. Economic, social and environmental assessment of bamboo for infrastructure development. *International Conference on Infrastructure Development in Africa*, 1–15.

Anbarashan, M., Parthasarthy, N. and Padmavathy, A. 2011. Ethno-floristic survey in sacred groves, Pudukottai district, Tamil Nadu- India. *Journal of Medicinal Plant Research 5*: 439–443.

Anjum, A., Tabassum, A. and Ambar, K. 2019. Tabasheer (*Bambusa arundinaceae* Retz.) a plant origin drug of Unani medicine–A review. *Journal of Ayurvedic and Herbal Medicine 5*: 31–34.

Anokye, R., Bakar, E.S., Abare, A.Y., Kalong, R.M. and Muhammad, A. 2014. The difference in density along the bamboo culms of *Gigantochloa scortichinii* and *Bambusa vulgaris*. *International Journal of Emerging Technology and Advanced Engineering 4*(10): 3–8.

Astutik, S., Pretzsch, J. and Kimengsi, J.N. 2019. Asian medicinal plants' production and utilization potentials: A review. *Sustainability 11*(19): 1–33.

Atanda, J. 2015. Environmental impacts of bamboo as a substitute constructional material in Nigeria. *Case Studies in Construction Materials 3*: 33–39.

Awang-Kanak, F., Abu Bakar, M.F., Mohamed, M. and Norazlimi, N.A. 2018, April 10–12. Utilization of natural resources: preliminary study on ethnopharmacological application of "ulam" or traditional vegetables among Sama-Bajau of Kampung Menunggui, Kota Belud, Sabah. In: *Proceedings of the International Conference on Applied Science and Technology*; Penang, Malaysia. Maryland: AIP Conference Proceedings.

Ayyanar, M. and Ignacimuthu, S. 2011. Ethnobotanical survey of medicinal plants commonly used by Kani tribals in Tirunelveli hills of western Ghats, India. *Journal of Ethnopharmacology 134*: 851–64.

Azeem, M.W., Hanif, M.A. and Khan, M.M. 2020. Chapter 3 – Bamboo. In: Hanif, M.A., Nawaz, H., Khan, M.M., Byrne, H.J.. (eds) *Medicinal Plants of South Asia*. Elsevier, United Kingdom, 29–45. https://doi.org/10.1016/B978–0-08–102659-5.00003–3

Bahru, T. and Ding, Y. 2021. A review on bamboo resource in the African region: A call for special focus and action. *International Journal of Forestry Research* 10–12. https://doi.org/10.1155/2021/8835673

Balkrishna, A., Katiyar, P., Singh, S., Mishra, R. and Arya, V. 2023. Vanshlochan substitution and adulteration: Discussions on the controversy over its original source. *Ayushdhara 10*: 1–4.

Banerjee, S., Basak, M., Dutta, S., Chanda, C., Dey, A. and Das, M. 2021. Ethnobamboology: Traditional uses of bamboos and opportunities to exploit genomic resources for better exploitation. In: Ahmad, Z., Ding, Y. & Shahzad, A. (eds) *Biotechnological Advances in Bamboo*. Springer, Singapore. https://doi.org/10.1007/978-981-16-1310-4_14

Banik, R.L. 2016. *Silviculture of South Asian Priority Bamboos*. Springer, Singapore, 21–115.

Bartram, T. 2015. *Bartram's Encyclopedia of Herbal Medicine: The Definitive Guide to Herbal Treatment of Diseases*. Hachette, United Kingdom.

Baskar, A.V., Bolan, N., Hoang, S.A., Sooriyakumar, P., Kumar, M., Singh, L., Jasemizad, T., Padhye, L.P., Singh, G., Vinu, A. and Sarkar, B. 2022. Recovery, regeneration and sustainable management of spent adsorbents from wastewater treatment streams: A review. *Science of the Total Environment 822*: 153555.

Basumatary, A., Middha, S.K., Usha, T., Basumatary, A.K., Brahma, B.K. and Goyal, A.K. 2017. Bamboo shoots as a nutritive boon for Northeast India: An overview. *3 Biotech 7*: 1–6.

Benjamin, M.A.Z., Ng, S.Y., Saikim, F.H. and Rusdi, N.A. 2021. Ethnobotany and traditional knowledge of bamboos (*Poaceae: Bambusoideae*) in Asia and their applications in the complementary and alternative medicine: A review. *Pharmacognosy Journal 13*: 1751–1762.

Bhatt, B. P., Singh, K. and Singh, A. 2005. Nutritional values of some commercial edible bamboo species of the North Eastern Himalayan region, India. *Journal of Bamboo and Rattan* 4(2), 111–124.

Bhujbal, S.K., Ghosh, P., Vijay, V.K. et al., 2022. Biotechnological potential of rumen microbiota for sustainable bioconversion of lignocellulosic waste to biofuels and value-added products. *Science of the Total Environment* 814: 152773

Bishnu Maya, K.C., Lamichhane, J., Khanal, S.N. and Gauchan, D.P. 2024. Traditional utilization of bamboo in the Central Siwalik region, Nepal. *PLoS One 19*(1): e0296886.

Canavan, S., Richardson, D.M., Visser, V., Le Roux, J.J., Vorontsova, M.S. and Wilson, J.R. 2017. The global distribution of bamboos: Assessing correlates of introduction and invasion. *AoB Plants 9*(1): lw078.

Chassagne, F., Deharo, E., Punley, H. and Bourdy, G. 2017. Treatment and management of liver diseases by Khmer traditional healers practicing in Phnom Penh area, Cambodia. *Journal of Ethnopharmacology 202*: 38–53.

Chaturvedi, K., Singhwane, A., Dhangar, M., Mili, M., Gorhae, N., Naik, A., Prashant, N., Srivastava, A.K. and Verma, S. 2023. Bamboo for producing charcoal and biochar for versatile applications. *Biomass Conversion and Biorefinery 14*: 15159–15185.

Cheng, Y., Wan, S., Yao, L., Lin, D., Wu, T., Chen, Y., Zhang, A. and Lu, C. 2023. Bamboo leaf: A review of traditional medicinal property, phytochemistry, pharmacology, and purification technology. *Journal of Ethnopharmacology 306*: 116166.

Chongtham, N. and Bisht, M.S. 2021, October 25. *Bamboo Shoot: Superfood for Nutrition, Health and Medicine*. CRC Press, New York.

Chongtham, N., Bisht, M.S., Santosh, O., Bajwa, H.K. and Indira, A. 2021. Mineral elements in bamboo shoots and potential role in food fortification. *Journal of Food Composition and Analysis 95*: 103662.

Choudhury, D., Sahu, J.K. and Sharma, G.D. 2012. Bamboo shoot: Microbiology, biochemistry and technology of fermentation - A review. *Indian Journal of Traditional Knowledge 11*: 242–249.

Christian, A., Knott, K., Vance, C. et al., 2015. Nutrient and mineral composition during shoot growth in seven species of Phyllostachys and *Pseudosasa bamboo* consumed by giant panda. *Journal of Animal Physiology and Animal Nutrition 99*: 1172–1183.

Cruz, J., Trombley, J., Carrington, L. and Cheng, X. 2021. Properties of the novel Chinese herbal medicine formula Qu du Qiang Fei i Hao Fang warrant further research to determine its clinical efficacy in COVID-19 treatment. *Medical Acupuncture 33*(1): 71–82. https://doi.org/10.1089/acu.2020.1466

Darabant, A., Rai, P.B., Staudhammer, C.L. and Dorji, T. 2016. Designing and evaluating bamboo harvesting methods for local needs: Integrating local ecological knowledge and science. *Environmental Management 58*(2): 312–322. https://doi.org/10.1007/s00267-016-0702-6

Das, A.K., Nanda, P.K., Chowdhury, N.R., Dandapat, P., Gagaoua, M., Chauhan, P., Pateiro, M. and Lorenzo, J.M. 2021. Application of pomegranate by-products in muscle foods: Oxidative indices, colour stability, shelf life and health benefits. *Molecules 26*(2): 467.

Das, M. 2019. Bamboo: Inherent source of nutrition and medicine. *Journal of Pharmacognosy and Phytochemistry 8*: 1338–1344.

de Boer, H.J., Lamxay, V. and Björk, L. 2011. Steam sauna and mother roasting in Lao PDR: Practices and chemical constituents of essential oils of plant species used in postpartum recovery. *BMC Complement and Alternative Medicine 11*: 1–10.

Del Rosario, A., McDermott, M.M. and Panee, J. 2012. Effects of a high-fat diet and bamboo extract supplement on anxiety- and depression-like neurobehaviours in mice. *British Journal of Nutrition 108*: 1143–1149.

Devi, Y.R., Chakma, A. and Yenkokpam, S. 2017. Traditional wisdom of Meitei community regarding elimination of cyanogenic glycosides in bamboo shoot food products *Indian Journal of Traditional Knowledge 16*: 470–475.

Emamverdian, A., Ding, Y., Ranaei, F. and Ahmad, Z. 2020. Application of bamboo plants in nine aspects. *The Scientific World Journal* 7284203. https://doi.org/10.1155/2020/7284203

Esakkimuthu, S., Mutheeswaran, S., Arvinth, S., Paulraj, M.G., Pandikumar, P. and Ignacimuthu, S. 2016. Quantitative ethnomedicinal survey of medicinal plants given for cardiometabolic diseases by the non-institutionally trained Siddha practitioners of Tiruvallur district, Tamil Nadu, India. *Journal of Ethnopharmacology 186*: 329–342.

Gabriel-Ajobiewe, R.A., Gabriel, B.M., Cyril-Olutayo, M.C., Olorunmola, F.O. and Adegbehingbe, K.T. 2021. Dosage determination of ethnopharmacologically used fermented *Zea mays* subsp. *mays* liquor, extract of *Bambusa vulgaris* Schrad. Ex. J.C. Wendel leaf against *Salmonella typhi. Jundishapur Journal of Natural Pharmaceutical Products 15*(4): 0–7. https://doi.org/10.5812/JJNPP.97866

Gagliano, J., Anselmo-Moreira, F., Sala-Carvalho, W.R. and Furlan, C.M. 2022. What is known about the medicinal potential of bamboo? *Advances in Traditional Medicine 22*(3): 467–495.

Gao, Q., Wang, D., Shao, S., Xue, Y., Zhang, Y., Chen, C., Tang, F., Sun, J., Li, Y. and Guo, Q. 2020. Identification and quantitation of the actual active components in bamboo juice and its oral liquid by NMR and UPLC-Q-TOF-MS. *Scientific Reports 10*(1): 19664.

Girija, P.L.T. and Sivan, N. 2022. Ayurvedic treatment of COVID-19/SARS-CoV-2: A case report. *Journal of Ayurveda and Integrative Medicine 13*(1): 100329. https://doi.org/10.1016/j.jaim.2020.06.001

Guo, Y., Liu, F.Y., Shen, Y., Xu, J.Y., Xie, L.Z., Li, S.Y., Ding, D.N., Zhang, D.Q. and Han, F.J. 2021. Complementary and alternative medicine for dysmenorrhea caused by endometriosis: A review of utilization and mechanism. *Evidence-Based Complementary and Alternative Medicine 1*: 6663602. https://doi.org/10.1155/2021/6663602

Gupta, D. and Ranjan, R. 2016. Role of bamboo in sustainable development. *International Journal of Advanced Science and Research 2*: 25–32.

Hanif, A., Hossan, M.S., Mia, M.M.K., Islam, M.J., Jahan, R. and Rahamatullah, M. 2009. Ethnobotanical survey of the Rakhain tribe inhabiting the Chittagong Hill Tracts region of Bangladesh. *American-Eurasian Journal of Sustainable Agriculture 3*: 172–180.

Ho, C.L., Lin, C.Y., Ka, S.M., Chen, A., Tasi, Y.L., Liu, M.L., Chiu, Y.C. and Hua, K.F. 2013. Bamboo vinegar decreases inflammatory mediator expression and NLRP3 inflammasome activation by inhibiting reactive oxygen species generation and protein kinase C-a/d activation. *PLoS One 8*: e75738.

Honfo, H., Tovissodé, F.C., Gnanglè, C., Mensah, S., Salako, V.K., Assogbadjo, A.E., Agbangla, C. and Kakaï, R.G. 2015. Traditional knowledge and use value of bamboo in southeastern Benin: Implications for sustainable management. *Ethnobotany Research and Applications 14*: 139–153.

Huang, L., Li, X. and Wang, Y. 2018. Bamboo and its potential medicinal uses. *Journal of Ethnopharmacology 216*: 1–15.

Iqbal, S.M., Hussain, L., Hussain, M., Akram, H., Asif, M., Jamshed, A., Saleem, A. and Siddique, R. 2022. Nephroprotective potential of a standardized extract of *Bambusa arundinacea*: In vitro and in vivo studies. *ACS Omega 7*(21): 18159–18167. https://doi.org/10.1021/acsomega.2c02047

Isukuru, E.J., Ogunkeyede, A.O., Adebayo, A.A. and Uruejoma, M.F. 2023. Potentials of bamboo and its ecological benefits in Nigeria. *Advances in Bamboo Science 4*: 100032.

Jadid, N., Kurniawan, E., Himayani, C.E.S., Andriyani, Prasetyowati, I., Purwani, K.I., Muslihatin, W., Hidayati, D. and Tjahjaningrum, I.T.D. 2020. An ethnobotanical study of medicinal plants used by the Tengger tribe in Ngadisari village, Indonesia. *PLoS One 15*: 1–16.

Jagtap, S.D., Deokule, S.S. and Bhosle, S.V. 2006. Some unique ethnomedicinal uses of plants used by the Korku tribe of Amravati district of Maharashtra, India. *Journal of Ethnopharmacology 107*: 463–469.

Jain, A., Katewa, S.S., Galav, P.K. and Sharma, P. 2005. Medicinal plant diversity of Sitamata wildlife sanctuary, Rajasthan, India. *Journal of Ethnopharmacology 102*: 143–157.

Jain, D.P.S., Jadav, D.D. and Awati, D.S. 2023. Reviewing the therapeutic value of *Bambusa arundinacea*: Bamboo. *International Journal of Homoeopathic Sciences 7*(1): 354–357. https://doi.org/10.33545/26164485.2023.v7.i1f.777

Jain, R. and Jain, S.K. 2010. Traditional medicinal plants as anticancer agents from Chhattisgarh, India: An overview. *International Journal of Phytomedicine 2*: 186–196.

Jaiswal, Y.S. and Williams, L.L. 2017. A glimpse of ayurveda – The forgotten history and principles of Indian traditional medicine. *Journal of Traditional and Complementary Medicine 7*(1): 50–53. https://doi.org/10.1016/j.jtcme.2016.02.002

Jaya, A.M., Musa, Y., Iswoyo, H., Asmi, N. and Siregar, L.F. 2019, August 3–4. Ethnobotanical study and identification of medicinal plants based on local knowledge. In: *Proceedings of the 1st International Conference of Interdisciplinary Research on Green Environmental Approach for Sustainable Development*; Universitas Muhammadiyah Buton, Indonesia. Bristol: IOP Conference Series: Rosario Earth and Environmental Science.

Jeeva, S. and Femila, V. 2012. Ethnobotanical investigation of Nadars in Atoor village, Kanyakumari district, Tamil Nadu, India. *Asian Pacific Journal of Tropical Biomedicine 2*: S593–S600.

Kikon, D. 2021. Bamboo shoot in our blood: Fermenting flavors and identities in Northeast India. *Current Anthropology 62*: S376–S387.

Kim, H. and Song, M.J. 2011. Analysis and recordings of orally transmitted knowledge about medicinal plants in the southern mountainous region of Korea. *Journal of Ethnopharmacology 134*: 676–696.

Kiruba, S., Jeeva, S., Das, S. and Kannan, D. 2007. Bamboo seeds as a means to sustenance of the indigenous community. *Indian Journal of Traditional Knowledge 6*: 199–203.

Kong, C.K., Tan, Y.N., Chye, F.Y. and Sit, N.W. 2020. Nutritional composition and biological activities of the edible shoots of *Bambusa vulgaris* and *Gigantochloa ligulata*. *Food Bioscience 36*: 100650.

Kumar, A., Jain, S. and Sharma, A. 2017. Bamboo fibers as a source of cellulose for pharmaceutical industry. *International Journey of Green Pharmacy 11*: 1–4.

Kumar, S., Rawat, D., Singh, B. and Khanduri, V.P. 2023. Utilization of bamboo resources and their market value in the western Himalayan region of India. *Advances in Bamboo Science 3*: 100019.

Lans, C., Georges, K. and Brown, G. 2007. Non-experimental validation of ethnoveterinary plants and indigenous knowledge used for backyard pigs and chickens in Trinidad and Tobago. *Tropical Animal Health and Production 39*: 375–385.

Lee, M.J., Kim, M.J., Song, Y.S., Song, Y.O. and Moon, G.S. 2008. Bamboo culm extract supplementation elevates HDL-cholesterol and ameliorates oxidative stress in C57BL/6 mice fed atherogenic diet. *Journal of Medicinal Food 11*: 69–77.

Li, X., Wang, Y. and Zhang, X. 2020. Bamboo leaf extract produces anti-viral activity against influenza virus. *Journal of Ethnopharmacolgy 273*: 1–8.

Libman, A., Bouamanivong, S., Southavong, B., Sydara, K. and Soejarto, D.D. 2006. Medicinal plants: An important asset to health care in a region of Central Laos. *Journal of Ethnopharmacology 106*: 303–311.

Liu, M.H., Ko, C.H., Ma, N., Tan, P.W., Fu, W.M. and He, J.Y. 2016. Chemical profiles, antioxidant and anti-obesity effects of extract of *Bambusa textilis* McClure leaves. *Journal of Functional Foods 22*: 533–546.

Loukas, M., Lanteri, A., Ferrauiola, J., Tubbs, R.S., Maharaja, G., Shoja, M.M., Yadav, A. and Rao, V.C. 2010. Anatomy in ancient India: A focus on the Susruta Samhita. *Journal of Anatomy 217*(6): 646–650. https://doi.org/10.1111/j.1469-7580.2010.01294.x

Lu, B., Cai, H., Huang, W., Wu, X., Luo, Y., Liu, L. and Zhang, Y. 2011. Protective effect of bamboo shoot oil on experimental nonbacterial prostatitis in rats. *Food Chemistry 124*: 1017–1023.

Luo, B., Ahmed, S. and Long, C. 2020. Bamboos for weaving and relevant traditional knowledge in Sansui, Southwest China. *Journal of Ethnobiology and Ethnomedicine 16*(1): 9.

Ma, D., Wang, S., Shi, Y., Ni, S., Tang, M. and Xu, A. 2021. The development of traditional Chinese medicine. *Journal of Traditional Chinese Medical Sciences 8*: S1–S9. https://doi.org/10.1016/j.jtcms.2021.11.002

Ma, Y., Zhang, T., Zhu, P., Cai, H., Jin, Y., Gao, K. and Li, J. 2022. Fabrication of $Ag_3PO_{(4)}$ polyaniline-activated biochar photocatalyst for efficient triclosan degradation process and toxicity assessment. *Science of the Total Environment 821*: 153453.

Masic, I. 2024. History of social medicine and public health - A review. *International Journal on Biomedicine and Healthcare 12*(1): 25. https://doi.org/10.5455/ijbh.2024.12.25-52

Maulana, S., Gumelar, Y., Fatrawana, A., Maulana, M.I., Hidayat, W., Sumardi, I., Wistara, N.J., Lee, S.H., Kim, N.H. and Febrianto, F. 2019. Destructive and non-destructive tests of bamboo oriented strand board under various shelling ratios and resin contents. *Journal of the Korean Wood Science and Technology 47*(4): 519–532. https://doi.org/10.5658/WOOD.2019.47.4.519

Medhi, P., Borthakur, S.K. and Hore, D.K. 2010. Phytoresources from north Cachar hills of Assam, India-IV: Bamboos and rattans. *Journal of Bamboo and Rattan 9*: 115–125.

Minagawa, T., Saitou, T., Suzuki, T., Domen, T., Yokoyama, H., Ishikawa, M., Hirakata, S., Nagai, T., Nakazawa, M., Ogawa, T. and Ishizuka, O. 2016. Impact of ao-dake-humi, Japanese traditional bamboo foot stimulator, on lower urinary tract symptoms, constipation and hypersensitivity to cold: A single-arm prospective pilot study. *BMC Complementary and Alternative Medicine 16*(1): 8.

Mizuki, I., Sato, A., Matsuo, A., Suyama, Y., Suzuki, J.I. and Makita, A. 2014. Clonal structure, seed set, and self-pollination rate in mass-flowering bamboo species during off-year flowering events. *PLoS One 9*(8). https://doi.org/10.1371/journal.pone.0105051

Mobeen, A. and Ahmad, A.K. 2021. The efficacy and safety of herbal combination of Unani medicine in chronic urticaria: A randomized, controlled study. *Journal of Traditional and Complementary Medicine 11*(4): 303–310. https://doi.org/10.1016/j.jtcme.2020.12.003

Mustafa, A.A., Derise, M.R., Yong, W.T.L. and Rodrigues, K.F. 2021. A concise review of *Dendrocalamus asper* and related bamboos: Germplasm conservation, propagation and molecular biology. *Plants 10*(9). https://doi.org/10.3390/plants10091897

Neamsuvan, O., Phumchareon, T., Bunphan, W. and Kaosaeng, W. 2016. Plant materials for gastrointestinal diseases used in Chawang district, Nakhon Si Thammarat province, Thailand. *Journal of Ethnopharmacology 194*: 179–187.

Nirmala, C. 2018. The healing touch of bamboo: Soul, mind and body. In: *11th World Bamboo Congress*, Farrelly 1984.

Nirmala, C., Bisht, M.S., Bajwa, H.K. and Santosh, C. 2018. Bamboo: A rich source of natural antioxidants and its applications in the food and pharmaceutical industry. *Trends in Food Science and Technology 77*: 91–99.

Nirmala, C., Bisht, M.S. and Laishram, M. 2014. Bioactive compounds in bamboo shoots: Health benefits and prospects for developing functional foods. *International Journal of Food Science and Technology 49*: 1425–1431.

Nongdam, P. and Tikendra, L. 2014. The nutritional facts of bamboo shoots and their usage as important traditional foods of Northeast India. *International Scholarly Research Notices*: 1–17. https://doi.org/10.1155/2014/679073

Noremylia, M.B., Aufa, A.N., Ismail, Z. and Hassan, M.Z. 2023. An overview of the potential usage of bamboo plants in medical field. In: Palombini, F.L. & Nogueira, F.M. (eds) *Bamboo Science and Technology*. Environmental Footprints and Eco-design of Products and Processes. Springer, Singapore. https://doi.org/10.1007/978-981-99-0015-2_3

Novaryatiin, S. and Indah, I. 2019. The medicinal plants used in Anjir Pulang Pisau, central Kalimantan-Indonesia. *Pharmacognosy Journal 11*: 1572–1579.

Ong, H.G. and Kim, Y.D. 2017. The role of wild edible plants in household food security among transitioning hunter-gatherers: Evidence from the Philippines. *Food Security 9*: 11–24.

Pan, S.Y., Litscher, G., Gao, S.H., Zhou, S.F., Yu, Z.L., Chen, H.Q., Zhang, S.F., Tang, M.K., Sun, J.N. and Ko, K.M. 2014. Historical perspective of traditional indigenous medical practices: The current renaissance and conservation of herbal resources. *Evidence-Based Complementary and Alternative Medicine*. Article ID 525340:1–20. https://doi.org/10.1155/2014/525340

Pandey, A.K. and Ojha, V. 2013. Standardization of harvesting age of bamboo shoots with respect to nutritional and anti-nutritional components. *Journal of Forestry Research 24*: 83–90.

Pandey, M.M., Rastogi, S. and Rawat, A.K.S. 2013. Indian traditional ayurvedic system of medicine and nutritional supplementation. *Evidence-Based Complementary and Alternative Medicine 1*: 376327. https://doi.org/10.1155/2013/376327

Panee, J. 2015. Potential medicinal application and toxicity evaluation of extracts from bamboo plants. *Journal of Medicinal Plant Research 9*(23): 681–692.

Park, H.S., Lim, J.H., Kim, H.J., Choi, H.J. and Lee, I.S. 2007. Antioxidant flavone glycosides from the leaves of *Sasa borealis*. *Archives of Pharmacal Research 30*: 161–166.

Patel, M., Mehta, P., Bakshi, S. and Tewari, S. 2016. Effect of *Phyllostachys parvifolia* leaf extract on ionizing radiation-induced genetic damage: A preliminary in vitro cytogenetic study. *Journal of Ayurveda and Integrative Medicine 7*: 138–140.

Patwardhan, K. and Kumar, M. 2021. Surgical practice and ayurveda: A realistic analysis of the current debate. *Journal of Ayurveda and Integrative Medicine 12*(1): 195–197. https://doi.org/10.1016/j.jaim.2020.12.008

Priyanka, D., Apeksha, N. and Sakshi, K. 2024. *Bambusa arundinaceae* : A review on phytochemistry and pharmacological activity. *International Journal of Novel Research and Development 9*(2): 266–273. https://www.ijnrd.org/papers/IJNRD2402232.pdf

Rahmatullah, M. and Biswas, K.R. 2012. Traditional medicinal practices of a Sardar healer of the Sardar (Dhangor) community of Bangladesh. *The Journal of Alternatives and Complementary Medicine 18*: 10–19.

Rasheed, M., Jawaid, M., Parveez, B., Zuriyati, A. and Khan, A. 2020. Morphological, chemical and thermal analysis of cellulose nanocrystals extracted from bamboo fibre. *International Journal of Biological Macromolecules 160*: 183–191.

Rawat, K., Sharma, V., Saini, N. et al., 2016. Impact of different boiling and soaking treatments on the release and retention of antinutrients and nutrients from the edible shoots of three bamboo species. *American Journal of Food Science and Nutrition Resources* 3: 31–41.

Ren, Y., Ma, Y., Zhang, Z., Qiu, L., Zhai, H., Gu, R. and Xie, Y. 2019. Total alkaloids from bamboo shoots and bamboo shoot shells of *Pleioblastus amarus* (Keng) Keng f. and their anti-inflammatory activities. *Molecules 24*: 2699.

Richa, Sharma, M.L. and Sharma, N.K. 2012. Bamboos of Himachal Pradesh. *Indian Forester 138*(12): 1141–1145.

Ritonga, M.A., Syamsuardi, N., Putu, I. and Damayanto, G.P. 2023. Ethnobotany of bamboo on Weh Island, Aceh, Indonesia. *Ethnobotany Research and Applications 26*: 1–19.

Rudra, S., Islam, K.N., Rahman, M.M. and Uddin, S.B. 2021. Medicinal plant diversity and their therapeutic uses in selected village common forests in Chittagong Hill Tracts, Bangladesh. *Journal of Herbs, Spices and Medicinal Plants 27*: 83–107.

Saini, N., Nirmala, C. and Bisht, M.S. 2015, September 17–22. Bamboo resource of Himachal Pradesh (India) and potential of shoots in Socio-economic development of the state. In: *Proceedings of 10th World Bamboo Congress*, theme: Community and Economic Development, Damyang, Korea.

Sangeetha, R., Diea, Y., Chaitra, C., Malvi, P. and Shinomol, G. 2015. The amazing bamboo: A review on its medicinal and pharmacological potential. *Indian Journal of Nutrition 2*: 1–7.

Santosh, O., Bajwa, H.K., Bisht, M.S. and Nirmala, C. 2018, September: 1–12. *Bamboo Shoot Fortified Cookies as a Healthy Snack* (11th ed.). World Bamboo Congress, Mexico.

Santosh, O., Bajwa, H.K., Bisht, M.S. and Chongtham, N. 2019. Functional biscuits from bamboo shoots: Enrichment of nutrients, bioactive compounds and minerals in bamboo shoot paste fortified biscuits. *International Journal of Food Science and Nutrition 4*: 89–94.

Santosh, O., Bajwa, H.K., Bisht, M.S. and Chongtham, N. 2021. Application of bamboo in the food and pharmaceutical industry. In: Ahmad Z. et al. (ed) *Biotechnological Advances in Bamboo.* Springer, Singapore, 401–429.

Sarkar, D., Chandra, A.K. Chakraborty, A., et al., 2020. Effects of bamboo shoots (*Bambusa balcooa*) on thyroid hormone synthesizing regulatory elements at cellular and molecular levels in thyrocytes. *Journal of Ethnopharmacology*, 250: 112463.

Sathiyaraj, K., Sivaraj, A., Thirumalai, T. and Senthilkumar, B. 2012. Ethnobotanical study of antifertility medicinal plants used by the local people in Kathiyavadi village, Vellore district, Tamilnadu, India. *Asian Pacific Journal of Tropical Biomedicine 2*: S1285–S1288.

Savithramma, N., Sulochana, C. and Rao, K.N. 2007. Ethnobotanical survey of plants used to treat asthma in Andhra Pradesh, India. *Journal of Ethnopharmacology 113*: 54–61.

Sawarkar, A.D., Shrimankar, D.D., Kumar, M., Kumar, P. and Singh, L. 2023. Bamboos as a cultivated medicinal grass for industries: A systematic review. *Industrial Crops and Products 203*: 117210.

Seki, T. and Maeda, H. 2010. Cancer preventive effect of Kumaizasa bamboo leaf extracts administered prior to carcinogenesis or cancer inoculation. *Anticancer Research 30*: 111–118.

Sertse, D., Disasa, T., Bekele, K., Alebachew, M., Kebede, Y., Eshete, N. and Eshetu, S. 2011. Mass flowering and death of bamboo: A potential threat to biodiversity and livelihoods in Ethiopia. *Journal of Biodiversity and Environmental Sciences 1*(5): 16–25.

Sharkar, P., Rahman, M.M., Haque Masum, G.Z., Nayeem, M.A., Hossen, M.M. and Azad, A.K. 2013. Ethnomedicinal importance of the plants in villages in Kushtia Sador and Mirpur Upozila, Bangladesh. *Journal of Herbs, Spices and Medicinal Plants 19*: 401–417.

Sharma, T.P. and Borthakur, S.K. 2008. Ethnobotanical observations on bamboos among Adi tribes in Arunachal Pradesh *Indian Journal of Traditional Knowledge 7*: 594–597.

Shukla, R., Sumit, G., Sajal, S., Dwivedi, P.K. and Mishra, A. 2012. Medicinal importance of bamboo. *International Journal of Biopharm and Phytochemical Research 1*: 9–15.

Silambarasan, R. and Ayyanar, M. 2015. An ethnobotanical study of medicinal plants in Palamalai region of Eastern Ghats, India. *Journal of Ethnopharmacology 172*: 162–178.

Singh, H.B., Kumar, B. and Singh, R.S. 2003. Bamboo resources of Manipur: An overview for management and conservation. *Journal of Bamboo and Rattan 2*: 43–55.

Singh, M.K. 2019. Ethno-medico-botanical observations of bamboos among indigenous people of Manipur. *Journal of Medicinal Plants 7*: 159–162.

Singh, R.A.J., Upadhyay, S.K., Rani, A., Kumar, P. and Kumar, A. 2020. Ethnobotanical study of Subhartipuram, Meerut, Uttar Pradesh, India. II. Diversity and pharmacological significance of shrubs and songclimbers. *International Journal of Pharmaceutical Research 12*: 383–393.

Singh, S.R., Singh, R., Kalia, S., Dalal, S., Dhawan, A.K. and Kalia, R.K. 2013. Limitations, progress and prospects of application of biotechnological tools in improvement of bamboo-a plant with extraordinary qualities. *Physiology and Molecular Biology of Plants 19*(1): 21–41. https://doi.org/10.1007/s12298-012-0147-1

Singh, V. and Pandey, R.P. 1998. *Ethnobotany of Rajasthan* (India). Scientific Publishers, Jodhpur.

Singhal, P., Bal, L.M., Satya, S., Sudhakar, P. and Naik, S.N. 2013. Bamboo Shoots: A novel source of nutrition and medicine. *Critical Reviews in Food Science and Nutrition 53*: 517–534.

Singhal, P., Satya, S. and Naik, S.N. 2021. Fermented bamboo shoots: A complete nutritional, anti-nutritional and antioxidant profile of the sustainable and functional food to food security. *Food Chemistry: Molecular Sciences 3*: 100041.

Song, M.-J., Kim, H., Lee, B.-Y., Brian, H., Park, C.H. and Hyun, C.W. 2014. Analysis of traditional knowledge of medicinal plants from residents in Gayasan National Park (Korea). *Journal of Ethnobiology and Ethnomedicine 10*(1): 47.

Soni, V., Jha, A.K., Dwivedi, J. and Soni, P. 2013. Traditional uses, phytochemistry and pharmacological profile of *Bambusa arudinacea* Retz. Tang. *Humanitas Medicine 3*(3): 20.1–20.6. https://doi.org/10.5667/tang.2013.0011

Soumya, V., Indira Muzib, Y. and Venkatesh, P. 2016. A novel method of extraction of bamboo seed oil (*Bambusa bambos* Druce) and its promising effect on metabolic symptoms of experimentally induced polycystic ovarian disease. *Indian Journal of Pharmacology 48*: 162–167.

Srisawat, T., Suvarnasingh, A. and Maneenoon, K. 2016. Traditional medicinal plants notably used to treat skin disorders nearby Khao Luang mountain hills region, Nakhon Si Thammarat, southern Thailand. *Journal of Herbs, Spices and Medicinal Plants 22*: 35–56.

Sujarwo, W. 2018. Bamboo resources, cultural values, and ex-situ conservation in Bali, Indonesia. *Reinwardtia 17*: 67–75.

Sundriyal, R.C., Upreti, T.C. and Varuni, R. 2002. Bamboo and cane resource utilisation and conservation in the Apatani plateau, Arunachal Pradesh, India: Implications for management. *Journal of Bamboo and Rattan 1*: 205–246.

Tang, Y., Li, Z., Yang, D., Fang, Y., Gao, S., Liang, S. and Liu, T. 2021. Research of insomnia on traditional Chinese medicine diagnosis and treatment based on machine learning. *Chinese Medicine (United Kingdom) 16*(1): 1–21. https://doi.org/10.1186/s13020-020-00409-8

Tripathi, Y.C. 2011. Bamboo on the touchstone of modern phyto-therapy. In: *Productivity Enhancement and Value Addition of Bamboos*. Excel India, New Delhi, 315–326.

U.S. Department of Agriculture, Agricultural Research Service. 2022. USDA Food and Nutrient Database for Dietary Studies 2019–2020. Food Surveys Research Group Home Page, http://www.ars.usda.gov/nea/bhnrc/fsrg

Udayan, P.S., George, S., Tushar, K.V. and Balachandran, I. 2005. Some common medicinal plants used by the Nayaka community, Savandurga forest of Magadi taluk, Bangalore rural district, Karnataka, India. *Journal of Natural Remedies 5*: 35–40.

Virtanen, P.K., Apurinã, F., Ruokolainen, K. and Manchineri, L. 2022. The role of *Guadua bamboo* in land management and indigenous perspectives on bamboo ecosystems in Southwestern Amazonia. *Human Ecology 50*(6): 1077–1088. https://doi.org/10.1007/s10745-022-00376-8

Wang, J., Jiang, J., Wang, J., Wang, Z., Yang, X. and Jia, L. 2019. The influence of gamma irradiation on the storage quality of bamboo shoots. *Radiation Physics and Chemistry 159*: 124–130.

Wang, J., Yue, Y.D., Tang, F. and Sun, J. 2012. TLC screening for antioxidant activity of extracts from fifteen bamboo species and identification of antioxidant flavone glycosides from leaves of *Bambusa textilis* McClure. *Molecules 17*: 12297–12311.

Wang, P., Maliang, H., Wang, C. and Ma, J. 2015. Bamboo charcoal by-products as sources of new insecticide and acaricide. *Industrial Crops and Product 77*: 575–581.

Wang, R., Guo, Z., Cai, C., Zhang, J., Bian, F., Sun, S. and Wang, Q. 2021. Practices and roles of bamboo industry development for alleviating poverty in China. *Clean Technologies and Environmental Policy 23*: 1687–1699.

Wang, Y., Chen, J., Wang, D., Ye, F., He, Y., Hu, Z. and Zhao, G. 2020. A systematic review on the composition, storage, processing of bamboo shoots: Focusing the nutritional and functional benefits. *Journal of Functional Foods 71*: 104015.

Watanabe, L.B., de Sena Aquino, A.C.M. and Edna Regina Amante, E.R. 2021. Paste as a new product from Bamboo shoot (*Dendrocalamus asper*): Physicochemical and microbiological properties. *Research, Society and Development 10*: e27610313353.

Wróblewska, K.B., de Oliveira, D.C.S., Grombone-Guaratini, M.T. and Moreno, P.R.H. 2019. Medicinal properties of bamboos. In: Perveen S. & Al-Taweel A. (eds) *Pharmacognosy: Medicinal Plants*. IntechOpen, London, 159–176.

Xiong, X., Yang, X., Liu, Y., Zhang, Y., Wang, P. and Wang, J. 2013. Chinese herbal formulas for treating hypertension in traditional Chinese medicine: Perspective of modern science. *Hypertension Research 36*(7): 570–579. https://doi.org/10.1038/hr.2013.18

Xu, K., Zhang, C., Dou, X., Ma, W. and Wang, C. 2019. Optimizing the modification of wood waste biochar via metal oxides to remove and recover phosphate from human urine. *Environmental Geochemistry and Health 41*: 1767–1776.

Xu, Y., Zhang, W., Sun, H., Yue, X. and Zhang, D. 2016. Study on the dynamic viscoelasticity of bamboo kraft black liquor. *BioRes 11*: 2655–2664.

Yang, Y. 2002. *Chinese Herbal Medicines Comparisons and Characteristics*. Churchill Livingstone, London.

Zhang, X., Li, X. and Wang, Y. 2016. Bamboo shoot extract improves skin injury in rats. *Journal of Ethnopharmacology 191*: 1–8.

Zhang, Y., Yao, X., Bao, B. and Zhang, Y. 2006. Anti-fatigue activity of atriterpenoid-rich extract from Chinese bamboo shavings (*Caulis bamfusae* in Taeniam). *Phytotherapy Research 20*: 872–876.

3 Bamboo Products and Their Antibacterial Properties

Damini Maithani, Purnita Raturi, Anjali Tiwari, Hemant Dasila, Nikita Pundir, and Viabhav Kumar Upadhyay

3.1 INTRODUCTION

In recent years, research has been driven towards finding sustainable materials due to rising eco-consciousness in society. Nature offers an immense diversity of fundamental blueprints and prototypes to scientists, engineers and researchers for technological advancements and addressing socio-economic challenges. Then, humans have been blessed with functionally engineered natural materials like tree wood and bones that are biodegradable and environmentally friendly, demonstrating the versatility of these natural structures and their utilization for human benefit (Gupta and Kumar, 2008; Komatsu et al., 2012). Several bioactive substances like phenols, phytosterols, flavonoids and dietary fibers are present in young bamboo shoots and leaves which may have potential health benefits.

Bamboo is a fast-growing multifunctional plant that has been used in making furniture, paper, agriculture and architecture and thus has revolutionized various industries. It is a renewable and easily available resource having tremendous economic potential. There are several products made from bamboo that are utilized for their antimicrobial properties, soil conservation, biofuel, timber and non-timber products and carbon sequestration characteristics. It is a sustainable natural resource with great tensile strength, flexibility, antimicrobial properties and numerous environmental benefits. All these properties make it a useful resource in handicraft, construction, pharmaceutical, textiles and pharmaceutical industries (Kaur et al., 2016). Bamboo has an inherent ability to withstand biotic and abiotic stress. Its antimicrobial property protects it from fungi, bacteria, insects and other biotic agents (Ramful et al., 2022). Bamboo vinegar, bamboo charcoal, juice, beer and various other products derived from bamboo are used frequently in different countries. Bamboo has numerous therapeutic benefits and has been used in traditional medicine for years. Besides natural bamboo products, bamboo composites are gaining a lot more attention due to improved microbial resistance and antibacterial activity (Mai et al., 2022; Zhang et al., 2017).

Increasing concerns about antimicrobial resistance and the outbreak of several infectious diseases have driven the interest of the research community towards finding sustainable, non-toxic and eco-friendly materials/polymers with antimicrobial properties. Data scattered across the literature makes it evident that bamboo has antimicrobial properties which has increased its popularity in manufacturing certain value-added products (Table 3.1). Afrin et al.'s (2012) study reported extracts from *Phyllostachys pubescens* displaying antibacterial properties against *Escherichia coli*. In another study, a bamboo charcoal and silver composite powder was prepared that displayed great antibacterial performance (Yang et al., 2009). Bamboo vinegar (BV) is a reddish-brown liquid that contains chemical compounds such as phenols, alcohols, acetic acid, etc. that impart antimicrobial, fungicidal and insecticidal activity to this fluid (Kartal et al., 2004). In a study conducted by Wang et al. (2018), an antibacterial bio-board was prepared using bamboo pretreated with BV. Treating bamboo with BV for 30 minutes improved the antibacterial property of the bio-board. Several studies demonstrated lignin as a major antimicrobial compound in bamboo which makes it an eco-friendly and sustainable solution. Phenolic and alcohol groups in lignin might be responsible for this antimicrobial activity. Afrin et al. (2012) reported water insoluble fraction of *P. pubescens* that

DOI: 10.1201/9781003485933-3

TABLE 3.1
Antimicrobial Properties of Some Bamboo Species

Bamboo spp.	Antimicrobial Properties	Pathogen Against	Bioactive Compound Responsible for the Activity	References
P. heterocycla var. *pubescens (Pradelle) Ohwi*	Bacteria	*Escherichia coli, Bacillus subtilis, Salmonella typhimurium, Sarcina lutea, Pseudomonas aeruginosa, Staphylococcus aureus*	2,6-Dimethoxy-pbenzoquinone	Nishina et al. (1991)
P. heterocycla var. *pubescens (Pradelle) Ohwi*	Fungus	*Candida albicans Saccharomyces cerevisiae, Aspergillus niger*	2,6-Dimethoxy-pbenzoquinone	Nishina and Uchibori (1991)
Japanese bamboo shoots *(Phyllostachys pubescens)*	Bacteria	*Erwinia carotovora, Agrobacterium radiobacter, Agrobacterium rhizogenes, Clavibacter michiganensis, Curtobacterium flaccumfaciens*	Chitin-binding peptides (Pp-AMP 1 and Pp-AMP 2)	Fujimura et al. (2005)
Japanese bamboo shoots *(Phyllostachys pubescens)*	Fungi	*Fusarium oxysporum, Geotrichum candidum*	Chitin-binding peptides (Pp-AMP 1 and Pp-AMP 2)	Fujimura et al. (2005)
Phyllostachys heterocycla cv. Pubescens, P. kwangsiensis, P. heterocycla fo. Gracilis, P. heterocycla (Carrière) Mitford cv heterocycla	Bacteria	*Escherichia coli, Staphylococcus aureus*	Essential oil	Jin et al. (2011)
Moso bamboo *(Phyllostachys pubescens)*	Bacteria	*Staphylococcus aureus*	Dichloromethane extract from shoot skins	Tanaka et al. (2011)
P. heterocycla var. *pubescens (Pradelle) Ohwi*	Bacteria	*Escherichia coli, Bacillus subtilis, Salmonella typhimurium, Sarcina lutea*	Chloroform/methanol extract (bark)	Valentino ct al. (2015)
P. heterocycla var. *pubescens (Pradelle) Ohwi*	Bacteria	*Pseudomonas aeruginosa, Staphylococcus aureus*	Chloroform/methanol extract (bark)	Valentino et al. (2015)
Natural bamboo fiber samples (CP-II, CP-IV, CP-I, CPF-IV, CPE-XI), Natural bamboo fibers (CP-1), *Phyllostachys rubromarginata*	Bacteria	*Staphylococcus aureus*	CP-II, CP-IV, CP-I, CPF-IV, CPE-XI CP-II, CP-IV, CP-I, CPF-IV, CPE-XI	Rocky and Thompson (2019)
Japanese bamboo *(Kyoto-Moso, Kyushu-Madake, Kyushu-Moso)*	Bacteria	*Staphylococcus aureus, Escherichia coli*	Lignin (C-O and C-H group)	Ramful et al. (2022)
Bambusa vulgaris	Bacteria	*Staphylococcus aureus, Escherichia coli*	Saponins, flavonoids, glycosides tannins, terpernoids	Musyimi et al. (2023)
Bambusa blumeana	Bacteria	*Escherichia coli, Staphylococcus aureus*	Apigenin, luteolin, *p*-coumaric acid	Saducos (2022)
Phyllostachys pubescens	Fungi	*Phytophthora capsici, Fusarium graminearum, Botryosphaeria dothidea, Venturia nashicola, Botrytis cinerea*	4-hydroxycinnamic acid	Liao et al. (2021)
Sasa veitchii	Bacteria	*Aeromonas salmonicida*	Fatty acids	Asakura et al. (2019)

primarily consists of lignin being responsible for antimicrobial activity. The use of antimicrobial fabrics and fibers has recently gained attention, especially in hospitals and biomedical applications, for improving hygiene and prevention of nosocomial infections. These can be used for preparing dressings that can help in healing infected wounds. There are commercially available fibers with antibacterial properties that are either acrylic polymers or consist of silver and copper salts. Some of these are BIOKRIL, COURTEX, Amicor and Microsafe celanesse, etc. (Raru et al., 2021). Yang et al. (2012) reported the antimicrobial effect of bamboo charcoal-supported silver having antimicrobial activity against *Staphylococcus aureus*, *Bacillus subtilis*, *E. coli* and *Pseudomonas aeruginosa*. Bamboo-based antimicrobial products have natural antimicrobial properties and, unlike synthetic antimicrobial agents, cannot be washed off easily. Using synthetic antimicrobial products negatively impacts human health and if not disposed of properly may cause toxicity to other life forms (Gokarneshan and Khan, 2020). Thus, using natural products with inherent antimicrobial properties is a suitable alternative that can be used in developing antibacterial products such as textiles, bio-boards, herbal toothpaste and various personal care products. Bamboo-based multifunctional material can also help in maintaining a bacteria-free environment which can help improve the indoor air quality in hospitals and industries (Rutala and Weber, 2013).

This chapter delves into a variety of bamboo-based products and the scientific basis for the antimicrobial properties of bamboo plants. Besides, some of the bamboo-derived products with proven antimicrobial/antibacterial properties against a variety of pathogenic microorganisms and their prospects in the future to develop sustainable and cost-effective solutions in various sectors have also been discussed.

3.2 PHYTOCHEMICALS IMPARTING ANTIBACTERIAL PROPERTIES TO BAMBOO

Bamboo can grow at a rapid rate and show remarkable resilience towards stressful environmental conditions. Antimicrobial properties are crucial in protecting bamboo against various pathogenic microorganisms such as bacteria, fungi, insects, etc. The inherent antibacterial characteristics of bamboo are primarily due to its unique chemical composition and structural characteristics, which provide natural immunity against phytopathogens (Ramful et al., 2022).

Extracts prepared from the leaves and roots of bamboo reveal the presence of various bioactive compounds including alkaloids, tannins, glycosides, saponins, phenols, flavonoids and terpenoids, which impart antimicrobial characteristics to the plant (Coffie et al., 2014; Kebede et al., 2021; Musyimi et al., 2023; Owolabi and Lajide, 2015). Alkaloids have negative effects on microbes and are reported to have antimicrobial, antifungal and anti-inflammatory properties. They interfere with growth and metabolic processes, causing antimicrobial effects (Okwu and Okwu, 2004). Similarly, tannins penetrate bacterial cell walls and interfere with cellular metabolism resulting in cell destruction and thus showing strong antibacterial activity. They also inhibit bacterial growth by forming complexes with bacterial proteins and interfering with their function (Dabbaghi et al., 2019). Phenols have hydroxyl groups attached to the benzene ring; these groups possess high antioxidant activities that allow them to neutralize free radicals, protecting cells from oxidative damage, which contributes to their antimicrobial efficacy (Duraiswamy et al., 2018). Glycosides enhance the resilience of bamboo against microbes, insects and herbivores (Sharma et al., 2017). Saponins are known for their action against hypocholesterolemia and hypertension, and are also used as cardiac depressants. These also exhibit antimicrobial activity by disrupting microbial cell membranes and causing cell lysis (Sharma et al., 2017). Flavonoids, another important secondary metabolite, are recognized for their antibacterial, antiviral, anti-inflammatory, enzyme-inhibiting and antioxidant properties. They enhance bamboo's antimicrobial capabilities by interfering with bacterial cell wall components and metabolic processes (Cushnie and Lamb, 2005).

In traditional medicine, bamboo has been utilized for its therapeutic properties. Bamboo leaves are used in traditional Asian medicine to treat a wide range of ailments, such as hypertension, fever, arteriosclerosis, cardiovascular diseases and some cancers. This is probably because bamboo species absorb physiologically active substances like polyphenols and other secondary plant metabolites (Wróblewska et al., 2018). Moreover, their phenolic components are utilized in China to create an antioxidant of bamboo (AOB) leaves preparation, which is employed as a food antioxidant and whose usage is approved by the regional Health Ministry. The primary constituents of the AOB are phenolic acids, lactones and flavonoids. The flavone C-glycosides, which include orientin, homoorientin, vitexin and isovitexin, are the primary flavonoids present in AOB (Zhang et al., 2005). Bamboo is useful in treating lung and stomach heat as well as fever disorders because it is said to have relaxing, cooling and phlegm-resolving characteristics (Panee, 2015). Bamboo shoots also have antiviral, anticancer and antibacterial qualities due to their high fiber content (Fujimura et al., 2005). Bamboo is considered a heart-protective vegetable and helps regulate blood pressure because of its high potassium content. Its relatively high cellulose content (up to 4%), promotes peristaltic movement in the intestines and helps with digestion. Additionally, it prevents constipation and lowers body fat (Ambika and Rajagopal, 2017). According to Akao et al. (2004), the lignophenol derivatives derived from a wood combination comprising bamboo *P. bambusoides* showed neuroprotective effects in cells affected by hydrogen peroxide-induced apoptosis. *Bambusa rutila* polysaccharide-rich extract exhibited hydroxyl radical scavenging efficacy comparable to that of vitamin C (Gao et al., 2012). Due to their high content of oligosaccharides and polysaccharides, prebiotic activity was assessed in bamboo shoots, and they could also slow down the growth of harmful bacteria, specifically *Salmonella* sp. (He et al., 2016). In terms of their antioxidant properties, bamboo extracts exhibit outstanding biological performance. This action may potentially have implications for the treatment of a variety of illnesses, including cancer, cardiovascular protection against neurodegenerative disorders, resistance to free radicals and many others (Wróblewska et al., 2018).

The antibacterial property of both natural and thermally modified Japanese bamboo powder was identified by Ramful et al. (2022) as being attributed to distinctive bands allocated to the C-O and C-H functional groups in lignin, as determined by FTIR (Fourier Transform Infrared Spectroscopy) analysis. This suggests that it works well with the other phenolic parts (*p*-coumaric acid, ferulic acid, caffeic acid, homoorientin and orientin) that are in the raw extract (Liu et al., 2012). The antibacterial activity of *P. pubescens* may be due to the presence of aromatic and phenolic functional groups in lignin (Afrin et al., 2012). Many species have demonstrated significant antioxidant activity, indicating that they may be used to treat a variety of illnesses, including anti-inflammatory, anticancer and other conditions involving oxidative processes. Furthermore, bamboo extracts may include physiologically active peptides and polysaccharides in addition to the typical secondary metabolites. According to Wróblewska et al. (2018), these macromolecules, when combined with polyphenols and other metabolites, may have a variety of biological effects, including anti-free radical, antiaging, antifatigue, antibacterial and antivirus properties. They can also be used as a functional dietary supplement, cosmetic ingredient, or food additive. The phytochemical composition of the plants varies depending on the part used, the location, the season, the climate and other factors, which can be the cause of reported variations in antimicrobial activity between researchers (Barral-Martinez et al., 2022).

3.3 ADVANTAGES OF BAMBOO-BASED ANTIMICROBIAL PRODUCTS

Pathogenic microorganisms are abundantly present in the environment around us. These are responsible for causing diseases, and infections and are also responsible for bad breath and odor. Besides increasing use of antimicrobial agents has also caused resistance in some of the pathogenic bacterial strains. For instance, methicillin-resistant *S. aureus* and other antibiotic-resistant gram-negative bacteria have raised the concerns in research community (Dadgostar, 2019). All this calls for sustainable solutions and new antimicrobial products to combat the threat posed by these pathogenic strains.

There are certain risks associated with antibiotics, including longer persistence, slow degradation, increasing resistance among pathogenic microorganisms and toxicity towards non-targeted organisms in aquatic and terrestrial ecosystems, etc. Plant-based antimicrobial compounds have low toxicity to the environment, humans and animals (Li et al., 2023). Developing polymeric materials with inherent antimicrobial potential has gained popularity these days in different fields. There are chemically diverse polymeric materials with antibacterial activity such as polymers with quaternary nitrogen atoms, cationic polymers including polystyrene and polyvinyl pyridine, acrylic compounds and their derivatives, positively charged conjugated polyelectrolytes, polysiloxanes, antimicrobial peptides, phenol derivatives and branched polyethyleneimine, etc. (Muñoz-Bonilla and Fernández-García, 2012). Biopolymers, however, have gained attention these days due to increasing awareness towards the environment and human health. Some of the natural polymers have inherent antimicrobial potential that makes them suitable choice for medical and pharmacological applications.

Plant-based materials are low-cost, renewable resources with no toxicity and broad-spectrum antibacterial activity that makes them a suitable choice over synthetic antibiotics (Al-Jumaili et al., 2018). Bamboo is the world's fastest-growing woody plant with therapeutic potential. Bamboo shoots are also used medicinally to cure and manage diabetes and cholesterol through the production of various bamboo shoot–derived products including vinegar and salt (Singhal et al., 2013).

Bamboo possesses natural antibacterial properties which makes it an excellent choice for promoting hygiene and reducing bacterial growth (Xu et al., 2008). Bamboo fibers and other derived products are used in food, clothing and textiles, hygiene and personal care products, etc. Antimicrobial and therapeutic benefits of bamboo have been exploited for years. Bamboo plants also have inherent larvicidal and antiseptic properties, mainly due to the presence of hydrocyanic acid (Singhal et al., 2013). Bamboo cellulose is the sustainable choice over synthetic polymers and is used as a packaging material in various industries due to its antimicrobial and UV protective characteristics (Rashid et al., 2023). Bamboo composites have better antimicrobial performance than natural bamboo materials. For instance, in a study conducted by Mai et al. (2022), bamboo material was *in situ* filled with hydrogel, chitosan and zinc oxide which led to improved antibacterial activity and resistance to microbial corrosion up to 100% for 56 days. Similarly, Zhang et al. (2017) reported improved antibacterial performance of bamboo material by impregnation with zinc oxide and graphene oxide against *E. coli* and *B. subtilis*. Figure 3.1 presents the applications of antimicrobial characteristics of bamboo for various value-added products.

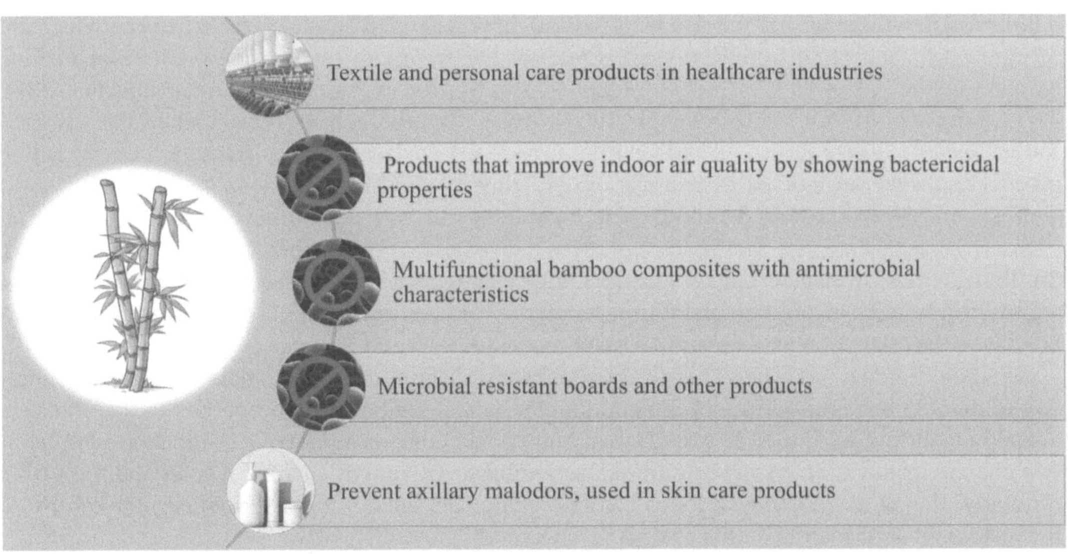

FIGURE 3.1 Application of bamboo-based antimicrobial products.

The biodegradable nature and antibacterial and anti-odor attributes of bamboo fibers make them a formidable choice in textile industries over synthetic fibers (Iroegbu and Ray, 2021). Bamboo-based composites are non-toxic in comparison to synthetic chemical sterilants used for sterilizing indoor environments such as formaldehyde and ethylene oxide (Rutala and Weber, 2013). Thus, bamboo-based antimicrobial products can be helpful in medical, healthcare and personal care products with inherent antibacterial action and unlike synthetic antimicrobial compounds, these are biodegradable, non-toxic, stable eco-friendly, sustainable and cannot be washed off easily.

3.4 BAMBOO PRODUCTS AND THEIR ANTIBACTERIAL CHARACTERISTICS

Bamboo has inherent antibacterial properties, which helps it to be sustainable in a variety of applications, especially in textiles and other items, where hygiene and cleanliness are primary considerations. Major constituents responsible for antibacterial properties of bamboo include polyphenolic chemicals. Their mode of action involves inhibiting enzymes such as phospholipase oxygenase by interacting with hydrosulfide groups and inactivating bacterial proteins (Cowan, 1999; Kim et al., 1995). According to numerous studies, there are significant biological benefits associated with bamboo plants including the prevention of cardiovascular diseases, antioxidant, anti-free radical and anti-aging properties (Lv et al., 2012). Bamboo has the potential to be an antibacterial due to the presence of bioactive compounds such as flavonoids, phenolics, tannins, anthraquinones, dietary fiber, polyphenols, carotenoids, anthocyanins, phytosterols and triterpenoids (Kalyan et al., 2023). This subsection covers some commercially available bamboo-based products with antibacterial activities.

The aforementioned properties of bamboo have made it popular in developing certain value-added products. For instance, hot-pressed bamboo-based particle boards with antibacterial properties is gaining popularity. In a study conducted by Yang et al. (2020), hot-pressed bamboo board developed by applying iron oxide was mildew resistant. Similarly, Wang et al. (2018) developed an antibacterial bio-board by treating bamboo with bamboo vinegar (BV) which has antibacterial, insecticidal and fungicidal properties.

Textile products made of bamboo fabric are usually popularized in the market for their eco-friendly and biodegradable nature (Hardin et al., 2009; Rodie, 2008). Bamboo fibers, derived primarily from the towering *Phyllostachys edulis* (Carrière) J.Houz., are indispensable for use in textile applications. Bamboo is being positioned as a stylish and long-lasting environmentally friendly building material because of the rise in popularity of regenerated bamboo viscose fibers, which are recognized for their high cellulose yield (Prakash, 2020). The most common method for making bamboo textiles involves transforming bamboo into viscose or rayon (Majumdar and Pol, 2014; Periyasamy and Militky, 2020; Xu et al., 2007). Apart from these, bamboo is a renewable resource for the textile industry due to its antibacterial properties, environmentally friendly fiber extraction process and wide range of textile applications (Kozłowski and Mackiewicz-Talarczyk, 2020).

Baby diapers made with bamboo fibers combined with wood pulp have natural antibacterial characteristics that are safe for the sensitive skin of babies and prevent infection (Malarvizhi, 2015). Similarly, female hygiene products such as sanitary napkins are also made with bamboo fibers. In a study conducted by Kathirvel and Ramachandran (2014), sanitary napkins were made using bamboo fibers that displayed antibacterial activity against *E. coli* and *S. aureus*.

Bamboo charcoal has inherent porosity and a large surface area which makes it a suitable additive to develop biocomposites and provide toughness and antibacterial properties to materials. In a study conducted by Liu et al. (2024), bamboo plastic composites after modification with bamboo charcoal/zinc oxide displayed remarkable antibacterial activity against *E. coli* and *S. aureus*. Similarly, Ren et al. (2020) developed bamboo charcoal loaded with silver-doped titanium oxide that displayed antibacterial action against *E. coli* and *S. aureus*.

Bamboo is increasingly used in the production of toothbrushes, toothpaste and various personal care products due to its sustainability, biodegradability and natural antimicrobial properties. These days, bamboo toothbrushes with plant-based bristles are available on the market. They are an eco-friendly substitute for standard nylon toothbrushes and can remove plaque just as well (Lodha et al., 2023). The culm, shoot and leaf of bamboo have been reported to possess antioxidation, anti-aging, antibacterial and antiviral properties (Azeez and Orege, 2018). Bamboo salt, a popular ingredient of traditional medicines in Korea, is known to possess antibacterial, antiviral and anti-inflammatory properties. It is an alkaline with a pH of 11.4 which makes it a powerful antioxidant. Recent studies have revealed reduced dental plaque, improved enamel strength and tooth demineralization prevented. Bamboo salt has antibacterial activity against *Streptococcus mutans* and other cariogenic microorganisms. In a study conducted by Biria et al. (2022), a herbal toothpaste with bamboo salt as a natural ingredient displayed antibacterial activity against *S. mutans* and *Lactobacillus* sp. A mixture of sodium fluoride and bamboo salt has been reported to reduce the growth of *S. mutans* and thus can help in preventing dental caries (Lee et al., 2019).

Some microorganisms are responsible for axillary malodor and create problems in the social interactions of human beings. Mostly deodorants are used as a management strategy to act against these microorganisms and inhibit the malodor (Bok et al., 2024). Bamboo salt has also been reported to have antibacterial activity against malodor-causing bacteria. Roh et al. (2009) evaluated the antibacterial effect of a body wash with 0.4% bamboo salt by weight. The body wash displayed antibacterial activity against *Corynebacterium glutamicum*, *C. xerosis*, and *C. jeikeium*. Bamboo salt has also been reported to act against *Helicobacter pylori* infections (Lee et al., 2022). Recently, natural ingredients have been popularized in skin care products due to various bioactive properties, including antioxidative, anti-inflammatory and antimicrobial activities. Essential oil from *Bambusa vulgaris,* due to its bioactive properties, is a better and cost-effective alternative for synthetic ingredients in skin care products (Alias et al., 2023).

3.5 CHALLENGES AND FUTURE PROSPECTS

Bamboo as a sustainable material offers diverse roles in different industries. It has gained popularity due to its inherent antibacterial characteristics and as a renewable resource. The presence of bioactive compounds with antimicrobial characteristics makes it a useful component of various textiles and personal hygiene products. It is being used in developing antibacterial products such as textiles, bio-boards, herbal toothpastes and various personal care products (Alias et al., 2023; Lee et al., 2022; Malarvizhi, 2015). Despite all the aforementioned benefits of bamboo-based antibacterial products, certain challenges prevent widespread acceptance and full exploration of this plant to develop antibacterial products. There are various bamboo species available across the world which vary in their physico-chemical and antibacterial activity, thus making it difficult to standardize the process globally and causing inconsistency in the performance of different products (Kaur et al., 2022). Thus, standardization procedures need to be established for the extraction and development of antibacterial products with consistent performance. Besides standardization and regulatory issues, market acceptance and risk assessment need to be done before commercializing antibacterial bamboo-based products. The future of bamboo-based antibacterial products seems promising as it provides sustainable, safe and eco-friendly solutions to the consumers.

3.6 CONCLUSION

Bamboo has natural antimicrobial properties that it uses for its defense against biotic stress and can be taken advantage of in manufacturing bamboo-derived products. Antibacterial properties of bamboo are a result of its unique chemical composition and structural characteristics. The presence of bioactive phytochemicals such as alkaloids, tannins, phenolics, glycosides, saponins and flavonoids, along with the complex organic polymer such as lignin, contribute to provide natural immunity against a variety of biotic threats to this plant. These properties not only enable bamboo to thrive in natural environments but also highlight its potential as a source of natural antimicrobial agents for various applications. Understanding the scientific basis of bamboo's antimicrobial/antibacterial properties provides valuable insights into its potential uses in medicine, agriculture and industry. If carefully exploited, it may also help to develop natural antibacterial bamboo-based value-added products; thus, limiting the use of synthetic chemical antibacterial agents.

REFERENCES

Afrin, T., Tsuzuki, T., Kanwar, R. K., & Wang, X. 2012. The origin of the antibacterial property of bamboo. *Journal of the Textile Institute 103*(8): 844–849.

Akao, Y., Seki, N., Nakagawa, Y., Yi, H., Matsumoto, K., Ito, Y., Ito, K., Funaoka, M., Maruyama, W., Naoi, M., & Nozawa, Y. 2004. A highly bioactive lignophenol derivative from bamboo lignin exhibits a potent activity to suppress apoptosis induced by oxidative stress in human neuroblastoma SH-SY5Y cells. *Bioorganic & Medicinal Chemistry 12*(18): 4791–4801.

Alias, A. H. D., Razak, N. Q., Yusoff, A., Chin, M. H. M., Kamal, K. H., Uyup, M. L., Abdullah, N. H., Ridzuan, S., Saaid, N. S., & Shafie, M. 2023. Antioxidant and antimicrobial activities of essential oils extracted from bamboo (*Bambusa vulgaris*) leaves and its application in skincare products: A review. *Biocatalysis and Agricultural Biotechnology 54*: 102930.

Raru, A., Florea-Burduja, E., Irovan, M., & Farima, D. 2021. Bamboo fiber antibacterial effect (a review). *Annals of the University of Oradea, Fascicle of Textiles, Leatherwork, 1*: 71–76.

Al-Jumaili, A., Kumar, A., Bazaka, K., & Jacob, M. V. 2018. Plant secondary metabolite-derived polymers: A potential approach to develop antimicrobial films. *Polymers 10*(5): 515.

Ambika, K., & Rajagopal, B. 2017. In vitro antimicrobial and antiproliferative activity of *Bambusa vulgaris*. *International Journal of Pharmacy and Pharmaceutical Research 9*(1): 10–22.

Asakura, H., Makino, S. I., Watanabe, K., Tuchida, Y., Kawabe, M., & Sakurai, D. 2019. Kuma bamboo grass (*Sasa veitchii*) extracts exhibit protective effects against atypical *Aeromonas salmonicida* infection in goldfish (*Carassius auratus*). *Biocontrol Science 24*(3): 145–154.

Azeez, M. A., & Orege, J. I. 2018. Bamboo, Its chemical modification and products. *Bamboo-Current and Future Prospects*: 25–48.

Barral-Martinez, M., Cassani, L., Carpena, M. G., Pereira, A., Garcia-Oliveira, P., Chamorro, F., & Simal-Gandara, J. 2022. Evaluation of the antimicrobial potential of extracts from plants of the family *Asteraceae*. *Medical Sciences Forum 12*(1): 17.

Biria, M., Rezvani, Y., Roodgarian, R., Rabbani, A., & Iranparvar, P. 2022. Antibacterial effect of an herbal toothpaste containing Bamboo salt: A randomized double-blinded controlled clinical trial. *BMC oral Health 22*(1): 193.

Bok, J., Choi, J., Lee, S., Lim, T. H., & Jang, Y. 2024. Antibacterial and deodorizing effects of cold atmospheric plasma-applied electronic deodorant. *Scientific Reports 14*(1): 3011.

Coffie, G. Y., Antwi-Boasiako, C., & Darkwa, N. A. 2014. Phytochemical constituents of the leaves of three bamboo (*Poaceae*) species in Ghana. *Journal of Pharmacognosy and Phytochemistry 2*(6): 34–38.

Cowan, M. M. 1999. Plant products as antimicrobial agents. *Clinical Microbiology Reviews 12*(4): 564–582.

Cushnie, T. T., & Lamb, A. J. 2005. Antimicrobial activity of flavonoids. *International Journal of Antimicrobial Agents 26*(5): 343–356.

Dabbaghi, A., Kabiri, K., Ramazani, A., Zohuriaan-Mehr, M. J., & Jahandideh, A. 2019. Synthesis of bio-based internal and external cross-linkers based on tannic acid for preparation of antibacterial superabsorbents. *Polymers for Advanced Technologies 30*(11): 2894–2905.

Dadgostar, P. 2019. Antimicrobial resistance: Implications and costs. *Infection and Drug Resistance 12*: 3903–3910.

Duraiswamy, B., Singanan, M., & Varadarajan, V. 2018. Physicochemical, phytochemicals and antioxidant evaluation of *Guazuma ulmifolia* fruit. *International Journal of Pharmacy and Pharmaceutical Sciences 10*(9): 87.

Fujimura, M., Ideguchi, M., Minami, Y., Watanabe, K., & Tadera, K. 2005. Amino acid sequence and antimicrobial activity of chitin-binding peptides, Pp-AMP 1 and Pp-AMP 2, from Japanese bamboo shoots (*Phyllostachys pubescens*). *Bioscience, Biotechnology, and Biochemistry 69*(3): 642–645.

Gao, Y. N., Tian, C. R., & Zhao, L. L. 2012. Extraction, purification and antioxidant activity of polysaccharides from bamboo leaves. *Journal of Forestry Research 23*: 139–143.

Gokarneshan, N., & Khan, A. 2020. Antimicrobial properties of bamboo fibres. *Biomedical Journal of Scientific & Technical Research 32*(5): 25299–25302.

Gupta, A., & Kumar, A. 2008. Potential of bamboo in sustainable development. *Asia Pacific Business Review 4*(3): 100–107.

Hardin, I. R., Wilson, S. S., Dhandapani, R., & Dhende, V. 2009. An assessment of the validity of claims for "Bamboo" fibers. *AATCC Review 9*(10): 33–36.

He, S., Wang, X., Zhang, Y., Wang, J., Sun, H., Wang, J., & Ye, Y. 2016. Isolation and prebiotic activity of water-soluble polysaccharides fractions from the bamboo shoots (*Phyllostachys praecox*). *Carbohydrate Polymers 151*: 295–304.

Iroegbu, A. O. C., & Ray, S. S. 2021. Bamboos: From bioresource to sustainable materials and chemicals. *Sustainability 13*(21): 12200.

Jin, Y. C., Yuan, K., & Zhang, J. 2011. Chemical composition, and antioxidant and antimicrobial activities of essential oil of *Phyllostachys* heterocycla cv. *Pubescens* varieties from China. *Molecules 16*(5): 4318–4327.

Kalyan, N., Santosh, O., Indira, A., Kuhad, A., & Chongtham, N. 2023. Therapeutic aspects of bamboo for wound healing. In: *Bamboo Science and Technology*. Springer Nature Singapore, Singapore, 237–264.

Kartal, S. N., Imamura, Y., Tsuchiya, F., & Ohsato, K. 2004. Preliminary evaluation of fungicidal and termiticidal activities of filtrates from biomass slurry fuel production. *Bioresource Technology 95*(1): 41–47.

Kathirvel, K. P., & Ramachandran, T. 2014. Development of antimicrobial feminine hygiene products using bamboo and aloevera fibers. *Journal of Natural Fibers 11*(3): 242–255.

Kaur, P. J., Pant, K. K., Satya, S., & Naik, S. N. 2016. Bamboo: The material of future. *International Journal Series in Multidisciplinary Research 2*(2): 27–34.

Kaur, P. J., Yadav, P., Gupta, M., Khandegar, V., & Jain, A. 2022. Bamboo as a source for value added products: Paving way to global circular economy. *BioResources 17*(3): 5437–5463.

Kebede, T., Gadisa, E., & Tufa, A. 2021. Antimicrobial activities evaluation and phytochemical screening of some selected medicinal plants: A possible alternative in the treatment of multidrug-resistant microbes. *PLoS One 16*(3): e0249253.

Kim, J., Marshall, M. R., & Wei, C. I. 1995. Antibacterial activity of some essential oil components against five foodborne pathogens. *Journal of Agricultural and Food Chemistry 43*(11): 2839–2845.

Komatsu, H., Onozawa, Y., Kume, T., Tsuruta, K., Shinohara, Y., & Otsuki, K. 2012. Canopy conductance for a Moso bamboo (*Phyllostachys pubescens*) forest in Western Japan. *Agricultural and Forest Meteorology 156*: 111–120.

Kozłowski, R. M., & Mackiewicz-Talarczyk, M. 2020. Introduction to natural textile fibres. In: *Handbook of Natural Fibres*. Woodhead Publishing, 1–13.

Lee, H. J., Park, A., & Oh, H. N. 2019. Effects of bamboo salt with sodium fluoride on the prevention of dental caries. *Journal of Dental Hygiene Science 19*(4): 288–293.

Lee, T. H., Jeong, H. Y., An, D. Y., Kim, H., Cho, J. Y., Hwang, D. Y., Lee, H. J., Ham, K. S., & Moon, J. H. 2022. Bamboo Salt and triple therapy synergistically inhibit helicobacter pylori-induced gastritis in vivo: A preliminary study. *International Journal of Molecular Sciences 23*(22): 13997.

Li, S., Jiang, S., Jia, W., Guo, T., Wang, F., Li, J., & Yao, Z. 2023. Natural antimicrobials from plants: Recent advances and future prospects. *Food Chemistry 432*: 137231.

Liao, M., Ren, X., Gao, Q., Liu, N., Tang, F., Wang, G., & Cao, H. 2021. Anti-fungal activity of Moso bamboo (*Phyllostachys pubescens*) leaf extract and its development into a botanical fungicide to control pepper phytophthora blight. *Scientific Reports 11*(1): 4146.

Liu, C., Zhang, S., Yan, S., Pan, M., & Huang, H. 2024. Mechanical and antibacterial properties of bamboo charcoal/ZnO-modified bamboo fiber/polylactic acid composites. *Forests 15*(2): 371.

Liu, L., Liu, L., Lu, B., Xia, D., & Zhang, Y. 2012. Evaluation of antihypertensive and antihyperlipidemic effects of bamboo shoot angiotensin converting enzyme inhibitory peptide in vivo. *Journal of Agricultural and Food Chemistry 60*(45): 11351–11358.

Lodha, V., Saraf, T., & Jawdekar, A. 2023. Comparison of three different solutions for the disinfection of bamboo toothbrushes with plant based bristle and nylon bristle toothbrushes: A randomized controlled trial. *JIDA: Journal of Indian Dental Association 17*(10):15–21.

Lv, Z. L., Lin, X., Miao, Z. H., Guo, H. X., Wang, J. A. H., Lei, M. L., Pan, Y., & Zhang, B. L. 2012. Antioxidant activity of bamboo-leaf extracts from the species *Dendrocalamopsis oldhami*. *Scientific Research and Essays 7*(44): 3789–3796.

Mai, X., Mai, J., Liu, H., Liu, Z., Wang, R., Wang, N., Li, X., Zhong, J., Deng, Q., & Zhang, H. 2022. Advanced bamboo composite materials with high-efficiency and long-term anti-microbial fouling performance. *Advanced Composites and Hybrid Materials* 5:1–8.

Majumdar, A., & Pol, S. B. 2014. Low stress mechanical properties of fabrics woven from bamboo viscose blended yarns. *Fibers and Polymers 15*: 1985–1991.

Malarvizhi, G. 2015. Development of herbal finished baby diapers with bamboo fiber. *International Journal of Humanities, Arts, Medicine and Sciences (BEST: IJHAMS)* *3*(2): 41–45.

Muñoz-Bonilla, A., & Fernández-García, M. 2012. Polymeric materials with antimicrobial activity. *Progress in Polymer Science* *37*(2): 281–339.

Musyimi, D. M., Khasabulli Buyela, D., Mbilu Kiema, F., & Chebii, T. K. 2023. Phytochemical compounds and antimicrobial activity of extracts of Bamboo against Staphylococcus aureus and *Escherichia coli*: Extracts of Bamboo leaves and roots. *Archives of Ecotoxicology* *5*(2): 45–51.

Nishina, A., Hasegawa, K., Uchibori, T., Seino, H., & Osawa, T. 1991. 6-dimethoxy-p-benzoquinone as an antibacterial substance in the bark of *Phyllostachys heterocycla* var. *Pubescens*, a species of thick-stemmed bamboo. *Journal of Agricultural and Food Chemistry* *39*(2): 266–269.

Nishina, A., & Uchibori, T. 1991. Antimicrobial activity of 2, 6-dimethoxy-p-benzoquinone, isolated from thick-stemmed bamboo, its analogs. *Agricultural and Biological Chemistry* *55*(9): 2395–2398.

Okwu, D. E., & Okwu, M. E. 2004. Chemical composition of *Spondias mombin* Linn plant parts. *Journal of Sustainable Agriculture and the Environment* *6*(2): 140–147.

Owolabi, M. S., & Lajide, L. 2015. Preliminary phytochemical screening and antimicrobial activity of crude extracts of *Bambusa vulgaris* Schrad. Ex JC Wendl.(*Poaceae*) from southwestern Nigeria. *American Journal of Essential Oils and Natural Products* *3*(1): 42–45.

Panee, J. 2015. Potential medicinal application and toxicity evaluation of extracts from bamboo plants. *Journal of Medicinal Plant Research* *9*(23): 681.

Periyasamy, A. P., & Militky, J. 2020. Sustainability in regenerated textile fibers. In: *Sustainability in the Textile and Apparel Industries.* Sourcing Synthetic and Novel Alternative Raw Materials. Springer nature, Singapore, 63–95.

Prakash, C. 2020. Bamboo fibre. In: *Handbook of Natural Fibres*. Woodhead Publishing, 219–229.

Ramful, R., Sunthar, T. P., Kamei, K., & Pezzotti, G. 2022. Investigating the antibacterial characteristics of Japanese bamboo. *Antibiotics* *11*(5): 569.

Rashid, M. M., Abir, N., Hossain, M. A., Iqbal, M. I., & Siddique, A. B. 2023. Bamboo cellulose: Structure, properties, and applications. In: *Cellulose Fibre Reinforced Composites*. Woodhead Publishing, 23–48.

Ren, Q., Zeng, Z., Jiang, Z., & Li, H. 2020. Functionalization of renewable bamboo charcoal to improve indoor environment quality in a sustainable way. *Journal of Cleaner Production* *246*: 119028.

Rocky, B. P., & Thompson, A. J. 2019. Production of natural bamboo fiber—2: Assessment and comparison of antibacterial activity. *AATCC Journal of Research* *6*(5): 1–9.

Rodie, J. B. 2008. Going green: Beyond marketing hype. *Textile World* *159*(6): 28–30.

Roh, S. S., Yun, W. S., Jung, J. Y., Yu, H. U., Hwang, D. S., Choi, S. M., & Chang, S. Y. 2009. The effect of body wash containing triclosan and bamboo salt on axillary malodor. *Journal of the Society of Cosmetic Scientists of Korea* *35*(3): 219–228.

Rutala, W. A., & Weber, D. J. 2013. Disinfection and sterilization: An overview. *American Journal of Infection Control* *41*(5): S2–S5.

Saducos, A. G. 2022. Potentials of Kawayang tinik (*Bambusa blumeana*) as new source antimicrobial agents. *Plant Science Today* *9*(3): 518–523.

Sharma, K. R., Kumar, S., Karna, L., Kalauni, S. K., & Pokharel, Y. R. 2017. Phytochemical analysis and in-vitro antimicrobial screening of some selected medicinal plants of Nepal. *International Journal of Recent Scientific Research* *8*: 14986–14994.

Singhal, P., Bal, L. M., Satya, S., Sudhakar, P., & Naik, S. N. 2013. Bamboo shoots: A novel source of nutrition and medicine. *Critical Reviews in Food Science and Nutrition* *53*(5): 517–534.

Tanaka, A., Kim, H. J., Oda, S., Shimizu, K., & Kondo, R. 2011. Antibacterial activity of Moso bamboo shoot skin (*Phyllostachys pubescens*) against Staphylococcus aureus. *Journal of Wood Science* *57*: 542–544.

Valentino, M. J. G., Ganado, L. S., Ganado, M. R., & Undan, J. R. 2015. Phytochemical screening and bio assay of the anti-microbial activity of three species of bamboo in Nueva Ecija, Philippines. *Advances in Environmental Biology* *9*(24): 389–396.

Wang, L., Ge, S., Liu, Z., Zhou, Y., Yang, X., Yang, W., Li, D., & Peng, W. 2018. Properties of antibacterial bioboard from bamboo macromolecule by hot press. *Saudi Journal of Biological Sciences* *25*(3): 465–468.

Wróblewska, K. B., de Oliveira, D. C., Grombone-Guaratini, M. T., & Moreno, P. R. H. 2018. Medicinal properties of bamboos. *Pharmacognosy-Medicinal Plants*: 1–18.

Xu, Q., Jiang, P., & Xu, Z. 2008. Soil microbial functional diversity under intensively managed bamboo plantations in southern China. *Journal of Soils and Sediments* *8*: 177–183.

Xu, Y., Lu, Z., & Tang, R. 2007. Structure and thermal properties of bamboo viscose, Tencel and conventional viscose fiber. *Journal of Thermal Analysis and Calorimetry* *89*(1): 197–201.

Yang, F. C., Wu, K. H., Huang, J. W., Horng, D. N., Liang, C. F., & Hu, M. K. 2012. Preparation and characterization of functional fabrics from bamboo charcoal/silver and titanium dioxide/silver composite powders and evaluation of their antibacterial efficacy. *Materials Science and Engineering: C 32*(5): 1062–1067.

Yang, F. C., Wu, K. H., Liu, M. J., Lin, W. P., & Hu, M. K. 2009. Evaluation of the antibacterial efficacy of bamboo charcoal/silver biological protective material. *Materials Chemistry and Physics 113*(1): 474–479.

Yang, L., Lou, Z., Han, X., Liu, J., Wang, Z., Zhang, Y., Wu, X., Yuan, C., & Li, Y. 2020. Fabrication of a novel magnetic reconstituted bamboo with mildew resistance properties. *Materials Today Communications 23*: 101086.

Zhang, J., Zhang, B., Chen, X., Mi, B., Wei, P., Fei, B., & Mu, X. 2017. Antimicrobial bamboo materials functionalized with ZnO and graphene oxide nanocomposites. *Materials 10*(3): 239.

Zhang, Y., Bao, B., Lu, B., Ren, Y., Tie, X., & Zhang, Y. 2005. Determination of flavone C-glucosides in antioxidant of bamboo leaves (AOB) fortified foods by reversed-phase high-performance liquid chromatography with ultraviolet diode array detection. *Journal of Chromatography A 1065*(2): 177–185.

4 The Medicinal and Nutraceutical Properties of Bamboo Shoots

Aanika Gupta and Pallavi Singh

4.1 INTRODUCTION

Widely recognised as "Green Gold" and "Poor Man's Timber", bamboo belongs to the *Poaceae* (*Gramineae*) family (subfamily Bambusoideae) and is regarded as the largest arborescent grass in the world (Das et al., 2021; Tewari et al., 2019). Since the dawn of time, bamboo has been of immense significance as it acts as a treasure of bioactive constituents. Different countries like India, China, Korea, and Japan have age-old traditional systems where bamboo is utilised to fulfil the deficiencies of vital nutrients and as a therapeutic agent to cure a wide array of disorders (Wróblewska et al., 2019). As per the estimations, approximately 80% of residents of developing nations rely extensively on herbal medications for the fulfilment of their primary healthcare needs and nearly about 25% of the pharmaceuticals prescribed by medical practitioners are sourced from plant species in developed nations (Hamilton, 2004). Furthermore, the estimations provided by the Food and Agricultural Organisation (FAO) show nearly 800 million inhabitants of the globe are experiencing significant deprivation of food and nutrition (Da Silva, 2014). As per the reports stipulated by the FAO, a spike of 60% is necessary for the agricultural produce yield to replenish the global food shortages by the year 2050. As bamboo is an economical, multifunctional, natural resource that exhibits resilience to a diverse range of climatic conditions, it can be utilised as a sustainable alternative to mitigate global food shortages by acting as a potent nutraceutical and as an active raw material in pharmaceuticals owing to its curative properties (Scherr and McNeely, 2012).

The juvenile shoots of bamboo act as rich repositories of phytochemicals possessing high nutritive value and health-promoting properties (Nirmala et al., 2014). The prevalence of substantial amounts of fibres enabled bamboo shoots to alleviate colon carcinoma and further maintain the concentration of cholesterol in the blood (Sohgaura et al., 2018). The nutritional composition of bamboo shoots exhibited the prevalence of appreciable quantities of various minerals including sodium, potassium, iron, selenium, magnesium, silicon, phosphorus and calcium, enabling its suitability to combat micronutrient deficiency prevailing in the human population worldwide (Bajwa et al., 2019; Chongtham et al., 2011). The consumption of bamboo shoots as delicacies dates to the tribal era and is currently devoured either as fermented or roasted vegetables by the Asian population (Choudhury et al., 2012; Van Dung et al., 2024). Additionally, about 2 million tonnes of nutritious bamboo shoots are eaten across the globe, encompassing Asia, Europe, Africa, North America and Oceania (Choudhury et al., 2011).

The current chapter focuses primarily on the various bioactive constituents prevailing in the bamboo shoots which provide the plant with robust nutraceutical and medicinal properties.

4.2 DISTRIBUTION AND BOTANICAL CHARACTERISTICS

Bambusoideae incorporates approximately 1575 species belonging to 116 diverse genera found worldwide (Ohrnberger, 1999; Sawarkar et al., 2021). The indigenous population of bamboo can be observed extensively in the tropical, sub-tropical and mild temperate zones encompassing Asia,

DOI: 10.1201/9781003485933-4

Africa, Europe and North, South and Central America (Lobovikov et al., 2007). Roughly 148 bamboo species within 33 genera and 6 diverse types are found to prevail profusely in India, where the higher incidence of *Bambusa*, *Dendrocalamus*, *Arundinaria*, *Chimonobambusa*, *Dinochloa* and *Ochlandra* genera have been reported by the India State of Forest Report (ISFR, 2017) (Kumari, 2019; Nirala et al., 2017). Bamboos are pivotal in mitigating soil erosion, protecting biodiversity, safeguarding riverbanks and capturing carbon, thereby, sustaining the health of forest ecosystems (Du et al., 2018; Han et al., 2013; Van der Lugt et al., 2018; Yuen et al., 2017).

Bamboo is characterised as an evergreen perennial woody grass with height varying from 30 cm to 4000 cm (Das et al., 2021). Further, the plant possesses rhizome-based roots which develop into a stem-like cylindrical structure known as culm which is completely devoid of the insides (Ahmad et al., 2021). Based on the type of rhizome, the juvenile culms tend to develop either from lateral buds in the case of leptomorph rhizomes or from apical buds in the case of pachymorph rhizomes. Culm has a diameter extending from 0.25 to 12 inches and is represented as a series of nodes segregated by internodes which bear petiolate scale leaves. As bamboo is a monocot, therefore it is distinguished by the prevalence of dispersed vascular bundles throughout its culms. Additionally, the culms are enclosed at the internodal region by the transformed triangular leaves known as the culm sheath to ensure the protection of juvenile shoots (Banik, 2000). The distinctive attribute of bamboo plants is the possession of compound inflorescence represented as an aggregation of spikelets (Keng, 1983). Moreover, the plant experiences three diverse types of flowering cycles which include annual flowering, gregarious flowering and sporadic flowering (Brandis, 1899).

4.3 ETHNOPHARMACOLOGICAL SIGNIFICANCE OF BAMBOO SHOOTS

Bamboos have captured global interest and hold immense significance in healthcare and food industries due to their prevalence of exceptional nutritional and therapeutic properties (Nirmala et al., 2018). Bamboos have been extensively utilised in folk medicines from ancient times, predominantly in Ayurveda, Unani, Chinese, Tibetan and Korean traditional medicine systems (Nirmala et al., 2018; Nirmala and Bisht, 2015; Sangeetha et al., 2015). Elaborate information on the application of bamboo shoots as a traditional medicine in the diverse nations of the Asian continent has been summarised in Table 4.1.

4.4 BIOACTIVE CHEMICAL CONSTITUENTS

To enhance the survival rates in scenarios of anthropogenic and natural crisis, plants often synthesise tonnes of bioactive compounds to promote their health. In the recent past, global research has identified that plants which were often neglected are endowed with voluminous amounts of phytoconstituents which exhibit a diverse array of medicinal properties. Bamboo shoots, being one of the underutilised and neglected plants, have revealed their prevalence of dietary fibres, sterols, phenols, flavonoids and glycosides which demonstrate tumour suppressive, hypoglycaemic, lipid-lowering, antimicrobial, antiarthritic, antifatigue, cardioprotective and antioxidant properties (Bajwa et al., 2020). Moreover, these bioactive compounds are now immensely valued for their low toxicity and environmental friendliness, rendering them a preferred choice over synthetically made pharmaceuticals. Table 4.2 depicts the nature of diverse bioactive chemical constituents commonly found in the juvenile shoots of bamboo. The chemical structure of the phytoconstituent was retrieved from PubChem (nih.gov).

4.5 MINERAL-RICH BAMBOO SHOOTS: CARDINAL FOR FOOD ENRICHMENT

Bamboo shoots are considered nutritional treasures, but their concentration of minerals experiences extensive fluctuations based on climatic conditions, soil type, season of cultivation and harvesting and maturity of plants and species (Bajwa et al., 2019; Saini et al., 2017). Studies highlight that

TABLE 4.1
Utilisation of Bamboo Shoots in Complementary and Alternative Medicinal Therapies

Species	Disease	Mode of Administration	Country	References
Bambusa arundinacea (Retz.) Willd.	Indigestion	Consumption of curry made from tender shoots.	India	(Sukumaran et al., 2014)
Bambusa bambos (L.) Voss	Malaria and rheumatism	Consumption of cooked juvenile shoots with rice.	India	(Silambarasan and Ayyanar, 2015)
Dendrocalamus giganteus Munro	Wounds and cuts	The paste of wool obtained by scrapping the juvenile culms is topically applied on skin.	India	(Medhi et al., 2010)
Dendrocalamus strictus	Skin injuries and Wounds	Concoction prepared from soft shoots is topically smeared on wounds and bruises.	India	(Singh et al., 2003)
Bambusa tulda	Skin injuries and Wounds	Paste prepared from soft shoots is topically smeared on wounds and bruises.	India	(Singh et al., 2003)
Dendrocalamus asper (Schult.f.) Backer	Muscle soreness	Consumption of crushed stems.	Indonesia	(Jadid et al., 2020)
Bambusa vulgaris Schrad. ex J.C. Wendl.	Malaria	Oral consumption of concoction prepared from stems.	Indonesia	(Novaryatiin and Indah, 2019)
Bambusa vulgaris	Leukoderma	-	Indonesia	(Setiawati et al., 2017)
Sasa borealis (Hack.) Makino	Hypoglycaemia	Consumption of decoction of stem.	Korea	(Kim and Song, 2011)
Bambusa bambos (L.) Voss	Lipoma and leprosy	Consumption of concoction of juvenile stems.	Thailand	(Srisawat et al., 2016)
Bambusa spp.	Hepatic diseases	Oral consumption of decoction prepared from woody stems.	Cambodia	(Chassagne et al., 2017)
Bambusa vulgaris Schrad. ex J.C. Wendl.	Allergic reactions	Topical application of burnt stems mixed with coconut oils on the affected area.	Philippines	(Abe and Ohtani, 2013)
Schizostachyum lumampao (Blanco) Merr.	Navel cut in infants	Topical application of burnt stem on wound.	Philippines	(Ong and Kim, 2015)
Bambusa spp.	Usage as postpartum dietary supplement	Consumption of boiled juvenile shoots.	Malaysia	(Awang-Kanak et al., 2018)
Dendrocalamus hamiltonii Nees & Arn. ex Munro	Gastroenteritis and oedema	Oral consumption of boiled culms with garlic and ginger.	China	(Luo et al., 2020)
Indosasa pingbianensis McClure	Cephalalgia and upper respiratory infections	Consumption of shoots.	China	(YuMing et al., 2004)
Phyllostachys heterocycla (Carrière) Matsum.	Cough and throat inflammation	Consumption of sap extracted from juvenile culms.	China	(YuMing et al., 2004)

bamboo shoots are packed with an abundance of essential macro- and microelements, rendering them a potential superfood. The predominant macronutrients comprise potassium, phosphorus, magnesium and calcium while trace micronutrients include zinc, nickel, manganese, cobalt, iron, selenium and copper (Christian et al., 2015; Chogantham et al., 2011).

Potassium has emerged as the dominant macronutrient. For instance, juvenile shoots of *Phyllostachys mannii* were found to have an exceptionally high potassium abundance around 6660 ± 70 mg/100 g (dry weight) (Saini et al., 2017; Premlata, 2020). *Bambusa arundinaria* tops the list with 1400 mg/100 g (wet weight basis) (Bhargava et al., 1996). On the other hand, *Melocanna baccifera*, *Bambusa tulda*, *Dendrocalamus hookeri*, *D. longispathus* and *D. sikkimensis* possess much

TABLE 4.2
Chemical Structures of Diverse Bioactive Compounds Found in Bamboo Shoots

S. No.	Phytochemical Name	Chemical Structure	PubChem ID	Molecular Weight	Phytochemical Family
1.	β-sitosterol	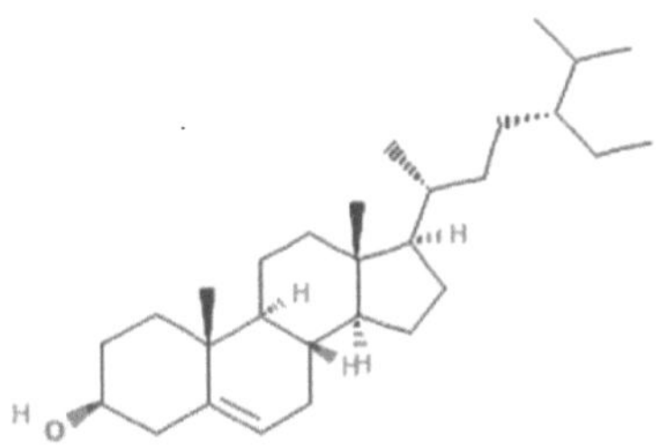	222284	414.7 g/mol	Sterol
2.	Campesterol	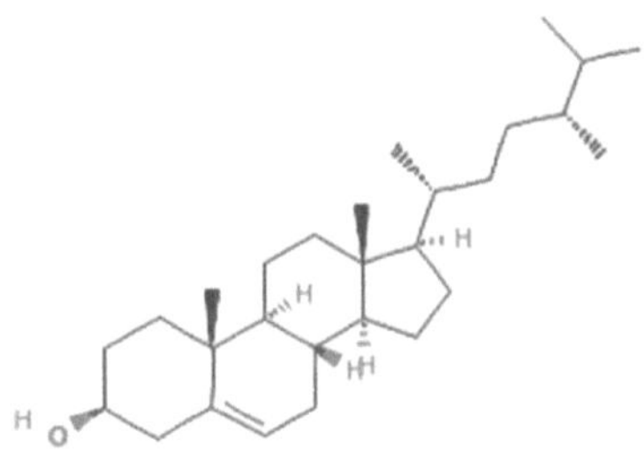	173183	400.7 g/mol	Sterol
3.	Stigmasterol	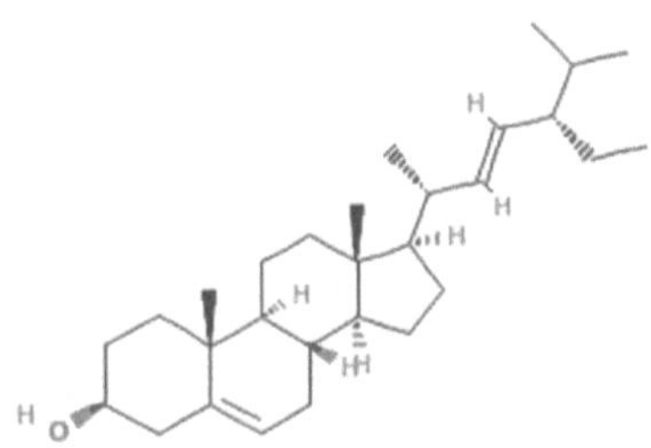	5280794	412.7 g/mol	Sterol
4.	Cholesterol	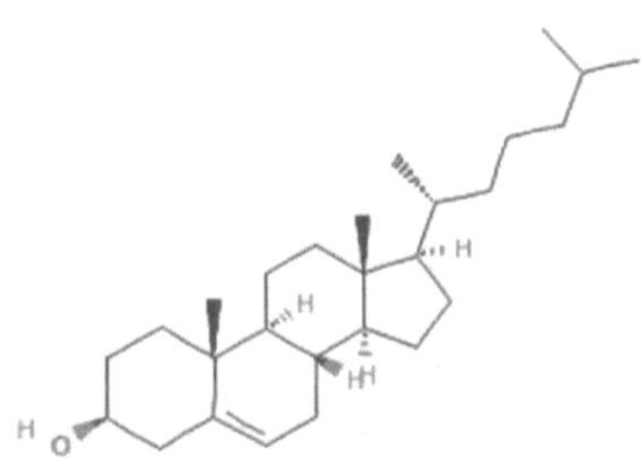	5997	386.7 g/mol	Sterol
5.	Ergosterol	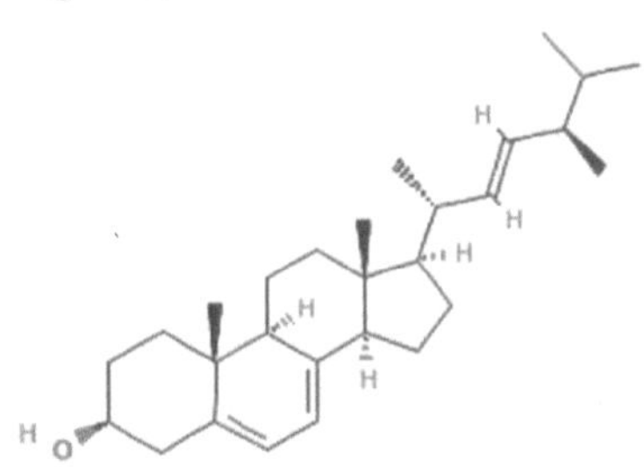	444679	396.6 g/mol	Sterol

(Continued)

TABLE 4.2 *(Continued)*
Chemical Structures of Diverse Bioactive Compounds Found in Bamboo Shoots

S. No.	Phytochemical Name	Chemical Structure	PubChem ID	Molecular Weight	Phytochemical Family
6.	Stigmastanol	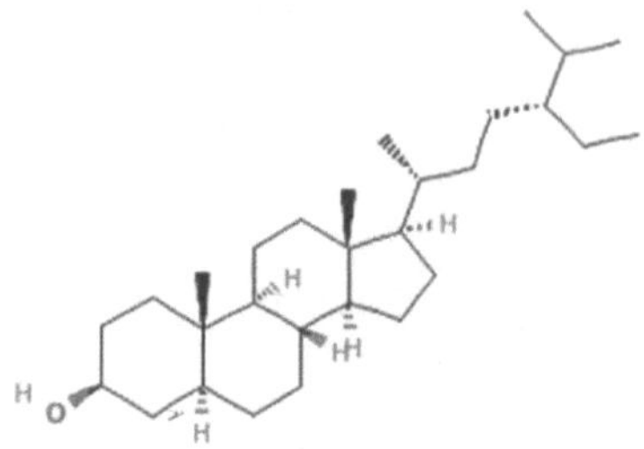	241572	416.7 g/mol	Sterol
7.	Protocatechuic acid	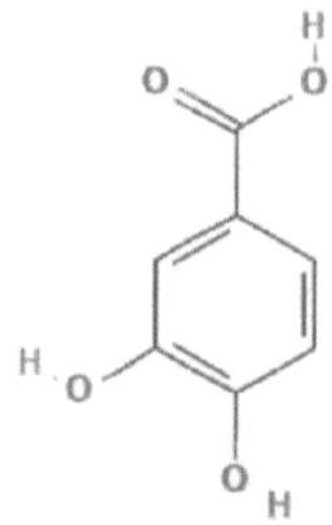	72	154.12 g/mol	Phenol
8.	p-hydroxybenzoic acid	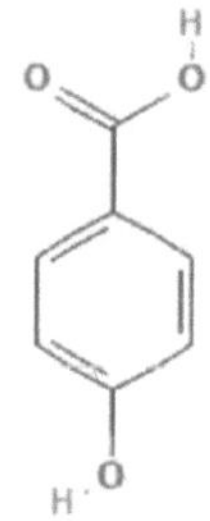	135	138.12 g/mol	Phenol
9.	Catechin	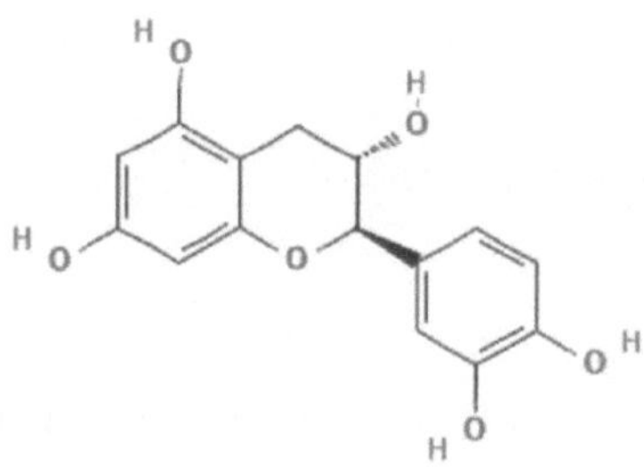	9064	290.27 g/mol	Phenol
10.	Caffeic acid	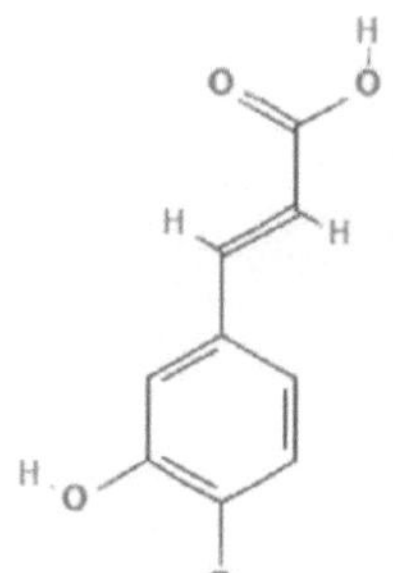	689043	180.16 g/mol	Phenol

(Continued)

TABLE 4.2 *(Continued)*
Chemical Structures of Diverse Bioactive Compounds Found in Bamboo Shoots

S. No.	Phytochemical Name	Chemical Structure	PubChem ID	Molecular Weight	Phytochemical Family
11.	Chlorogenic acid	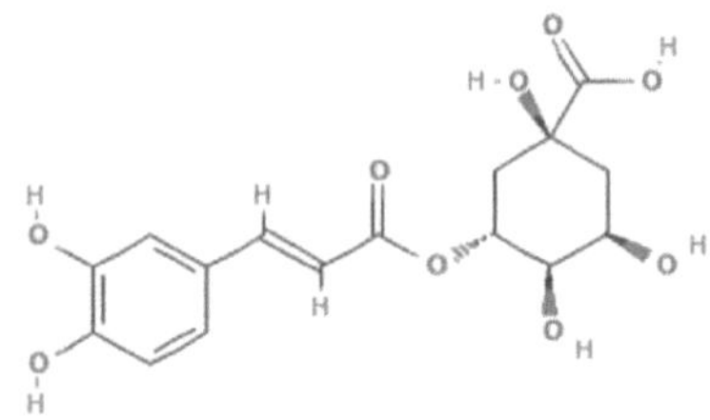	1794427	354.31 g/mol	Phenol
12.	Syringic acid	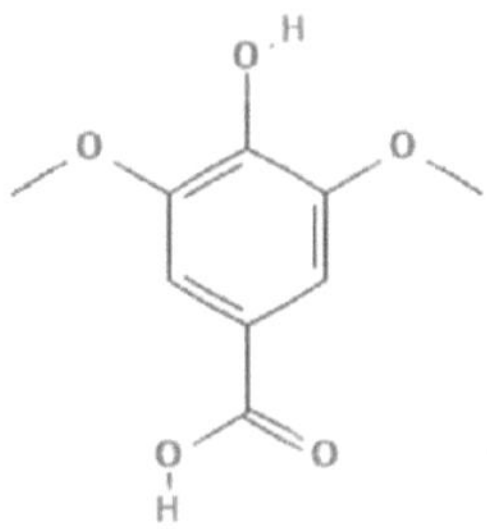	10742	198.17 g/mol	Phenol
13.	p-coumaric acid	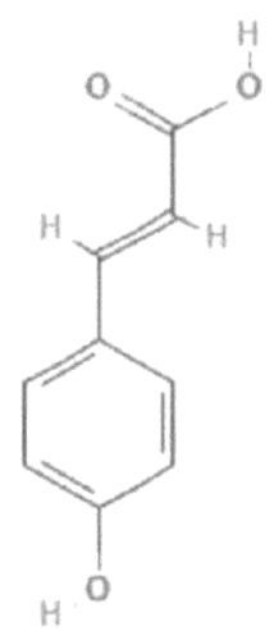	637542	164.16 g/mol	Phenol
14.	Ferulic acid	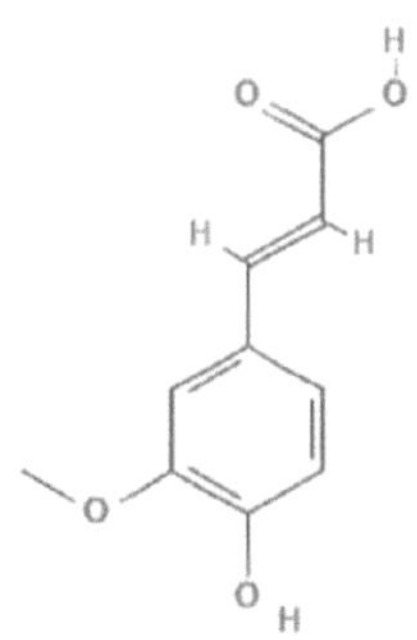	445858	194.18 g/mol	Phenol
15.	Gallic Acid	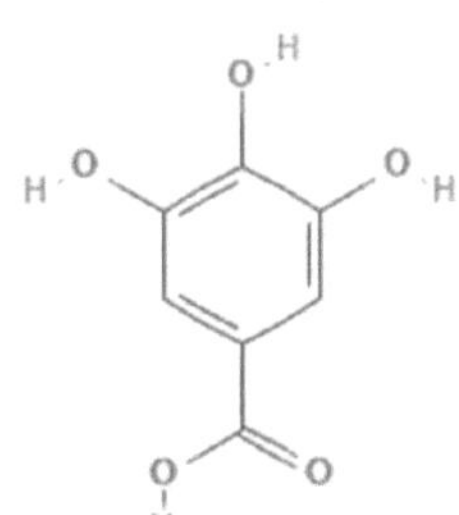	370	170.12 g/mol	Phenol

(Continued)

TABLE 4.2 *(Continued)*
Chemical Structures of Diverse Bioactive Compounds Found in Bamboo Shoots

S. No.	Phytochemical Name	Chemical Structure	PubChem ID	Molecular Weight	Phytochemical Family
16.	Vanillic Acid	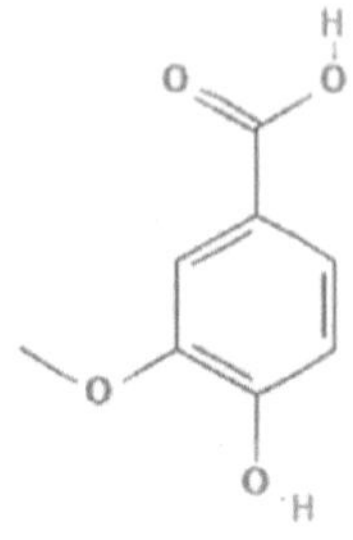	8468	168.15 g/mol	Phenol
17.	Isovitexin	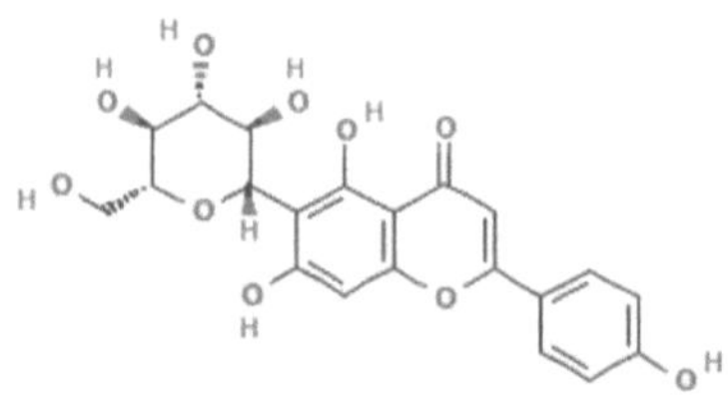	162350	432.4 g/mol	Flavonoid
18.	Isoorientin	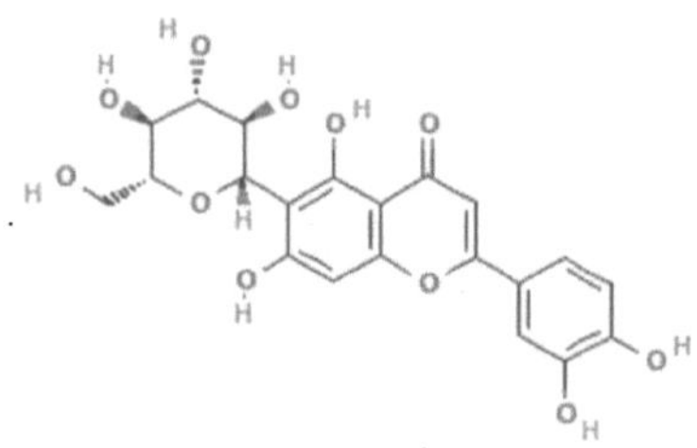	114776	448.4 g/mol	Flavonoid
19.	Tricin	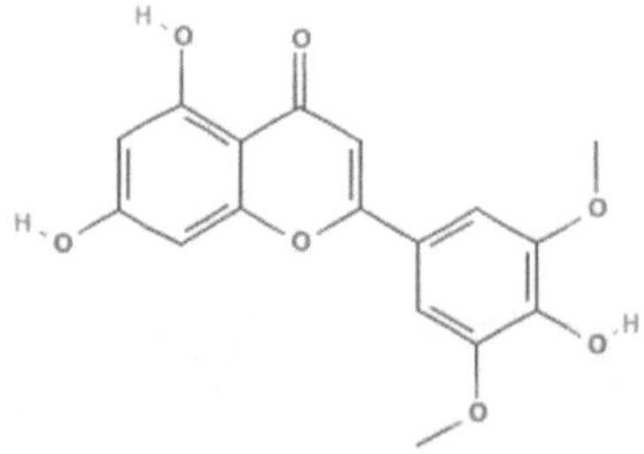	5281702	330.29 g/mol	Flavonoid
20.	Orientin	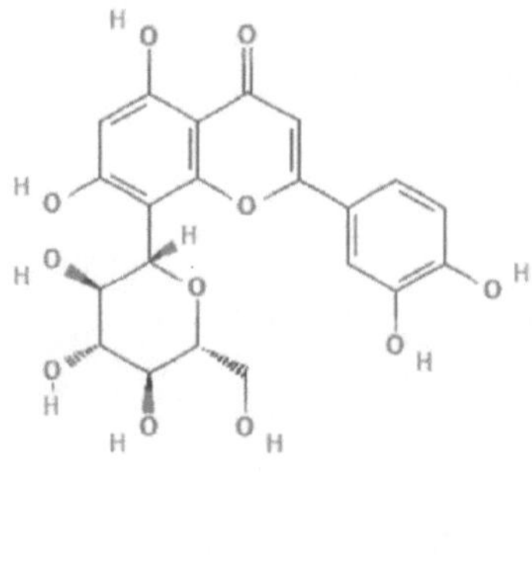	5281675	448.4 g/mol	Flavonoid

(Continued)

TABLE 4.2 *(Continued)*
Chemical Structures of Diverse Bioactive Compounds Found in Bamboo Shoots

S. No.	Phytochemical Name	Chemical Structure	PubChem ID	Molecular Weight	Phytochemical Family
21.	Vitexin	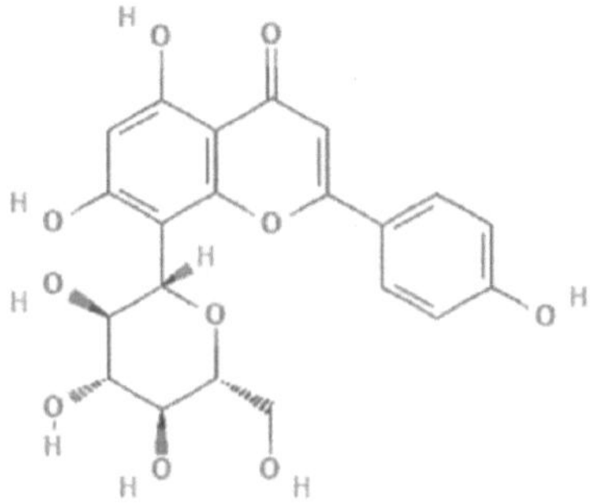	5280441	432.4 g/mol	Flavonoid
22.	Rutin	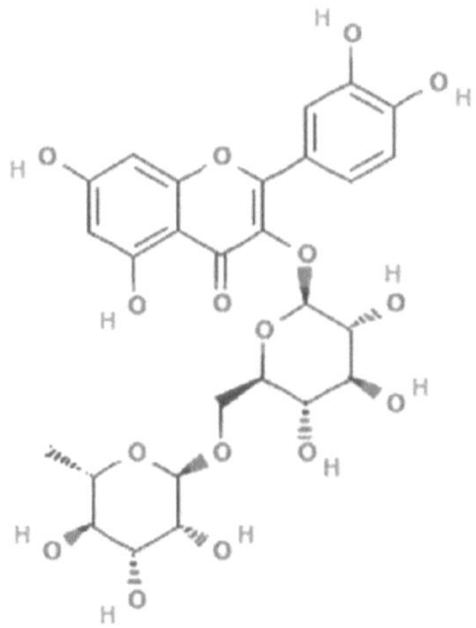	5280805	610.5 g/mol	Flavonoid
23.	Quercetin	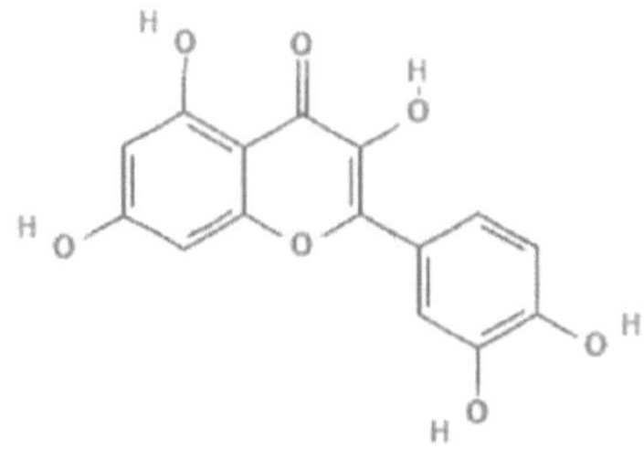	5280343	302.23 g/mol	Flavanoid
24.	Taxiphyllin	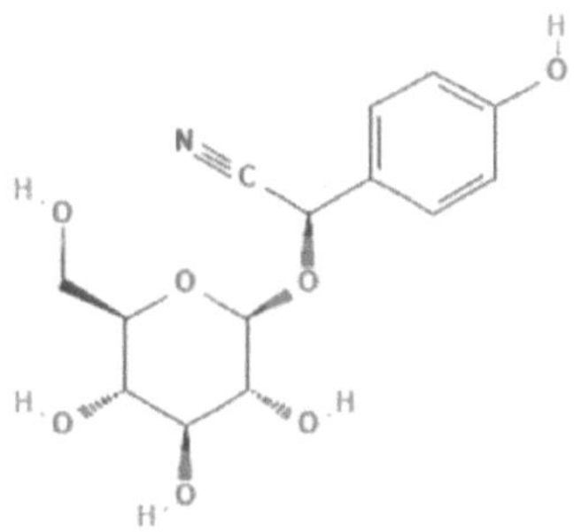	107721	311.29 g/mol	Cyanogen Glycoside

lower potassium levels around 20 mg/100 g (wet weight basis) (Bhargava et al., 1996; Nirmala et al., 2007). Interestingly, in 10-day-old fresh shoots of *B. bamboos*, *B. tulda*, *D. asper* and *D. giganteus*, a noticeable drop in potassium levels was recorded (Bhatt et al., 2005). This suggests that with the advancing age of bamboo shoots, the content of potassium largely decreases. Furthermore, potassium levels in bamboo shoots vary significantly based on storage duration, growing location and altitude (Bajwa et al., 2019; Saini et al., 2017). Additionally, a substantial reduction in potassium is often observed when processing techniques such as fermentation, boiling and brining are applied. However, the potassium content can be effectively preserved through freeze-drying the bamboo shoots, thereby maintaining their nutritional value.

Phosphorus is the second most prevalent macronutrient whose content ranges from 0.46 g to 0.93 g/100 g in freshly harvested bamboo shoots (Bajwa et al., 2019; Saini et al., 2017). Notably, the content of phosphorus varies within the different parts of the bamboo shoots, as observed in *Yushania alpina*, where the upper region possessed 763 mg/100 g, while the lower portion had 481 mg/100 g (Karanja et al., 2015). On a dry-weight basis, *D. hookerii* recorded the highest phosphorus content of 1000 mg/100 g, followed by *M. baccifera* (800 mg/100 g) and *D. giganteus*, *D. sikkimensis* and *P. bambusoides* (700 mg/100 g each) (Bhatt et al., 2005). Furthermore, sprouts of *B. bamboos* and *P. mannii* revealed the prevalence of 750 ± 50 and 930 ± 60 mg/100 g (dry weight) of phosphorus, respectively (Saini et al., 2017; Premlata, 2020).

Magnesium has emerged as one of the vital nutrients in freshly harvested bamboo shoots, with impressive concentrations recorded in *M. baccifera* (300 ± 20 mg/100 g of dry weight), followed by *D. sikkimens* and *D. manipureanus* (240 ± 20 mg/100 g of dry weight) (Saini et al., 2017; Premlata, 2020). Interestingly, the upper region of *Y. alpina* shoots exhibit even higher magnesium content, measuring 430 mg/100 g (Karanja et al., 2015). When the bamboo shoots are subjected to the process of sun drying and fermentation, the magnesium concentration is enhanced to a slight extent while boiling and treatment with a brine solution significantly reduces its concentration (Bajwa et al., 2019; Saini et al., 2017). However, such magnesium-packed bamboo shoots can significantly boost immune functions and the muscular function and contribute towards the maintenance of heart health and neural function.

Among several species, *D. sikkimensis* emerges as a powerhouse of calcium, with an impressive concentration of 1900 mg/100 g of dry weight, followed by *D. hookerii* with 1600 mg/100 g of dry weight (Bhatt et al., 2005). The abundance of calcium in bamboo shoots highlights their potential to promote bone strength and support muscular functions. The shoot tip of *P. edulis* exhibits particularly high levels of iron with a concentration of 12.92 mg/100 g of dry weight while the entire shoot averages an iron content of 9.25 mg/100 g (Huang et al., 2014). Likewise, the shoots of *Pleioblastus maculatus* contributes significantly to iron intake with a concentration of 6.75 mg/100 g of dry weight (Wang et al., 2014). These iron-loathed bamboo shoots can significantly enhance haemoglobin levels, thereby supporting overall energy metabolism.

A study revealed that bamboo shoots have exceptionally high amounts of micronutrients when compared with regularly consumed vegetables, including tomatoes, peppers, potatoes and spinach (Huang and Yang, 2006). As the freshness of bamboo shoots diminishes, a notable depletion in the quantities of both macro- and micronutrients is observed (Nirmala et al., 2007). Furthermore, with the advancing age of the bamboo shoots, a marked decline in the mineral content is revealed (Nirmala and Sharma, 2008). Such dynamic alterations in mineral composition substantially enhance the nutritional profile of bamboo shoots and emphasize their importance in addressing specific dietary requirements.

4.6 THERAPEUTIC POTENTIAL OF BAMBOO SHOOTS

Juvenile bamboo shoots have always been an integral part of folk medicine and traditional diets, predominantly on the Asian continent. Scientific research conducted worldwide in recent years has consequently brought significant attention to bamboo shoots due to their phenomenal therapeutic

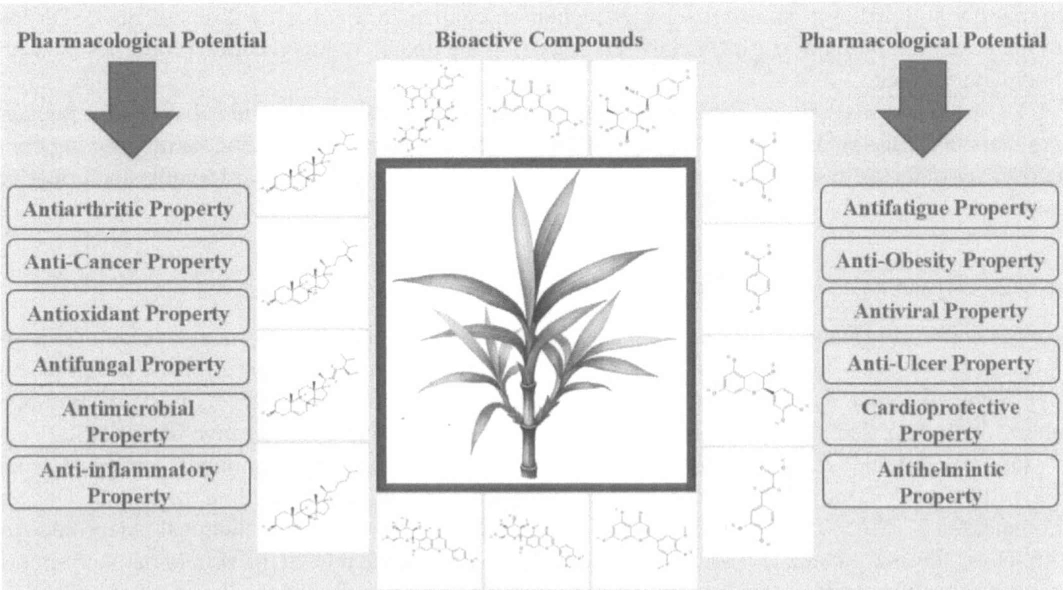

FIGURE 4.1 Depiction of therapeutic significance of bamboo and its different bioactive compounds.

potential, attributed to their rich composition of bioactive constituents, essential nutrients and functional peptides (Figure 4.1). These attributes have established bamboo shoots as an emerging superfood with considerable potential in preventing and managing diverse health conditions. The elaborate account of some of the various medicinal properties, such as hypoglycaemic, tumour suppressive, anti-inflammatory and cholesterol-lowering potential, depicted by bamboo shoots have been discussed in the following sections.

4.6.1 Hypoglycaemic Potential

Diabetes mellitus is a growing health crisis worldwide, with over 830 million inhabitants affected in 2022 and 1.6 million mortalities reported in 2021 (WHO factsheet). Alarmingly, speculations suggest that beyond 750 billion people across the globe will be affected with diabetes by 2045 (Shoaib et al., 2022). Managing hyperglycaemia, a major concern in diabetes, often involves pharmaceuticals that inhibit enzymes like α-amylase and α-glucosidase, which are pivotal in the metabolism of starch and release of glucose (Baenas et al., 2016).

Interestingly, recent research underlined the promising hypoglycaemic potential of dietary fibres extracted from bamboo shoots. A pharmacological analysis revealed that soluble dietary fibres derived from bamboo shoots exhibited the IC_{50} value of 0.76 mg/mL and 0.65 mg/mL against α-amylase and α-glucosidase, respectively (Zheng et al., 2019). Notably, the inhibitory effect on α-amylase was comparable to that of Acarbose, a well-known antidiabetic drug, which has an IC_{50} value of 0.78 mg/mL. In an *in-vivo* experiment involving male Kunming mice, administration of a 5% w/w dose of total dietary fibres for four weeks resulted in a remarkable 38.2% reduction in blood glucose levels. These findings demonstrated that bamboo dietary fibres can inhibit key enzymes and promote insulin secretion from β-cells and glucose absorption, thus acting as a natural solution for blood sugar control.

Another study further validated the antidiabetic effects of enzymatic bamboo shoot dietary fibres (EMBSDF) through *ex-vivo* and *in-vivo* approaches (Yao et al., 2024). The results revealed that administering 312.8 mg/kg of EMBSDF significantly improved insulin sensitivity in hyperglycaemic Sprague-Dawley mice and boosted glucagon-like peptide (GLP-1) levels by 57.79%. GLP-1 plays a vital role in facilitating insulin secretion, rendering this an important finding in diabetes management. Furthermore, *in-vitro* experimentation confirmed that higher doses of EMBSDF led to increased inhibition of α-glucosidase, emphasizing the potent hypoglycaemic properties of bamboo dietary fibres.

These studies suggest that bamboo shoots could be a revolutionary natural substitute for managing diabetes, offering dual benefits of insulin regulation and enzyme inhibition. With growing concerns over diabetes worldwide, bamboo shoot dietary fibres may provide a sustainable and effective approach to control blood sugar levels.

4.6.2 Tumour-Suppressive Potential

As per the World Health Organization 2018 reports, carcinoma is the predominant reason for mortalities around the world. Therefore, natural sources are extensively explored for the development of effective therapeutics to prevent carcinoma in the global population (Gogoi et al., 2024). Bamboo shoots with their rich bioactive profile demonstrated remarkable antiproliferative potential.

One exceptional study evaluated the antiproliferative effect of compounds extracted from *Phyllostachys heterocycla* var. pubescens shoots on the breast cancer line Michigan Cancer Foundation-7 (MCF-7) (Abdelhameed et al., 2020). A novel phytochemical identified as "(6′-O-octadeca-8″,11″-dienoyl)-sitosterol-3-O-d-glucoside" derived from the skin of bamboo shoots exhibited an IC_{50} value equivalent to 25.8 μM, closely matching that of the standard chemotherapeutic drug Fluorouracil (IC_{50} value: 26.98 μM). Notably, this compound stimulated apoptosis in cancer cells through the elevated production of p53, Bax and caspase 3, 9 and 8 while suppressing the BCL2 gene, a known antiapoptotic marker.

Further reinforcing the tumour-suppressive potential of bamboo shoots, the pharmacological activity of bioactive constituents has been investigated. Quercetin, a flavonoid found in the juvenile shoots of bamboo, has shown potent antiproliferative effects by facilitating the expression of p53 and activating the NF-κB signalling pathway (Priyadarsini et al., 2010). Quercetin induced apoptosis in lung cancer cells (A549) by upregulating p53 and halting cell division (Chan et al., 2013). Additionally, quercetin inhibits cancer progression by preventing STAT-3 tyrosine phosphorylation, a crucial step in inflammation-driven carcinogenesis (Liao and Lin, 2014). By modulating key cell cycle regulators such as p21WAF1, cyclin B and p27KIP1, quercetin has demonstrated its efficacy in suppressing tumour growth in breast, lung, hepatoma and squamous cell carcinoma (NavaneethaKrishnan et al., 2019).

Another tumour-suppressive flavonoid in bamboo shoots is rutin. A study on HCT116 human colon carcinoma cells revealed that rutin upregulates the expression of caspase-3 and β-Actin, inducing apoptosis and arresting cancer cell proliferation at the G_0–G_1 phase (Jayameena et al., 2018). The IC_{50} value for rutin was reported equivalent to 679.858 μM, indicating its significant anticancer effect.

Additionally, caffeic acid, a phenolic compound present in bamboo shoots, plays a pivotal role in preventing cancer by inhibiting the progression of hepatocarcinoma (Espíndola et al., 2019). Caffeic acid achieves this by minimising the expression of matrix metalloproteinases (MMP2 and MMP9), suppressing STAT signalling and preventing angiogenesis. Furthermore, it reduces oxidative stress by limiting the production of reactive oxygen species and enhancing the oxidation of DNA in tumour cells, thereby reducing cancer cell survival (Espíndola et al., 2019).

These findings emphasise the immense potential of bamboo shoots as a natural, multi-faceted approach to cancer therapy, providing a promising substitute or a complement to conventional treatments.

4.6.3 Anti-Inflammatory Activity

Inflammation characterised as a protective mechanism is adopted by the body against a diverse array of potential pathogens and in healing tissue injuries (Ahmed, 2011). While inflammation plays a vital role in restoring cellular homeostasis, it also manifests as vasodilation, erythema, pain, swelling and heat at the affected site (Medzhitov, 2007). However, chronic or excessive inflammation is often associated with several diseases, encompassing diabetes, rheumatoid arthritis and carcinoma (Shi et al., 2009).

The latest research has highlighted the anti-inflammatory potential of bamboo shoots. One study, focussed on *Pleioblastus amarus* (Keng), revealed that both the shoot and its shells exhibited notable anti-inflammatory effects (Ren et al., 2019). The shoot extract suppressed nitric oxide synthesis in murine macrophages, with an IC_{50} value of 78 μm/mL, while the shell extract was even more potent, with an IC_{50} value of 55 μm/mL. Furthermore, both extracts at 100 μm/mL effectively suppressed the activity of cyclooxygenase-2 (COX-2) and inducible nitric oxide synthase (iNOS), reducing the synthesis of key inflammatory markers like interleukin 1 (IL-1) and tumour necrosis factor α (TNF-α) at the transcription level. This anti-inflammatory action was linked to the attenuation of the Extracellular signal regulated kinase signalling pathway, as demonstrated by the diminished phosphorylation of ERK.

Another study explored the anti-inflammatory properties of bamboo shoots from *Gigantochloa albocillata*, specifically in the context of gout arthritis (Mustaffa et al., 2022). Extracts prepared with different solvents—distilled water, methanol and 1:1 distilled water-methanol blend—were tested for their ability to reduce inflammation. The water-methanol extract, after ultrasonication for 30 minutes, exhibited the highest inhibition of 74.15%, comparable to the pharmaceutical anti-inflammatory drug Allopurinol. Active phenolic compounds such as ellagic acid and ferulic acid were identified in the extract, which contributed significantly to its anti-inflammatory effects.

Additionally, gallic acid, found in bamboo shoots, is vital in reducing inflammation by suppressing the production of chemokines, cytokines and adhesion factors while also minimising cell infiltration (Bai et al., 2021). Gallic acid is also crucial in regulating Mitogen activated protein kinase pathway and NF-κB signalling pathways, further supporting its anti-inflammatory activity.

These studies suggest the potential of bamboo shoots to act as a natural remedy for preventing inflammation, providing an alternative to conventional pharmaceutical treatments.

4.6.4 Cholesterol Reducing and Antihypertensive Property

About 17.9 million fatalities across the globe in the year 2019 were the direct consequence of cardiovascular diseases, as per the statistics provided by World Health Organisation. Hypertension is the major risk factor linked to cardiovascular disorders and related mortality (Turner and Hooper, 2002). The blood pressure regulation is predominantly controlled by the angiotensin converting enzyme (ACE).

Exciting research has highlighted the cholesterol-lowering and antihypertensive potential of bamboo shoots. One study investigated the effects of bamboo shoot peptides obtained from *Phyllostachys pubescens* on the high-fat diet of Sprague Dawley rats and spontaneously hypertensive rats (Liu et al., 2012). The results were impressive: the bamboo shoot peptide significantly lowered the levels of triglycerides, low-density lipoprotein cholesterol and total cholesterol in mice along with downregulation of fatty acid synthase activity. Moreover, it substantially lowered systolic blood pressure and suppressed ACE activity when administered at 50 mg/day/kg body weight for 30 days.

In a related study, dietary fibres extracted from bamboo shoots of *Fargesia spathacea* demonstrated remarkable cholesterol-lowering properties (Wu et al., 2020). The ability of both soluble and insoluble dietary fibres to adsorb cholesterol was notably high: 10.82 mg/g and 9.59 mg/g at a pH equivalent to 7.0, exceeding the cholesterol adsorption capacities of dietary fibres of other bamboo species such as *P. pubescens*, *P. iridescens*, *P. edulis* and *P. praecox*.

These findings suggest the therapeutic potential of bamboo shoots in controlling hypertension and blood cholesterol levels, acting as a natural alternative for enhancing heart health (Table 4.3).

TABLE 4.3
Therapeutic Effects of Bamboo Shoots

S. No.	Therapeutic Potential	Species	Active Constituent	References
1.	Hypoglycaemic property	*Dendrocalamus hamiltonii, D. latiforus, Chimonobambusa quadrangularis, Bambusa vulgaris, Gigantochloa ligulata*	Dietary fibres	(Imdad et al., 2022; Kong et al., 2020; Li et al., 2016; Li et al., 2018; Tang et al., 2022a; Tang et al., 2022b; Yao et al., 2024; Zheng et al., 2019)
2.	Tumour suppressive and anticancer property	*Phyllostachys heterocycla* var. pubescens, *Dendrocalamus asper*	Sterol, flavonoids	(Abdelhameed et al., 2020; Ontaha et al., 2021)
3.	Anti-inflammatory property	*Pleioblastus amarus* (Keng), *Gigantochloa albocillata, Phyllostachys pubescens, Bambusa arundinacae*	Alkaloids, phenolics	(Lu et al., 2011; Mustaffa et al., 2022; Ren et al., 2019; Vanitha et al., 2016)
4.	Antihypertensive property	*P. pubescens, P. nigra, Gigantochloa apus, Bambusa blumeana*	Peptide, phenolics, flavonoids, tannins, saponins, tannins	(Alvarez et al., 2023; Liu et al., 2012; Liu et al., 2013; Octaviani, 2020a; Octaviani, 2020b; Park and Jhon, 2010)
5.	Antiobesity property	*P. pubescens, Fargesia spathacea, P. edulis*	Dietary fibres	(Li et al., 2021; Liu et al., 2012; Lu et al., 2010; Tang et al., 2022; Wu et al., 2020; Zhou et al., 2023)
6.	Antioxidant property	*P. pubescens, P. nigra, B. vulgaris, G. ligulate, B. polymorpha, Chusquea bambusoides, C. meyeriana, C. capituliflora Trin. var.* pubescens, *Merostachys pluriflora*	Phenolics	(Kong et al., 2020; Milani et al., 2020; Park and Jhon, 2010; Thomas et al., 2016; Wróblewska et al., 2019)
7.	Prebiotic property	*Fargesia spathacea, Chimonobambusa quadrangularis, P. praecox*	Dietary fibres, polysaccharides	(Chen et al., 2019; He et al., 2016; Liu et al., 2012; Tang et al., 2022)
8.	Antimicrobial property	*B. polymorpha, P. pubescens, B. balcooa, D. hamiltonii*	Peptides, sterols, tannins, phenolics, glycosides, flavonoids, saponins	(ES et al., 2021; Fujimura et al., 2005; Singh et al., 2012; Tanaka et al., 2013; Thomas et al., 2016)

4.7 NUTRACEUTICAL POTENTIAL

The tender bamboo shoots not only serve as an exotic Asian delicacy, widely known for their crispy texture and luscious flavour, but also as a treasury exhibiting prolific amounts of carbohydrates, proteins, vitamins, minerals and gut-friendly dietary fibres offering substantial health-promoting benefits, thereby rendering bamboo shoots as an emerging nutraceutical (Shi and Yang, 1992). Here is a closer look into a thorough exploration of bamboo shoots' noteworthy nutritional composition.

4.7.1 Vitamin Profile

Freshly harvested bamboo shoots are often loaded with appreciable quantities of essential vitamins including vitamin B1, B3, B6, E and A which significantly contribute towards the maintenance of health (Visuphaka, 1985; Xia, 1989). Notably, 152.31 mg/100 g (dry weight) of vitamin C has been found in the shoots of *D. hookerii,* followed by *M. baccifera* with 108.57 mg/100 g of (dry weight) (Bhatt et al., 2005). Interestingly, juvenile shoots of *D. oldhami* showcased the existence of high levels of vitamin C equivalent to 7.3 mg/100 g, significantly higher than *B. nutans* which reported 1.19 mg/100 g of vitamin C (Huang and Yang, 2006). Consequently, the prevalence of extensive ascorbic acid content imparts effective antiradical activity, both *in vivo* and *ex vivo,* rendering bamboo shoots a natural source of antioxidants, which can contribute to minimizing oxidative stress and strengthening immune responses (Nirmala et al., 2018). Moreover, the shoots of *P. amarus* possess considerable levels of riboflavin (0.04 mg/100 g), pyridoxine (0.38 mg/100 g), thiamine (0.09 mg/100 g) and β-carotene (0.65 μg/100 g) (dry weight), while the shoots of *D. oldhami* revealed the presence of riboflavin (0.7 mg/1000 g), thiamine (1.7 mg/1000 g) and niacin (3.5 mg/1000 g) (wet weight), all of which are immensely crucial for maintaining skin health, energy metabolism and vision, consequently contributing towards the nutraceutical potential of bamboo shoots (Dang et al., 2017; Zhu and Luo, 2012).

4.7.2 Proteins and Biofunctional Peptides

Beyond vitamin content, bamboo shoots exhibit a remarkable protein reservoir which can help in fulfilling the daily requirements of essential amino acids in the human diet. Bamboo shoots are known to possess substantial amounts of 17 major amino acids, of which 8 are considered essential for the maintenance of homeostasis within the human body (Nongdam and Tikendra, 2014). In addition, the quantity of protein is significantly determined by the maturity and species of bamboo shoots, varying from 21.1 g to 33.4 g/100 g (dry weight) and 1.49 g to 4.04 g/100 g (wet weight) (Chongtham et al., 2011; Karanja et al., 2015; Nirmala et al., 2007). For instance, the fresh shoots of *D. giganteus* contain 3.86 g/100 g (wet weight) of protein while the dried shoots of *Y. alpine* exhibit protein content equivalent to 33.4 g/100 g (dry weight) (Karanja et al., 2015; Nirmala et al., 2007).

Apart from protein, the nutritious shoots of bamboo also possess certain bioactive peptides which display therapeutic potential. A protein named "Dendrocin", obtained from the shoots of *D. latiflora,* revealed antifungal potential with IC_{50} values equivalent to 5.1 μM, 1.8 μM and 1.4 μM against *Mycosphaerella arachidicola, Botrytis cincerea,* and *Fusarium oxysporum,* thereby preventing the proliferation in their mycelia (Wang and Ng, 2003). Further, two peptides, Pp-AMP2 and Pp-AMP1, isolated from *P. pubescens* shoots, were effective against *Geotrichum candidum* and *F. oxysporum* fungi at an IC_{50} value of 2 μg/mL and *Agrobacterium radiobacter, Erwinia carotovora, Clavibacter michiganensis, A. rhizogenes* and *Curtobacterium flacccumfaciens* bacteria (Fujimura et al., 2005). The antimicrobial effects of the bamboo shoots can be highly advantageous in preventing the incidence of disease and food preservation. Additionally, aspartic acid-tyrosine dipeptide extracted from the shoots of *P. pubescens* revealed antihypertensive potential by suppressing the activity of ACE, a vital target in managing high blood pressure (Liu et al., 2012). The presence of such biofunctional peptides contributes towards the cardioprotective property of bamboo shoots.

4.7.3 Carbohydrates and Dietary Fibres

Bamboo shoots are a rich repository of diverse carbohydrates, including glycoprotein, fructose, glucose, sucrose, starch and polysaccharides like xyloglucan, β-D-glucan, arabinogalactan, arabinoxylan trisaccharide and tetra-saccharide and xyloglucan disaccharide, arabinoxylan and glucomannan (Edashige and Ishii, 1998; Ishii and Hiroi, 1990a; Ishii and Hiroi, 1990b; Karanja, 2017; Tanabe et al., 2017). These monosaccharides function as readily available energy substrates while the polysaccharides serve as prebiotics which stabilize the gut microbiome equilibrium, thereby promoting the overall well-being of gastrointestinal health. Furthermore, the content of carbohydrates within the shoots varies from 2.0 g to 9.94 g/100 g, depending upon the species of bamboo (Singhal et al., 2013).

One of the unparallelled attributes of bamboo shoots is the prevalence of high dietary fibres in them, which constitute 60% to 90% of total carbohydrate content (Felisberto et al., 2017;

Karnjanapratum et al., 2019). Such immense quantities of fibres can actively improve digestion and may help in regulating blood glucose levels and reducing blood cholesterol levels (Imdad et al., 2022; Kong et al., 2020; Li et al., 2016; Li et al., 2018; Tang et al., 2022; Tang et al., 2022). The medicinal applications of dietary fibres and polysaccharides extracted from the shoots of bamboo can be seen as mentioned in the table above (Table 4.1).

4.8 CONCLUSION

The past few years have experienced a substantially increased exploration of bamboo shoots as potent nutraceutical and pharmaceutical agents, especially due to the prevalence of an impressive array of phytocompounds and their extensive role in promoting overall health. As an immensely renewable natural resource, bamboo shoots not only provide an option for enhancing human health by the utilisation of traditional knowledge in the modern pharmaceutical industry but can also sustain the ecology. Bamboo shoots are treasures full of vital nutrients encompassing amino acids, vitamins, dietary fibres and minerals like potassium, magnesium and calcium among many others, rendering bamboo a suitable functional food to combat growing malnutrition among the global population. Additionally, the capacity of bamboo shoots to demonstrate several therapeutic effects encompassing hypoglycaemia, cardioprotective, anti-inflammatory, tumour suppressive along with lipid-lowering activity has made it a significant medicinal plant. With further research and innovation, bamboo shoots can be globally recognised as a promising functional food, promoting both human health and the plant's ecosystem.

REFERENCES

Abdelhameed, R. F., Nafie, M. S., Ibrahim, A. K., Yamada, K., Abdel-Kader, M. S., Ibrahim, A. K., … Habib, E. S. 2020. Cytotoxic, apoptosis-inducing activities, and molecular docking of a new sterol from bamboo shoot skin *Phyllostachys heterocycla* var. pubescens. *Molecules 25*(23): 5650.

Abe, R., & Ohtani, K. 2013. An ethnobotanical study of medicinal plants and traditional therapies on Batan Island, the Philippines. *Journal of Ethnopharmacology 145*(2): 554–565.

Ahmad, Z., Upadhyay, A., Ding, Y., Emamverdian, A., & Shahzad, A. 2021. Bamboo: Origin, habitat, distributions and global prospective. *Biotechnological Advances in Bamboo: The "Green Gold" on the Earth*: 1–31.

Ahmed, A. U. 2011. An overview of inflammation: Mechanism and consequences. *Frontiers in Biology 6*(4): 274–281.

Alvarez, M. L., Uy, L. Y., & Torio, M. A. 2023. Antihypertensive, antioxidative, and hypocholesterolemic activities of bioactive peptides from bamboo (*Bambusa blumeana*) shoots. *The Philippine Journal of Crop Science 48*(3): 1–13.

Awang-Kanak, F., Bakar, M. F. A., Mohamed, M., & Norazlimi, N. A. 2018 September. Utilization of natural resources: Preliminary study on ethnopharmacological application of 'ulam'or traditional vegetables among Sama-Bajau of Kampung Menunggui, Kota Belud, Sabah. In AIP Conference Proceedings (Vol. 2016, No. 1). AIP Publishing.

Baenas, N., Piegholdt, S., Schloesser, A., Moreno, D. A., García-Viguera, C., Rimbach, G., & Wagner, A. E. 2016. Metabolic activity of radish sprouts derived isothiocyanates in drosophila melanogaster. *International Journal of Molecular Sciences 17*(2): 251.

Bai, J., Zhang, Y., Tang, C., Hou, Y., Ai, X., Chen, X., … Meng, X. 2021. Gallic acid: Pharmacological activities and molecular mechanisms involved in inflammation-related diseases. *Biomedicine & Pharmacotherapy 133*: 110985.

Bajwa, H. K., Santosh, O., & Chongtham, N. 2020. Bioactive compounds in bamboo shoot. *Bioactive Compounds in Underutilized Vegetables and Legumes*: 1–22.

Bajwa, H. K., Santosh, O., Koul, A., Bisht, M. S., & Nirmala, C. 2019. Quantitative determination of macroelement and microelement content of fresh and processed bamboo shoots by wavelength dispersive X-ray fluorescence spectrometry. *X-Ray Spectrometry 48*(6): 637–643.

Banik, R. L. 2000. Silviculture and field guide to priority bamboos of Bangladesh and South Asia. Government of the People's Republic of Bangladesh, Bangladesh Forest Research Institute.

Bhargava, A., Kumbhare, V., Srivastava, A., & Sahai, A. 1996. Bamboo parts and seeds for additional source of nutrition. *Journal of Food Science and Technology (Mysore) 33*(2): 145–146.

Bhatt, B. P., Singh, K., & Singh, A. 2005. Nutritional values of some commercial edible bamboo species of the North Eastern Himalayan region, India. *Journal of Bamboo and Rattan 4*(2): 111–124.

Brandis, D. (1899). Biological Notes on Indian Bamboos. *Indian Forester, 25*(1): 1–25..

Chan, S. T., Yang, N. C., Huang, C. S., Liao, J. W., & Yeh, S. L. 2013. Quercetin enhances the antitumor activity of trichostatin A through upregulation of p53 protein expression in vitro and in vivo. *PLoS One 8*(1): e54255.

Chassagne, F., Deharo, E., Punley, H., & Bourdy, G. 2017. Treatment and management of liver diseases by Khmer traditional healers practicing in Phnom Penh area, Cambodia. *Journal of Ethnopharmacology 202*: 38–53.

Chen, G., Chen, X., Yang, B., Yu, Q., Wei, X., Ding, Y., & Kan, J. 2019. New insight into bamboo shoot (*Chimonobambusa quadrangularis*) polysaccharides: Impact of extraction processes on its prebiotic activity. *Food Hydrocolloids 95*: 367–377.

Chongtham, N., Bisht, M. S., & Haorongbam, S. 2011. Nutritional properties of bamboo shoots: Potential and prospects for utilization as a health food. *Comprehensive Reviews in Food Science and Food Safety 10*(3): 153–168.

Choudhury, D., Sahu, J. K., & Sharma, G. D. 2011. Bamboo shoot based fermented food products: A review. *Journal of Scientific and Industrial Research 70*(03): 199–203.

Choudhury, D., Sahu, J. K., & Sharma, G. D. 2012. Value addition to bamboo shoots: A review. *Journal of Food Science and Technology 49*: 407–414.

Christian, A. L., Knott, K. K., Vance, C. K., Falcone, J. F., Bauer, L. L., Fahey, G. C., Jr, ... Kouba, A. J. 2015. Nutrient and mineral composition during shoot growth in seven species of *Phyllostachys* and *Pseudosasa* bamboo consumed by giant panda. *Journal of Animal Physiology and Animal Nutrition 99*(6): 1172–1183.

Dang, L. M., Chen, D. H., & Yin, A. W. 2017. Analysis of nutritional components in bamboo shoots of *Pleioblastus Amarus. Guangzhou Chemical Industry 45*(14): 133–152.

Das, B., Singh, P. K., & Munna, J. 2021. Miraculous medicinal properties of bamboo. *Journal of Medicinal Plants 9*(5): 28–30.

Da Silva, J. G. 2014. Don't just produce more food-produce better food. *National Geographic Special Series Future Food 5*: 1–2.

Du, H., Mao, F., Li, X., Zhou, G., Xu, X., Han, N., ... Zhou, Y. 2018. Mapping global bamboo forest distribution using multisource remote sensing data. *IEEE Journal of Selected Topics in Applied Earth Observations and Remote Sensing 11*(5): 1458–1471.

Edashige, Y., & Ishii, T. 1998. Hemicellulosic polysaccharides from bamboo shoot cell-walls. *Phytochemistry 49*(6): 1675–1682.

ES, C. E., Nonglait, R., & Suting, B. 2021. Antimicrobial and antioxidant activity of fermented bamboo shoot *Dendrocalamus hamiltonii. Current Trends in Biotechnology and Pharmacy 15*(5): 425–436.

Espíndola, K. M. M., Ferreira, R. G., Narvaez, L. E. M., Silva Rosario, A. C. R., Da Silva, A. H. M., Silva, A. G. B., ... Monteiro, M. C. 2019. Chemical and pharmacological aspects of caffeic acid and its activity in hepatocarcinoma. *Frontiers in Oncology 9*: 541.

Felisberto, M. H. F., Miyake, P. S. E., Beraldo, A. L., & Clerici, M. T. P. S. 2017. Young bamboo culm: Potential food as source of fiber and starch. *Food Research International 101*: 96–102.

Forest Survey of India. (2017). *India State of Forest Report 2017.* Ministry of Environment, Forest and Climate Change, Government of India. https://fsi.nic.in/forest-report-2017

Fujimura, M., Ideguchi, M., Minami, Y., Watanabe, K., & Tadera, K. 2005. Amino acid sequence and antimicrobial activity of chitin-binding peptides, Pp-AMP 1 and Pp-AMP 2, from Japanese bamboo shoots (*Phyllostachys pubescens*). *Bioscience, Biotechnology, and Biochemistry 69*(3): 642–645.

Gogoi, I., Dowara, M., & Chetia, P. 2024. Traditional medicinal plants and their ethnomedicinal values. In: *Traditional Resources and Tools for Modern Drug Discovery: Ethnomedicine and Pharmacology.* Springer Nature Singapore, Singapore, 377–399.

Hamilton, A. C. 2004. Medicinal plants, conservation and livelihoods. *Biodiversity & Conservation 13*: 1477–1517.

Han, N., Du, H., Zhou, G., Xu, X., Cui, R., & Gu, C. 2013. Spatiotemporal heterogeneity of Moso bamboo aboveground carbon storage with Landsat Thematic Mapper images: A case study from Anji County, China. *International Journal of Remote Sensing 34*(14): 4917–4932.

He, S., Wang, X., Zhang, Y., Wang, J., Sun, H., Wang, J., ... Ye, Y. 2016. Isolation and prebiotic activity of water-soluble polysaccharides fractions from the bamboo shoots (*Phyllostachys praecox*). *Carbohydrate Polymers 151*: 295–304.

Huang, C. L., & Yang, Y. F. 2006. A study on the major nutrient components and trace element of bamboo shoots of *Pleioblastus amarus*. *Journal of Bamboo Research 25*(3): 32–35.

Huang, X., Li, J., Zhang, W., Peng, S., Zhu, Y., & Lin, L. 2014. Nutritional components and active ingredient analysis of different parts of winter Moso bamboo shoots. *Food Science & Technology 39*: 59–63.

Imdad, L., Shah, M. A. A., Hassan, J., Humayun, F., Sadiq, A., & Javed, M. 2022. Probing the therapeutic effects of bamboo shoots supplementation on diabetic male and female patients: Therapeutic effects of bamboo shoots in diabetic subject. *Journal of Aziz Fatimah Medical & Dental College 4*(1): 26–30.

Ishii, T., & Hiroi, T. 1990a. Isolation and characterization of feruloylated arabinoxylan oligosaccharides from bamboo shoot cell-walls. *Carbohydrate Research 196*: 175–183.

Ishii, T., & Hiroi, T. 1990b. Linkage of phenolic acids to cell-wall polysaccharides of bamboo shoot. *Carbohydrate Research 206*(2): 297–310.

Jadid, N., Kurniawan, E., Himayani, C. E. S., Andriyani, Prasetyowati, I., Purwani, K. I., … Tjahjaningrum, I. T. D. 2020. An ethnobotanical study of medicinal plants used by the Tengger tribe in Ngadisari village, Indonesia. *PLoS One 15*(7): e0235886.

Jayameena, P., Sivakumari, K., Ashok, K., & Rajesh, S. 2018. Rutin: A potential anticancer drug against human colon cancer (HCT116) cells. *International Journal of Biology, Pharmacy and Allied Sciences 7*(9): 1731–1745.

Karanja, P. N. 2017. *Physicochemical properties of bamboo shoots of selected species grown in Kenya and utilization as human food* (Doctoral dissertation, Faculty of Agriculture, JKUAT).

Karanja, P. N., Kenji, G. M., Njoroge, S. M., Sila, D. N., Onyango, C. A., Koaze, H., & Baba, N. 2015. Variation of nutrients and functional properties within young shoots of a bamboo species (*Yushaniaalpina*) growing at Mt. Elgon Region in Western Kenya. *Journal of Food and Nutrition Research 3*(10): 675–680.

Karnjanapratum, S., Kaewthong, P., Takeungwongtrakul, S., Sae-leaw, T., Hong, J. H., & Nalinanon, S. 2019. Production of fiber hydrolysate from bamboo shoot with antioxidative properties by enzymatic hydrolysis. *Current Applied Science and Technology 19*(3): 225–234.

Keng, P. C. 1983. A revision of the genera of bamboos from the world IV. *Journal of Bamboo Research 2*(2): 1–18.

Kim, H., & Song, M. J. 2011. Analysis and recordings of orally transmitted knowledge about medicinal plants in the southern mountainous region of Korea. *Journal of Ethnopharmacology 134*(3): 676–696.

Kong, C. K., Tan, Y. N., Chye, F. Y., & Sit, N. W. 2020. Nutritional composition and biological activities of the edible shoots of *Bambusa vulgaris* and *Gigantochloa ligulata*. *Food Bioscience 36*: 100650.

Kumari, P. 2019. The *Bambusoideae* in India: An updated enumeration. *Plantae Scientia 1*(06): 99–117.

Li, Q., Fang, X., Chen, H., Han, Y., Liu, R., Wu, W., & Gao, H. 2021. Retarding effect of dietary fibers from bamboo shoot (*Phyllostachys edulis*) in hyperlipidemic rats induced by a high-fat diet. *Food & Function 12*(10): 4696–4706.

Li, X., Fu, B., Guo, J., Ji, K., Xu, Y., Dahab, M. M., & Zhang, P. 2018. Bamboo shoot fiber improves insulin sensitivity in high-fat diet-fed mice. *Journal of Functional Foods 49*: 510–517.

Li, X., Guo, J., Ji, K., & Zhang, P. 2016. Bamboo shoot fiber prevents obesity in mice by modulating the gut microbiota. *Scientific Reports 6*(1): 32953.

Liao, Y. R., & Lin, J. Y. 2014. Quercetin, but not its metabolite quercetin-3-glucuronide, exerts prophylactic immunostimulatory activity and therapeutic anti-inflammatory effects on lipopolysaccharide-treated mouse peritoneal macrophages ex vivo. *Journal of Agricultural and Food Chemistry 62*(13): 2872–2880.

Liu, L., Liu, L., Lu, B., Chen, M., & Zhang, Y. 2013. Evaluation of bamboo shoot peptide preparation with angiotensin converting enzyme inhibitory and antioxidant abilities from byproducts of canned bamboo shoots. *Journal of Agricultural and Food Chemistry 61*(23): 5526–5533.

Liu, L., Liu, L., Lu, B., Xia, D., & Zhang, Y. 2012. Evaluation of antihypertensive and antihyperlipidemic effects of bamboo shoot angiotensin converting enzyme inhibitory peptide in vivo. *Journal of Agricultural and Food Chemistry 60*(45): 11351–11358.

Lobovikov, M., Ball, L., & Guardia, M. (2007). World bamboo resources: *A thematic study prepared in the framework of the Global Forest Resources Assessment 2005* (Non-wood Forest Products No. 18). Food and Agriculture Organization of the United Nations, Rome, Italy

Lu, B., Cai, H., Huang, W., Wu, X., Luo, Y., Liu, L., & Zhang, Y. 2011. Protective effect of bamboo shoot oil on experimental nonbacterial prostatitis in rats. *Food Chemistry 124*(3): 1017–1023.

Lu, B., Xia, D., Huang, W., Wu, X., Zhang, Y., & Yao, Y. 2010. Hypolipidemic effect of bamboo shoot oil (*P. pubescens*) in Sprague–Dawley rats. *Journal of Food Science 75*(6): H205–H211.

Luo, B., Ahmed, S., & Long, C. 2020. Bamboos for weaving and relevant traditional knowledge in Sansui, Southwest China. *Journal of Ethnobiology and Ethnomedicine 16*: 1–9.

Medhi, P., Borthakur, S. K., & Hore, D. K. 2010. Phyto resources from North Cachar Hills of Assam, India-IV: Bamboos and Rattans. *Journal of Bamboo and Rattan 9*(3): 115–125.

Medzhitov, R. 2007. Recognition of microorganisms and activation of the immune response. *Nature 449*(7164): 819–826.

Milani, G., Curci, F., Cavalluzzi, M. M., Crupi, P., Pisano, I., Lentini, G., ... Corbo, F. 2020. Optimization of microwave-assisted extraction of antioxidants from bamboo shoots of *Phyllostachys pubescens*. *Molecules 25*(1): 215.

Mustaffa, N. A. Z., Mansur, S. A., & Kassim, A. S. M. 2022. Ultrasound-assisted-extraction (UAE) of phenolic compounds from bamboo shoot and its potential source of anti-inflammatory agents for gout treatment. *Progress in Engineering Application and Technology 3*(1): 74–83.

NavaneethaKrishnan, S., Rosales, J. L., & Lee, K. Y. 2019. ROS-mediated cancer cell killing through dietary phytochemicals. *Oxidative Medicine and Cellular Longevity 2019*(1): 9051542.

Nirala, D. P., Nirbhay Ambasta, N. A., & Phallo Kumari, P. K. 2017. A review on distribution of bamboos. *Lifesciences Leaflets 92*: 70–78.

Nirmala, C., & Bisht, M. S. (2017). Bamboo: A prospective ingredient for functional food and nutraceuticals. *Journal of Bamboo and Rattan*, 16(1-2), 99–112.

Nirmala, C., Bisht, M. S., Bajwa, H. K., & Santosh, O. 2018. Bamboo: A rich source of natural antioxidants and its applications in the food and pharmaceutical industry. *Trends in Food Science & Technology 77*: 91–99.

Nirmala, C., Bisht, M. S., & Laishram, M. 2014. Bioactive compounds in bamboo shoots: Health benefits and prospects for developing functional foods. *International Journal of Food Science & Technology 49*(6): 1425–1431.

Nirmala, C., David, E., & Sharma, M. L. 2007. Changes in nutrient components during ageing of emerging juvenile bamboo shoots. *International Journal of Food Sciences and Nutrition 58*(8): 612–618.

Nirmala, C., & Sharma, M. L. 2008. A comparative study of nutrient components of freshly harvested, fermented and canned bamboo shoots of *Dendrocalamus giganteus* Munro. *Bamboo Science & Culture 21*(1): 41.

Nongdam, P., & Tikendra, L. 2014. The nutritional facts of bamboo shoots and their usage as important traditional foods of northeast India. *International Scholarly Research Notices 2014*(1): 679073.

Novaryatiin, S., & Indah, I. 2019. The medicinal plants used in Anjir Pulang Pisau, Central Kalimantan-Indonesia. *Pharmacognosy Journal 11*(6): 1572–1579.

Octaviani, P. 2020a, February. Identification of chemical content and antihypertensive activity test of ethanol extract from Tali Bamboo Shoot (*Gigantochloa apus* (Schult.&Schult. F.)). In: *1st International Conference on Community Health (ICCH 2019)* (pp. 237–241). Atlantis Press.

Octaviani, P. 2020b, February. The effect of antihypertensive of Tali's bamboo shoot ethanol extract (*Gigantochloa apus* (Schult.&Schult. F.)) to male white rats *Sprague dawley*. In *1st International Conference on Community Health (ICCH 2019)* (pp. 178–181). Atlantis Press.

Ohrnberger, D. 1999. The Bamboos of the World: Annotated Nomenclature and Literature of the Species and the Higher and Lower Taxa. Elsevier, Amsterdam, 596.

Ong, H. G., & Kim, Y. D. 2015. Herbal therapies and social-health policies: Indigenous Ati Negrito Women's dilemma and reproductive healthcare transitions in the Philippines. *Evidence-Based Complementary and Alternative Medicine 2015*(1): 491209.

Ontaha, Y., Samal, N. S., Nurfardila, N., & Angria, N. 2021. Pemanfaatan ekstrak rebung bambu betung (*Dendrocalamus asper*) terhadap sel cancer mcf-7 secara in vitro. *Jurnal Riset Kesehatan Poltekkes Depkes Bandung 13*(2): 466–475.

Park, E. J., & Jhon, D. Y. 2010. The antioxidant, angiotensin converting enzyme inhibition activity, and phenolic compounds of bamboo shoot extracts. *LWT-Food Science and Technology 43*(4): 655–659.

Premlata, T., Sharma, V., Bisht, M. S., & Nirmala, C. 2020. Edible bamboo resources of Manipur: consumption pattern of young shoots, processing techniques and their commercial status in the local market.

Priyadarsini, R. V., Murugan, R. S., Maitreyi, S., Ramalingam, K., Karunagaran, D., & Nagini, S. 2010. The flavonoid quercetin induces cell cycle arrest and mitochondria-mediated apoptosis in human cervical cancer (HeLa) cells through p53 induction and NF-κB inhibition. *European Journal of Pharmacology 649*(1–3): 84–91.

Ren, Y., Ma, Y., Zhang, Z., Qiu, L., Zhai, H., Gu, R., & Xie, Y. 2019. Total alkaloids from bamboo shoots and bamboo shoot shells of *Pleioblastus amarus* (Keng) Keng f. and their anti-inflammatory activities. *Molecules 24*(15): 2699.

Saini, N., Rawat, K., Bisht, M. S., & Nirmala, C. 2017. Qualitative and quantitative mineral element variances in shoots of two edible bamboo species after processing and storage evaluated by wavelength dispersion x-ray fluorescence spectrometry. *International Journal of Innovative Research in Science, Engineering and Technology 6*(5): 8265–8270.

Sangeetha, R., Diea, Y. K. T., Chaitra, C., Malvi, P. G., & Shinomol, G. K. 2015. The amazing bamboo: A review on its medicinal and pharmacological potential. *Indian Journal of Nutrition 2*(1): 1–7.

Sawarkar, A. D., Shrimankar, D. D., Kumar, M., Kumar, P., Kumar, S., & Singh, L. 2021. Traditional system versus DNA barcoding in identification of bamboo species: A systematic review. *Molecular Biotechnology 63*(8): 651–675.

Scherr, S. J., & McNeely, J. A. (Eds.). 2012. *Farming With Nature: the Science and Practice of Ecoagriculture*. Island Press.

Setiawati, T., Mutaqin, A. Z., Irawan, B., Anamillah, A., & Iskandar, J. 2017. Species diversity and utilization of bamboo to support life's the community of Karangwangi Village, Cidaun Sub-District of Cianjur, Indonesia. *Biodiversitas Journal of Biological Diversity 18*(1): 58–64.

Shi, C., Yuan, S., Zhang, H., Zhang, T., Wang, L., & Xu, Z. 2009. Cell-mediated immune responses and protective efficacy against infection with mycobacterium tuberculosis induced by Hsp65 and hIL-2 fusion protein in mice. *Scandinavian Journal of Immunology 69*(2): 140–149.

Shi, Q. T., & Yang, K. S. 1992, December. Study on relationship between nutrients in bamboo shoots and human health. In: *Proceedings of the International Symposium on Industrial Use of Bamboo. International Tropical Timber Organization and Chinese Academy.* Bamboo and its Use, Beijing, China, 338–346.

Shoaib, A., Azmi, L., Pal, S., Alqahtani, S. S., Rahamathulla, M., Hani, U., … Shakeel, F. 2022. Integrating nanotechnology with naturally occurring phytochemicals in neuropathy induced by diabetes. *Journal of Molecular Liquids 350*: 118189.

Silambarasan, R., & Ayyanar, M. 2015. An ethnobotanical study of medicinal plants in Palamalai region of Eastern Ghats, India. *Journal of Ethnopharmacology 172*: 162–178.

Singh, H. B., Kumar, B., & Singh, R. S. 2003. Bamboo resources of Manipur: An overview for management and conservation. *Journal of Bamboo and Rattan 2*(1): 43–55.

Singh, S. A., Bora, T. C., & Singh, N. R. 2012. Preliminary phytochemical analysis and antimicrobial potential of fermented *Bambusa balcooa* shoots. *The Bioscan 7*(3): 391–394.

Singhal, P., Bal, L. M., Satya, S., Sudhakar, P., & Naik, S. N. 2013. Bamboo shoots: A novel source of nutrition and medicine. *Critical Reviews in Food Science and Nutrition 53*(5): 517–534.

Sohgaura, A., Bigoniya, P., & Shrivastava, B. 2018. Diuretic potential of *Cynodon dactylon, Emblica officinalis, Kalanchoe pinnata* and *Bambusa nutans. Journal of Pharmacognosy and Phytochemistry 7*(3): 2895–2900.

Srisawat, T., Suvarnasingh, A., & Maneenoon, K. 2016. Traditional medicinal plants notably used to treat skin disorders nearby Khao Luang mountain hills region, Nakhon Si Thammarat, Southern Thailand. *Journal of Herbs, Spices & Medicinal Plants 22*(1): 35–56.

Sukumaran, S., Brintha, T. S., Subitha, P., Sheebha, Y. A., & Jeeva, S. 2014. Usage of medicinal plants by two cultural communities of Kanyakumari district, Tamil Nadu, South India. *Journal of Chemical and Pharmaceutical Research 6*(8): 67–79.

Tanabe, C., Furuta, K., Maeda, M., & Kimura, Y. 2017. Structural feature of N-glycans of bamboo shoot glycoproteins: Useful source of plant antigenic n-glycans. *Bioscience, Biotechnology, and Biochemistry 81*(7): 1405–1408.

Tanaka, A., Shimizu, K., & Kondo, R. 2013. Antibacterial compounds from shoot skins of Moso bamboo (*Phyllostachys pubescens*). *Journal of Wood Science 59*: 155–159.

Tang, C., Wu, L., Zhang, F., Kan, J., & Zheng, J. 2022a. Comparison of different extraction methods on the physicochemical, structural properties, and in vitro hypoglycemic activity of bamboo shoot dietary fibers. *Food Chemistry 386*: 132642.

Tang, C., Yang, J., Zhang, F., Kan, J., Wu, L., & Zheng, J. 2022b. Insight into the physicochemical, structural, and in vitro hypoglycemic properties of bamboo shoot dietary fibre: Comparison of physical modification methods. *International Journal of Food Science & Technology 57*(8): 4998–5010.

Tewari, S., Negi, H., & Kaushal, R. 2019. Status of bamboo in India. *International Journal of Economic Plants 6*(Feb, 1): 030–039.

Thomas, R., Jebin, N., Saha, R., & Sarma, D. K. 2016. Antioxidant and antimicrobial effects of Kordoi (*Averrhoa carambola*) fruit juice and bamboo (*Bambusa polymorpha*) shoot extract in pork nuggets. *Food Chemistry 190*: 41–49.

Turner, A. J., & Hooper, N. M. 2002. The angiotensin-converting enzyme gene family: Genomics and pharmacology. *Trends in Pharmacological Sciences 23*(4): 177–183.

Van der Lugt, P., ThangLong, T., & King, C. 2018. Carbon sequestration and carbon emissions reduction through bamboo forests and products. *Biochemical and Biophysical Research Communications 307*(3): 750–755.

Van Dung, D., Thao, L. D., Ngoan, L. D., Phung, L. D., & Roubík, H. 2024. Effects of biochar produced from tropical rice straw, corncob, and bamboo tree at different processing temperatures on in vitro rumen fermentation and methane production. *Biomass Conversion and Biorefinery 14*(2): 2577–2584.

Vanitha, B., Rajesh Kumar, R., Duraiswamy, B., Gnanavel, G., & Neelamma, G. 2016. Estimation of total phenol and total flavanoid content with antioxidant and anti-inflammatory activity of *Bambusa arundinacae* (bamboo shoot) grown in Nilgiris. *Der Pharma Chem 8*(20): 307–16.

Visuphaka, K. (1985, October). The role of bamboo as a potential food source in Thailand. In G. Lessard & A. Chouinard (Eds.), Recent research on bamboos: *Proceedings of the International Bamboo Workshop* (pp. 301–303). International Development Research Centre & Chinese Academy of Forestry.

Wang, G. Y., Ma, S., & Gou, G. Q. 2014. Analysis of nutrient components of bamboo shoots of *Pleioblastus maculatus. Journal of Guizhou University (Natural Sciences) 31*(6): 42–44.

Wang, H. X., & Ng, T. B. 2003. Dendrocin, a distinctive antifungal protein from bamboo shoots. *Biochemical and Biophysical Research Communications 307*(3): 750–755.

Wróblewska, K. B., Baby, A. R., Guaratini, M. T. G., & Moreno, P. R. H. 2019. In vitro antioxidant and photoprotective activity of five native Brazilian bamboo species. *Industrial Crops and Products 130*: 208–215.

Wu, W., Hu, J., Gao, H., Chen, H., Fang, X., Mu, H., … Liu, R. 2020. The potential cholesterol-lowering and prebiotic effects of bamboo shoot dietary fibers and their structural characteristics. *Food Chemistry 332*: 127372.

Xia, N. H. 1989. Analysis of nutritive constituents of bamboo shoots in Guangdong. *Acta Botanica Austro Sinica 4*: 199–206.

YuMing, Y., KangLin, W., Pei, S., & Jiming, H. 2004. Bamboo diversity and traditional uses in Yunnan, China *24*(2): 157–165.

Yao, X., Wan, R., Li, C., Li, G., Zhang, B., Deng, Z., & Li, H. 2024. The hypoglycemic effect of enzymatic modified dietary fiber from bamboo shoot on type 2 diabetes rats. *Journal of Food Science 89*(9): 5900–5911.

Yuen, J. Q., Fung, T., & Ziegler, A. D. 2017. Carbon stocks in bamboo ecosystems worldwide: Estimates and uncertainties. *Forest Ecology and Management 393*: 113–138.

Zheng, Y., Wang, Q., Huang, J., Fang, D., Zhuang, W., Luo, X., … Cao, H. 2019. Hypoglycemic effect of dietary fibers from bamboo shoot shell: An in vitro and in vivo study. *Food and Chemical Toxicology 127*: 120–126.

Zhou, X., Ma, L., Dong, L., Li, D., Chen, F., & Hu, X. 2023. Bamboo Shoot dietary fiber alleviates gut microbiota dysbiosis and modulates liver fatty acid metabolism in mice with high-fat diet-induced obesity. *Frontiers in Nutrition 10*: 1161698.

Zhu, Y., & Luo, C. G. 2012. Analysis of nutrient components in *Dendrocalamopsis oldhami* bamboo shoot. *Nonwood Forest Research 30*(3): 103–105.

5 Bamboos as a Rich Source of Antioxidants

Aribam Indira, Babita Joshi, Nikita Kalyan, and Nirmala Chongtham

5.1 INTRODUCTION

Oxidative stress is a harmful physiological condition to human health. It is a negative imbalance between free radicals, which are highly reactive oxygen-containing molecules, and antioxidative molecules. Antioxidants are the molecules which inhibit the oxidation of biomolecules caused by free radicals. Free radicals can damage cells, proteins, and DNA, contributing to ageing and various disorders. Antioxidants can be categorized into two main types: endogenous and exogenous. Endogenous antioxidants are produced within the body and include enzymes like superoxide dismutase (SOD), glutathione transferase (GST), oxidized and reduced glutathione (GSH and GSSH), catalase (Cat), and glutathione peroxidase (GPx), which the body naturally produces to neutralize free radicals. Exogenous antioxidants are obtained from diet, i.e., foods and supplements, and include vitamins, minerals, and phytochemicals. Although our body naturally produces antioxidants, excessive free radical production from oxidative stress necessitates the intake of antioxidants from external sources to maintain balance (Indira et al., 2024b). Phytochemicals are naturally occurring compounds found in plants that serve as potent antioxidants. Vitamin C (ascorbic acid) and vitamin E (tocopherol) are well-known antioxidants. Vitamin C is water-soluble and found in fruits and vegetables, while vitamin E is fat-soluble and found in nuts, seeds, and oils. Minerals such as selenium, zinc, copper, manganese, and iron play a crucial role in maintaining the antioxidant defenses within the body by acting as cofactors for enzymes, participating in redox reactions and help stabilize the cellular structures against oxidative damage. Beyond their health benefits, these antioxidants are also valuable in food preservation, enhancing the shelf life and stability of food products by preventing oxidation-related spoilage. The growing interest in plant antioxidants underscores their significance not only in promoting human health but also in their applications in the food and pharmaceutical industries (Muscolo et al., 2024). Common sources of antioxidants are diet-rich that include fruits such as berries (blueberries, strawberries), citrus fruits (oranges, lemons), and grapes; vegetables namely, leafy greens (spinach, kale), cruciferous vegetables (broccoli, Brussels sprouts), and root vegetables (carrots, sweet potatoes); nuts and seeds like almonds, sunflower seeds, and flaxseeds; and beverages such as green tea, black tea, and red wine. Plant-based antioxidants help neutralize free radicals in the body, reducing oxidative stress and contributing to overall health and the prevention of chronic diseases (Indira et al., 2023b). The use of plants for the prevention and treatment of various diseases has garnered global interest due to their minimal side effects and lack of toxicity. The protective properties of natural products are attributed to their bioactive compounds, which can serve as prophylactic agents against several chronic human diseases. These bioactive compounds in edible plants play a crucial role in human health by influencing various aspects of well-being (Indira et al., 2024a). While plant-based treatments typically have lower potency compared to pharmaceutical drugs, they can be consumed regularly as part of a normal diet, potentially leading to significant long-term physiological benefits. Growing awareness of these health advantages has increased the demand for plant-based foods rich in bioactive compounds. Besides the health-benefiting properties, antioxidants also play a crucial role in industries such as petrochemicals, food, cosmetics, and medicine, where they help stabilize polymeric products in the food, and pharmaceutical sectors,

 DOI: 10.1201/9781003485933-5

antioxidants are essential for preventing oxidation-related issues such as deterioration, rancidity, and discoloration during processing and storage (Nirmala et al., 2018).

Bamboo, the arborescent and fastest-growing, multipurpose plant is an excellent source of bioactive compounds that offer numerous health benefits, namely improved digestion and the prevention of metabolic disorders, cardiovascular disease, and certain types of cancer (Behera and Balaji, 2021; Gagliano et al., 2022; Indira et al., 2023b). Nearly every part of the bamboo plant, including its leaves, seeds, culms, shoots, sheaths, and rhizomes, possesses therapeutic potential. Among these, the leaves and shoots are particularly noted for their nutritional and medicinal benefits. The health-benefiting properties of bamboo have been attributed to the presence of phytochemicals, vitamins, and minerals (Nirmala et al., 2018). Various macromolecules, such as cellulose lignin, volatile oils, phenolic compounds, flavonoids, flavon-C-glycosides, sterols, alkaloids, and terpenes, have been reported from different parts of bamboo plants (Kimura et al., 2022). Therefore, the versatile plant exemplifies the therapeutic potential of plant-based antioxidants. As awareness of the health benefits of plant-based antioxidants grows, so does the demand for foods rich in these bioactive compounds. This trend underscores the importance of integrating antioxidant-rich plants like bamboo into our diets to enhance overall health and well-being, while also highlighting the broader industrial applications of antioxidants in preserving product quality and stability.

5.2 BIOACTIVE COMPOUNDS IN BAMBOO

Bioactive compounds, well known for their diverse range of biological activities, constitute a diverse class of secondary metabolites produced naturally in plants through various pathways such as shikimate, phenylpropanoid, melonic, mevalonic, and methylerythriol pathways. Phenolic compounds prevent the oxidative damage of various biomolecules such as DNA, lipids, and proteins by chelating metal ions and scavenging various reactive species such as superoxide radical, hydroxyl radical, peroxyl radical, hypochlorous acid, and peroxynitrous acid (Indira et al., 2022).

The health benefits of bamboo can be largely attributed to its diverse phytochemical content (Figure 5.1). Research revealed the presence of different classes of phytochemicals, including tannins, polyphenols, saponins, general and cyanogenic glycosides, alkaloids, carotenoids, flavonoids, triterpenoids, steroids, anthraquinones, anthracene glycosides, and coumarins. Flavonols, flavones,

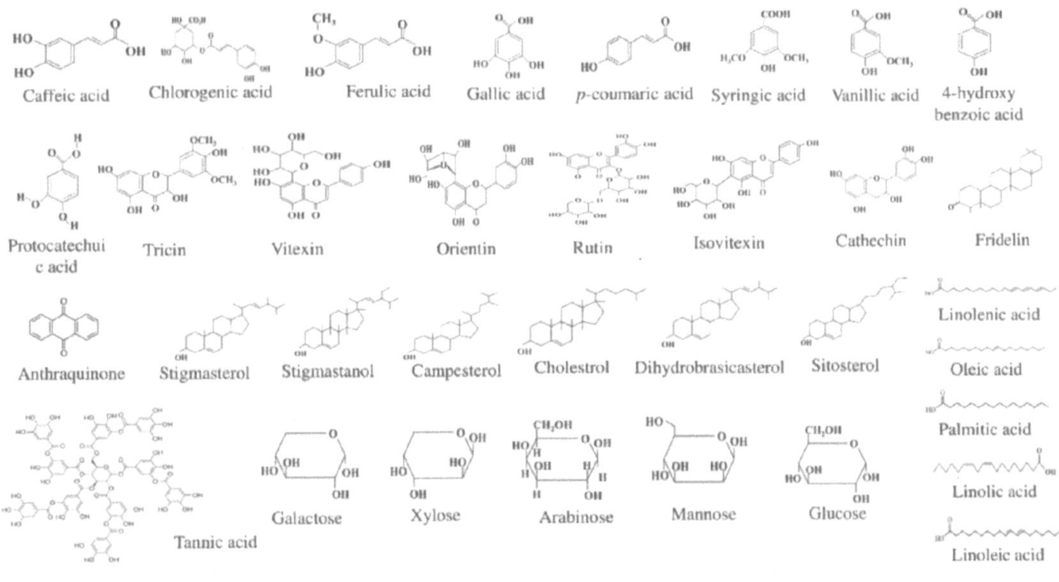

FIGURE 5.1 Major bioactive compounds in bamboo.

flavanones, flavanols, isoflavones, and antocyanidins are the most common flavonoids found in bamboo. A study by Coffie et al. (2014) analyzed the phytochemicals present in both dry and fresh leaf samples of *Bambusa vulgaris*, *B. ventricosa*, and *Oxytenanthera abyssinica* that revealed the presence of 11 different classes of phytochemicals, including tannins, polyphenols, saponins, general and cyanogenic glycosides, alkaloids, carotenoids, flavonoids, triterpenoids, steroids, anthraquinones, anthracene glycosides, and coumarins. Similar results were found in the presence of coumarins, flavonoids, phenols, phytosterols, saponins, tannins, and triterpenes in the shoots of *B. nutans* (Indira et al., 2023a). Phytochemical screening of methanolic and aqueous extracts from the leaves of *Gigantochloa levis* and *G. manggong* revealed the presence of saponins, diterpenes, triterpenes, phenols, phytosterols, tannins, and flavonoids (Supriyatin and Sukmawati, 2014; Tongco et al., 2016). The leaves of bamboo predominantly contain phenols, particularly flavonoids and flavone glycosides. Additionally, seeds of *B. arundinacea* were found to contain phenols, flavonoids, steroids, and tannins (Thamizharasan et al., 2015).

Volatile organic compounds such as sterols, fatty acids, esters, alcohols, aldehydes, and terpenes were identified in the shoots of *B. nutans* and *B. bambos* with the major compounds being anthraquinone (26.85%), β-sitosterol (16.57%), cyclooctenone (12.17%), palmitic acid (10.72%), benzoic acid (9.95%), hexadecanol (9.30%), 1-dodecanol (7.53%), phenol (6.15%), and decanal (5.48%) (Indira et al., 2024b; Saini et al., 2024).

Polyphenols are well known for their wide range of biological properties because of their molecular structure. Major phenolic compounds present in the bamboo shoots are gallic acid, ferulic acid, coumaric acid, p-coumaric acid, caffeic acid, protocatechuic acid, p-hydroxybenzoic acid, rutin, catechol, naringin, luteolin, catechin, synergic acid, and chlorogenic acid (Indira et al., 2024b; Park and Jhon, 2010). Park et al. (2007) identified two antioxidative flavonoid C-glycoside derivatives, isoorientin and isoorientin 2″-O-α-L-rhamnoside, as well as tricin 7-O-β-D-glucopyranoside and apigenin 6-C-β-D-xylopyranosyl-8-C-β-D-glucopyranoside, which demonstrated strong free radical scavenging activity against t-BOOH-induced oxidative damage in HepG2 cells. Mequinol is a phenol derivative that is found in appreciable higher quantity in pyrolyzed culm of *Guadua angustifolia* and is a topical solution that is used in case of hyperpigmentation skin problems (Gomez et al., 2021).

Bamboo shoots and sheaths are also rich in phytosterols, a class of bioactive compounds that belongs to the triterpene family (Nirmala et al., 2014). The main phytosterols present in bamboo shoots are β-sitosterol, campesterol, stigmasterol, cholesterol, ergosterol, and stigmastanol. Lu et al. (2009) evaluated the sterol content in two parts (shoot shell and body) of four bamboo species (*Dendrocalamus latiflorus*, *Pleioblastus amarus*, *P. pubescens*, and *P. praecox*) harvested in winter, spring, and summer seasons and reported the presence of β-sitosterol, stigmasta-3,5-dien-7-one, stigmast-4-en-3-one, stigmasterol, campesterol, and ergosterol. Similarly, Saini et al. (2024) identified three major phytosterols in the shoots of *B. bambos*, i.e., β-sitosterol, campesterol, and stigmasterol, in the amounts of 16.57%, 6.23%, and 4.89%, respectively.

Polysaccharides, another important class of phytochemicals, are polymers of aldoses and/or ketoses linked together by glycosidic linkages and occur naturally in bamboo. Water-soluble polysaccharides from shoots of *Bambusa rigida, G. levis, P. edulis, P. pubescens, P. praecox, P. reticulata, P. heterocycla,* and *Sasa borealis* have been characterized. The main polysaccharides identified in bamboo shoots are rhamnose, arabinose, xylose, mannose, glucose and galactose, beta-glucans, alpha glucan, arabinogalactan, acidic xylan, (4-O-methylglucurono)-arabinoxylan, 4-O-methyl-D-glucuronic acid (4-O-Me-D-GlcA), L-arabinose, and D-xylose, 4-O-methyl-D-glucurono-L-arabino-D-xylan, L-arabino-4-O-methyl-D-glucurono-D-xylan, glucuronic acid (Azmi et al., 2012; He et al., 2016; Huang et al., 2017; Kweon et al., 2004; Sun et al., 2016; Wen et al., 2011; Zhang et al., 2011; Zheng et al., 2016). Similarly, dietary fiber, is a key ingredient of a healthy diet, is present in copious amounts in bamboo shoots (Table 5.1).

The presence of these bioactive compounds in bamboo has helped in the activation of a host defense mechanism that may help to maintain a disease-free state. These bioactive compounds with the least toxicity and maximum selectivity can be used in pharmacology by optimizing their safety,

TABLE 5.1
Common Phenolic Compounds, Polysaccharides, Terpenes, Fatty Acids, and Alkaloids Reported in Bamboo

Species	Phenolic Compounds	Polysaccharides	Terpenes	Fatty Acids/ Fatty Alcohol	Alkaloids
Bambusa bambos	anthraquinone	–	β-sitosterol, campesterol, stigmasterol,	–	–
B. nutans	gallic acid, caffeic acid, 4-hydroxybenzoic acid, coumaric acid, ferulic acid, rutin, catechol, naringin, luteolin	–	β-sitosterol, stigmasterol, campesterol	palmitic acid, hexadecanol	–
B. rigida	*p*-hydroxybenzoic acid, syringic acid, vanillic acid, vanillin acid, vanillin, syringaldehyde, *p*-hydroxybenzyl aldehyde,	uronic acid, galactose, glucose, arabinose	lup-20(29)-en-3-one	–	–
B. tulda	gallic acid, caffeic acid, chlorogenic acid, vanillic acid	–	–	–	–
B. tuldoides	–	–	lup-20(29)-en-3- one, lup-20(29)-en-3-ol	–	–
Dendrocalamus asper	gallic acid, caffeic acid, chlorogenic acid, vanillic acid, methylphenol, 2-pentylfuran, 2-methoxyphenol	–	–	–	–
D. latiflorus	–	–	β-sitosterol, campesterol, stigmasterol, ergosterol, stigmastanol, cholesterol, stigmasta-3,5-dien-7-one, stigmast-4-en-3-one	–	–
D. strictus,	gallic acid, caffeic acid, chlorogenic acid, vanillic acid	–	–	–	–
Gigantochloa levis	glucan	–	–	–	–
Leleba oldhami	–	galactose, mannose, glucose, arabinose,	–	–	–
Phyllostachys bambusoides	luteolin	–	–	–	–

(Continued)

TABLE 5.1 *(Continued)*
Common Phenolic Compounds, Polysaccharides, Terpenes, Fatty Acids, and Alkaloids Reported in Bamboo

Species	Phenolic Compounds	Polysaccharides	Terpenes	Fatty Acids/ Fatty Alcohol	Alkaloids
P. amarus,	orientin, vitexin, isovitexin, tricin	–	β-sitosterol, campesterol, stigmasterol, ergosterol, stigmastanol, cholesterol, stigmasta-3,5-dien-7-one, stigmast-4-en-3-one	–	–
P. aureosulcata	orientin, vitexin, isovitexin, tricin	–	–	–	–
P. edulis	*p*-coumaric acid, ferulic acids, chlorogenic acid, rutin	O-(4-O-trans-feruloyl-α-D-xylopyranosyl)-(1→6)-D-glucopyranose,O-[5-O-(trans-p-coumaroyl)-α-L-arabinofuranosyl]-(1→3)-O-β-D-xylopyranosyl-(1→4)-D-xylopyranose	friedelin	–	–
P. nigra,	catechin, chlorogenic acid, luteolin	–	friedelin, friedelane-3-ol, lup-20(29)-en-3-ol	–	–
P. praecox	–	–	β-sitosterol, campesterol, stigmasterol, ergosterol, stigmastanol, cholesterol, stigmasta-3,5-dien-7-one, stigmast-4-en-3-one	–	–
P. propinqua	orientin, vitexin, isovitexin, tricin	–	–	–	–
P. pubescens	protocatechuic acid, *p*-hydroxybenzoic acid, caffeic acid, syringic acid, *p*-coumaric acid, ferulic acids	–	β-sitosterol, campesterol, stigmasterol, ergosterol, stigmastanol, cholesterol, dihydrobrassicasterol, stigmasta-3,5-dien-7-one, stigmast-4-en-3-one	linoleic acid, palmitic acid, oleic acid, linolenic acid	–
Pleioblastus amarus	–	–	–	–	betaine, arecaidine, cordycepin, arecolin, cephalotaxine, ligustrazine, N-feruloyltramine, berberine, colchiceine, monocrotaline, berberrubine

- Not reported

efficacy, and acceptability. Novel herbal drugs can be developed by using the active compounds that will help in the prevention and eradication of various contemporary illnesses such as diabetes, arthritis, cancer, cardiovascular diseases, etc. as immune checkpoint inhibitors via multiple targets (Sharma and Barooah, 2017). Therefore, bamboo offers a multifaceted health benefit due to the richness in phytochemicals that show immunomodulatory and anti-inflammatory properties, thereby improving the overall well-being.

5.3 ANTIOXIDANTS IN BAMBOO

Antioxidants derived from plants are essential compounds that play a crucial role in maintaining health and preventing disease. These natural substances protect cells from the damaging effects of oxidative stress caused by free radicals, which are unstable molecules that can cause cellular damage and contribute to aging and chronic diseases. They serve several critical functions in the body, such as neutralizing free radicals by donating electrons to free radicals and preventing cellular damage, reducing inflammation by combating oxidative stress, and preventing chronic diseases by reducing oxidative damage. Antioxidants help prevent diseases such as heart disease, cancer, diabetes, and neurodegenerative disorders like Alzheimer's and Parkinson's disease; supporting immune function contributes to the proper functioning of the immune system by protecting immune cells from oxidative damage. These antioxidants, including selected phenolic compounds, flavonoids, carotenoids, vitamins, and minerals, help in the neutralization of free radicals. They can reduce inflammation, support immune function, and lower the risk of various diseases such as heart disease, cancer, and neurodegenerative disorders.

Although high antioxidant activity of bamboo leaves and shoots has been demonstrated through ABTS, DPPH, ORAC, and FRAP tests, the assessment of antioxidant properties of various compounds and natural products by *in-vitro* and *in-vivo* experiments is essential for understanding the actual role of these substances outside and within the complex biological systems. The dual approach will assure the effectiveness of the antioxidants not only in the controlled *in-vitro* system but also in the *in-vivo* dynamic and multifaceted environment of a living organism. It is crucial for the development of effective and safe plant-derived food supplements and pharmaceuticals.

The leaves of *B. nutans*, *B. texilis*, *B. vulgaris,* and *D. oldhami* can serve as natural sources of antioxidants as an alternative to synthetic antioxidants (Nirmala et al., 2018). The antioxidant properties of bamboo leaf flavonoids have practical applications in the food processing industry as well as in animal husbandry. In China, up to 46% of daily antioxidant intake from vegetable sources is contributed by bamboo shoots (Yang et al., 2005). The incorporation of bamboo leaf flavonoids has been shown to enhance the antioxidant capacity of broilers (Shen et al., 2019).

Notable compounds isolated from the leaves and shoots of different bamboo species of *Bambusa*, *Dendrocalamus*, and *Sasa* (e.g., *B. nutans*, *B. balcooa*, *P. edulis*, *P. bambusoides*, *P. pubescens*, *S. albomarginata*) include hexadecanol, caffeic acid, coumaric acid, rutin, 3-O-(3′-methylcaffeoyl) quinic acid, 5-O-caffeoyl-4-methylquinic acid, 3-O-caffeoyl-1-methylquinic acid, isoorientin 4″-O-β-D-xylopyranoside, isoorientin 2″-O-α-L-rhamnoside, orientin, and isoorientin.

The DPPH free radical scavenging activity of the leaves of 11 bamboo species, *B. balcooa*, *B. nutans*, *B. polymorpha*, *B. tulda*, *B. vulgaris*, *D. asper*, *D. hamiltonii*, *D. strictus*, *D. membranaceus*, *D. giganteus*, and *Thyrostachys siamensis,* revealed the range of their IC_{50} inhibition values from 49.92% in *B. nutans* to 99.60% in *D. hamiltonii*. The FRAP assay activities of leaves of different bamboo species ranged from 48.95% in *T. siamensis* to 63.69% in *B. vulgaris* (Indira et al., 2022). The antioxidant potential of bamboo leaves may be variable depending on the seasons of the sample collection. Ni et al. (2012) found the highest DPPH free radical scavenging ability of leaves of *S. argenteastriatus* collected during the months from November to March, with IC_{50} values ranging from 284.08 to 457.42 μg/mL. The ferric ion–reducing power in FRAP assay was also higher during these months, with the values ranging from 234.57 to 422.87 μmol/L. Badwaik et al. (2014) reported 26.62% DPPH free radical scavenging activity of fresh shoots of *B. balcooa*.

In a subsequent study, the researchers found the DPPH free radical scavenging activities of shoots of *B. balcooa*, *B. pallida*, *B. tulda*, and *D. hamiltonii* ranged from 19.17% to 27.12% (Badwaik et al., 2015). Variation in the DPPH and ABTS after processing the *D. hamiltonii* shoots revealed the higher antioxidant activities after fermentation (Santosh et al., 2021). Joshi et al. (2024) assessed the DPPH radical scavenging activity and FRAP assay of *B. balcooa* extracts. The leaf extract exhibited the highest DPPH scavenging activity, with the lowest IC_{50} value of 73.92 ± 5.12 μg/mL, while boiled shoot extract (BSE) showed the lowest activity, indicated by the highest IC_{50} value of 351.48 ± 5.64 μg/mL. A similar result of higher scavenging activity by fermented shoot than fresh shoot extract was reported in the study. Moreover, in the FRAP assay, a similar trend was reported in the extracts. For *D. asper*, the DPPH free radical scavenging activity of fresh shoots was found to be 33.10 μg/mL with ABTS radical scavenging activities of 40.54 μmol/L (Sansenya et al., 2023). Gong et al. (2016) isolated acetic acid lignin from the shoot shell of *D. latiflorus*, which exhibited significant antioxidant activity (633.01 μM/g Trolox equivalent).

Khatun et al. (2013) assessed the antioxidant activities using colorimetric and electron spin resonance (ESR) assays for aqueous and 80% methanol (MeOH) Kumaisasa extracts; the DPPH radical scavenging activity was similar for both extracts. However, the ORAC value of the MeOH extract was nearly twice that of the aqueous extract. The MeOH extract also exhibited higher hydroxyl and superoxide radical scavenging activities. This correlated with the higher flavonoid and phenolic content, especially flavonoids, in the MeOH extract compared to the aqueous extract. Notably, both the polyphenol content and antioxidant activities were significantly reduced when the sample was extracted overnight compared to a 1-hour extraction. When measuring hydroxyl radical scavenging activity using the Fenton reaction and ultraviolet (UV) illumination, the activity was significantly higher for the Fenton system, indicating the metal-chelating nature of the Kumaisasa extract.

The butanol extract of *Phyllostachys edulis* leaves has demonstrated significant antioxidant activity, effectively scavenging both the stable DPPH free radical and the superoxide anion radical (O2-) in the xanthine/xanthine oxidase assay (Kweon et al., 2001). This potent antioxidant activity is attributed to three novel compounds: 3-O-(3′-methylcaffeoyl)quinic acid, 5-O-caffeoyl-4-methylquinic acid, and 3-O-caffeoyl-1-methylquinic acid. These compounds significantly inhibited superoxide generation in 12-O-tetradecanoylphorbol-13-acetate-induced human promyelocytic leukemia HL-60 cells. Similar results of improved *in-vivo* antioxidant activity by *B. nutans* shoot extract was reported in blood and liver tissue of LACA mice that exhibited a reduced amount of lipid peroxidation (LPO) and increased activity of GSH (Indira et al., 2024a). Bajwa et al. (2019) reported a reduction of glutathione reductase (GR), GPx, and Cat by 21%, 20%, and 69%, respectively. Additionally, the preventive effect of Kumaisasa intake was reported in mice subjected to restraint stress-induced oxidative stress, as evident from the TBARS and GSH levels of liver tissue.

The non-edible part of *Phyllostachys edulis,* i.e., sheath extract, decreased the reactive oxygen species (ROS) production without hampering cell viability in HepG2 liver cells (Tundis et al., 2023). Bamboo leaf extract reduced malondialdehyde (MDA) levels in the jejunum and liver. Supplementation with bamboo leaf extract also enhanced the activity of antioxidant enzymes and increased the expression of nuclear factor erythroid 2-related factor 2 (Nrf2) and several targeted genes in the jejunum and liver of suckling piglets (Yu et al., 2022).

5.4 HEALTH BENEFITS OF BAMBOO

Bamboo has been utilized in traditional medical practices for around 10,000 years (Kaur, 2017). Advances in scientific techniques and rigorous experimental studies have validated many of bamboo's traditional uses. Various therapeutic properties of bamboo have been substantiated, including cardioprotective, hepatoprotective, renal-protective, immune-modulating, antiaging, anticarcinogenic, anti-inflammatory, antihypertensive, antihyperglycemic, and antimicrobial effects. Almost all parts of the bamboo plant, such as the culm, shoot, leaf, resin, shoot shell, culm sheath, seeds, and roots, are used in clinical applications (Nirmala and Bisht, 2017). The polyphenols, phytosterols, and polysaccharides derived from bamboo are of special interest in the context of health-promoting benefits.

Polyphenols exert immunomodulatory effects by regulating the immune cells, impeding the synthesis of proinflammatory cytokines, influencing gene expression, and suppressing toll-like receptors (TLRs) and pro-inflammatory gene expression. Dietary polyphenols demonstrate immunomodulatory effects on macrophages; enhance the proliferation of B cells and T cells; exert impact on dendritic cells; and suppress the activity of Type 1 T helper (Th1), Th2, Th17, and Th9 cells. They deactivate NF-κB and regulate pathways such as mitogen-activated protein kinase (MAPk) and arachidonic acids. Additionally, key signaling pathways, including phosphatidylinositide 3-kinases/protein kinase B (PI3K/AkT), inhibitor of kappa kinase/c-Jun amino-terminal kinases (IKK/JNK), mammalian target of rapamycin complex 1 (mTORC1), and JAK/STAT are also inactivated. These polyphenols manifest immunomodulatory properties against allergic reactions and autoimmune diseases by inhibiting the proliferation of autoimmune T cells and downregulating pro-inflammatory cytokines, including interleukin-6 (IL-6), IL-1, and interferon-γ (IFN-γ) (Shakoor et al., 2021). Due to their diverse mechanisms of action, the polyphenols show promise for the prevention and treatment of autoimmune diseases such as type 1 diabetes, rheumatoid arthritis, and multiple sclerosis and these activities of polyphenols are due to their inhibition of eicosanoid-producing enzymes, modulation of ROS production, and their enhanced antioxidant activity. They downregulate enzymes like xanthine oxidase and NADPH oxidase (NOX) while enhancing the activity of endogenous antioxidant enzymes such as superoxide dismutase (SOD), catalase, and glutathione peroxidase (GPx). They also have an inhibitory effect against phospholipase A2 (PLA2), cyclooxygenase (COX), and lipoxygenase (LOX), leading to a reduction in prostaglandins (PGs) and leukotrienes (LTs) production. The benefit of polyphenols in inflammatory bowel disease is attributed to the mitigation of inflammation by inducing regulatory Treg cells (T cells) in the intestine, inhibition of tumor necrosis factor-alpha (TNF-α), promotion of apoptosis, and reduction of DNA damage. The activities of polyphenols are dependent on factors such as the sugar moiety, type, position, and extent of glycosylation in their structure and biotransformation (Sobhani et al., 2021). Therefore, it is essential to recognize the correlation between the structure and mode of action of each polyphenol for their use for the prevention and treatment of different inflammatory and autoimmune conditions.

Phytosterols or plant sterols are known to have many health-promoting properties, including immunomodulation, anti-inflammatory, anticancer, antiulcer, and lowering serum cholesterol (Chongtham and Bisht, 2020). Beta-sitosterol and its glycoside demonstrate anti-inflammatory, antineoplastic, antipyretic, and immune-modulating activities. These effects are evidenced by an augmentation in viable peripheral blood mononuclear cells; the activation of dendritic cells; and an elevation in plasma concentration of apolipoprotein A1, which regulates the secretion of pro-inflammatory cytokines and chemokines. Therefore, being a reservoir of phytosterols coupled with the abundant dietary fiber content in bamboo, they not only contribute to the improvement of lipid profiles and body weight but also showcase their remarkable potential in fostering immune modulation, anti-inflammatory responses, and overall health enhancement.

Polysaccharides are structurally diverse macromolecules that are reported to possess a wide range of bioactive properties that have biological functions, namely immunomodulatory, antitumor, antivirus, antioxidant, neuroprotective, and antidiabetic effects (Zhao et al., 2019). Most bioactivities of polysaccharides are assumed to be attributed to the modulation of innate immunity, specifically macrophage function. Polysaccharides stimulate macrophages, causing them to release ROS; nitric oxide (NO); and a variety of cytokines and chemokines; TNF-α; IL-1, IL-6, IL-8, and IL-12; and interferon (IFN)-γ connected to immunomodulation, antitumor activity, wound healing, and other therapeutic effects (Ramberg et al., 2010). Food low in dietary fiber has been associated with several disorders such as constipation, hemorrhoids, diverticulitis, obesity, heart problems, cancer, and diabetes. Being a rich source of dietary fiber, bamboo shoots have also been reported to play a key role in maintaining the lipid profile and body weight (Bajwa et al., 2021; Nirmala et al., 2018).

5.5 CONCLUSION

Many of the traditional health claims of bamboo have been validated by various *in-vitro* and *in-vivo* scientific studies which have proven to have protective effects on major organs and overall health. These beneficial effects have been attributed to the comprehensive bioactive compounds present in them. Therefore, bamboo is an underexplored plant with varied benefits; it requires further exploration and research of the mechanism behind its health benefits.

REFERENCES

Azmi, A.F.M.N., Mustafa, S., Hashim, D.M. & Manap, Y.A. 2012. Prebiotic activity of polysaccharides extracted from *Gigantochloa levis* (Buluh beting) shoots. *Molecules 17*(2): 1635–1651.

Badwaik, L.S., Borah, P.K. & Deka, S.C. 2014. Antimicrobial and enzymatic antibrowning film used as coating for bamboo shoot quality improvement. *Carbohydrate Polymers 103*: 213–220.

Badwaik, L.S., Borah, P.K. & Deka, S.C. 2015. Production and purification of anti-bacterial biometabolite from wild-type lactobacillus, isolated from fermented bamboo shoot: Future suggestions and a proposed system for secondary metabolite onsite recovery during continuous fermentation. *Applied Biochemistry and Biotechnology 175*: 1915–1925.

Bajwa, H.K., Santosh, O., Koul, A., Bisht, M.S. & Nirmala, C. 2019. Phytomodulatory effects of fresh and processed shoots of an edible bamboo *Dendrocalamus hamiltonii* Nees & Arn. Ex Munro on antioxidant defense system in mouse liver. *Journal of Food Measurement and Characterization 13*: 3250–3256.

Bajwa, H.K., Santosh, O., Koul, A., Bisht, M.S. & Nirmala, C. 2021. Phytomodulatory effects of *Dendrocalamus hamiltonii* (Nees & Arn. Ex Munro) shoots on the hepatic status of healthy male Balb/c mice. *Journal of Pharmacognosy and Phytochemistry 10*(1): 1860–1865.

Behera, P. & Balaji, S. 2021. Health benefits of fermented bamboo shoots: The twenty-first century green gold of northeast India. *Applied Biochemistry and Biotechnology 193*: 1800–1812.

Chongtham, N. & Bisht, M.S. 2020. *Bamboo Shoot: Superfood for Nutrition, Health and Medicine*. CRC Press, Boca Raton.

Coffie, G.Y., Antwi-Boasiako, C. & Darkwa, N.A. 2014. Phytochemical constituents of the leaves of three bamboo (*Poaceae*) species in Ghana. *Journal of Pharmacognosy and Phytochemistry 2*(6): 34–38.

Gagliano, J., Anselmo-Moreira, F., Sala-Carvalho, W.R. & Furlan, C.M. 2022. What is known about the medicinal potential of bamboo? *Advances in Traditional Medicine 22*(3): 467–495.

Gomez, J.P., Velez, J.P.A., Pinzon, M.A., Arango, J.A.M. & Muriel, A.P. 2021. Chemical characterization and antiradical properties of pyroligneous acid from a preserved bamboo, *Guadua angustifolia* Kunth. *Brazilian Archives of Biology and Technology 64*: e21190730.

Gong, J., Huang, J., Xiao, G., Chen, F., Lee, B., Ge, Q., You, Y., Liu, S. & Zhang, Y. 2016. Antioxidant capacities of fractions of bamboo shaving extract and their antioxidant components. *Molecules 21*(8): 996.

He, S., Wang, X., Zhang, Y., Wang, J., Sun, H., Wang, J., Cao, X. & Ye, Y. 2016. Isolation and prebiotic activity of water-soluble polysaccharides fractions from the bamboo shoots (*Phyllostachys praecox*). *Carbohydrate Polymers 151*: 295–304.

Huang, J.Q., Pang, M.R., Li, G.Y., Wang, N., Jin, L. & Zhang, Y. 2017. Alleviation of cyclophosphamide-induced immunosuppression in mice by naturally acetylated hemicellulose from bamboo shavings. *Food and Agricultural Immunology 28*(2): 328–342.

Indira, A., Joshi, B., Koul, A. & Chongtham, N. 2024a. Comparative hepato-ameliorative effects of *Bambusa nutans* fresh and fermented shoot extracts on STZ induced diabetic LACA mice. *Talanta 274*: 126035.

Indira, A., Joshi, B., Oinam, S., Koul, A. Chongtham, N. 2023a. Assessment of preliminary subacute oral toxicity of *Bambusa nutans* aqueous extract based on histoarchitectural and biochemical alterations in LACA mice. *Comparative Clinical Pathology 32*(2): 253–265.

Indira, A., Joshi, B., Oinam, S., Koul, A. & Chongtham, N. 2023b. Potential of bamboo in the prevention of diabetes-related disorders: Possible mechanisms for prevention. In: *Bamboo Science and Technology*. Springer Nature Singapore, Singapore, 89–124.

Indira, A., Santosh, O., Koul, A. & Nirmala, C. 2022. Comparative assessment of the antioxidant potential of bamboo leaves, along with some locally and commercially consumed beverages in India. *Advances in bamboo Science 1*: 100007.

Indira, A., Shahar, B., Joshi, B. & Chongtham, N. 2024b. Assessment of bioactive compound variations and in-vitro and in-vivo antioxidant activity in edible fresh and processed *Bambusa nutans* shoot through FTIR, GC/MS and HPLC analyses. *Food Chemistry 452*: 139552.

Joshi, B., Indira, A., Kalyan, N., Koul, A. and Chongtham, N. 2024. Assessment of in vitro antioxidant potential of *Bambusa balcooa* crude extracts and their safety evaluation based on biochemical alteration in C57bL/6 mice. *Food and Humanity 3*: 100335.

Kaur, H. 2017. Phytomodulatory effects of fresh and processed shoots of *Dendrocalamus hamiltonii* on lipid profile and antioxidant defense system. *A Ph. D.* (Doctoral dissertation, thesis submitted to Panjab University, Chandigarh, India).

Khatun, A., Nakajima, K., Saito, M., Giridharan, V.V., Hara, H. & Konishi, T. 2013. In vitro and in vivo antioxidant potential of Japanese kumaisasa bamboo leaf (*Sasa senanensis* Rehd). *Journal of Biochemical and Pharmacological Research 1*(4): 236.

Kimura, I., Kagawa, S., Tsuneki, H., Tanaka, K. & Nagashima, F. 2022. Multitasking bamboo leaf-derived compounds in prevention of infectious, inflammatory, atherosclerotic, metabolic, and neuropsychiatric diseases. *Pharmacology and Therapeutics 235*: 108159.

Kweon, M.H., Hwang, H.J. & Sung, H.C. 2001. Identification and antioxidant activity of novel chlorogenic acid derivatives from bamboo (*Phyllostachys edulis*). *Journal of Agricultural and Food Chemistry 49*(10): 4646–4655.

Kweon, M.H., Jung, M.J. & Sung, H.C. 2004. Cytoprotective effects of heme oxygenase-1 induction by 3-O-caffeoyl-1-methylquinic acid. *Free Radical Biology and Medicine 36*(1): 40–52.

Lu, B., Ren, Y., Zhang, Y. & Gong, J. 2009. Effects of genetic variability, parts and seasons on the sterol content and composition in bamboo shoots. *Food Chemistry 112*(4): 1016–1021.

Muscolo, A., Mariateresa, O., Giulio, T. & Mariateresa, R. 2024. Oxidative stress: The role of antioxidant phytochemicals in the prevention and treatment of diseases. *International Journal of Molecular Sciences 25*(6): 3264.

Ni, Q., Xu, G., Wang, Z., Gao, Q., Wang, S. & Zhang, Y. 2012. Seasonal variations of the antioxidant composition in ground bamboo *Sasa argenteastriatus* leaves. *International Journal of Molecular Sciences 13*(2): 2249–2262.

Nirmala, C. & Bisht, M.S. 2017. Bamboo: A prospective ingredient for functional food and nutraceuticals. *Bamboo Journal 1*(30): 82–99.

Nirmala, C., Bisht, M.S., Bajwa, H.K. & Santosh, O. 2018. Bamboo: A rich source of natural antioxidants and its applications in the food and pharmaceutical industry. *Trends in Food Science & Technology 77*: 91–99.

Nirmala, C., Bisht, M.S. & Laishram, M. 2014. Bioactive compounds in bamboo shoots: Health benefits and prospects for developing functional foods. *International Journal of Food Science and Technology 49*(6): 1425–1431.

Park, E.J. & Jhon, D.Y. 2010. The antioxidant, angiotensin converting enzyme inhibition activity, and phenolic compounds of bamboo shoot extracts. *LWT-Food Science and Technology 43*(4): 655–659.

Park, H.S., Lim, J.H., Kim, H.J., Choi, H.J. & Lee, I.S. 2007. Antioxidant flavone glycosides from the leaves of *Sasa borealis*. *Archives of Pharmacal Research 30*: 161–166.

Ramberg, J.E., Nelson, E.D. & Sinnott, R.A. 2010. Immunomodulatory dietary polysaccharides: A systematic review of the literature. *Nutrition Journal 9*(1): 22.

Saini, N., Nirmala, C. & Bisht, M.S. 2024. Isolation and quantification of phytosterols in fresh and processed shoots of an edible bamboo *Bambusa bambos* (L.) Voss using GC-MS. *Indian Journal of Natural Product and Resources 15*(1): 135–145.

Sansenya, S., Payaka, A. & Mansalai, P. 2023. Biological activity and inhibition potential against α-glucosidase and α-amylase of 2, 4-di-tert-butylphenol from bamboo shoot extract by *in-vitro* and in silico studies. *Process Biochemistry 126*: 15–22.

Santosh, O., Bajwa, H.K., Bisht, M.S. & Chongtham, N. 2021. Antioxidant activity and sensory evaluation of crispy salted snacks fortified with bamboo shoot rich in bioactive compounds. *Applied Food Research 1*(2): 100018.

Shakoor, H., Feehan, J., Apostolopoulos, V., Platat, C., Al Dhaheri, A.S., Ali, H.I., Ismail, L.C., Bosevski, M. & Stojanovska, L. 2021. Immunomodulatory effects of dietary polyphenols. *Nutrients 13*(3): 728.

Sharma, N. & Barooah, M. 2017. Microbiology of khorisa, its proximate composition and probiotic potential of lactic acid bacteria present in khorisa, a traditional fermented bamboo shoot product of Assam.

Shen, M.M., Zhang, L.L., Chen, Y.N., Zhang, Y.Y., Han, H.L., Niu, Y., He, J.T., Zhang, Y.L., Cheng, Y.F. & Wang, T. 2019. Effects of bamboo leaf extract on growth performance, meat quality, and meat oxidative stability in broiler chickens. *Poultry Science 98*(12): 6787–6796.

Sobhani, M., Farzaei, M.H., Kiani, S. & Khodarahmi, R. 2021. Immunomodulatory; anti-inflammatory/antioxidant effects of polyphenols: A comparative review on the parental compounds and their metabolites. *Food Reviews International 37*(8): 759–811.

Sun, J., Ding, Z.Q., Gao, Q., Xun, H., Tang, F. & Xia, E.D. 2016. Major chemical constituents of bamboo shoots (*Phyllostachys pubescens*): Qualitative and quantitative research. *Journal of Agricultural and Food Chemistry 64*(12): 2498–2505.

Supriyatin, S.R. & Sukmawati, D. 2014. Phytochemical analysis of Manggong bamboo leaf extract. In *Presented at the Proceeding of International Conference on Research, Implementation and Education* B-165.

Thamizharasan, S., Umamaheswari, S., Hari, R. & Ulagaratchagan, V. 2015. Quantitative phytochemical analysis of *Bambusa arundinacea* seeds. *International Journal of Pharmacognosy and Phytochemical Research 7*(5): 980–983.

Tongco, J.V., Rodriguez, E.B., Abasolo, W.P., Mun, S.P. & Razal, R.A. 2016. Mineral, nutritional, and phytochemical profile, total phenolic content, and radical scavenging activity of Philippine bamboo. *Natural Product Sciences 22*(1): 60–63.

Tundis, R., Augimeri, G., Vivacqua, A., Romeo, R., Sicari, V., Bonofiglio, D. & Loizzo, M.R. 2023. Anti-inflammatory and antioxidant effects of leaves and sheath from bamboo (*Phyllostacys edulis* J. Houz). *Antioxidants 12*(6): 1239.

Wen, J.L., Xiao, L.P., Sun, Y.C., Sun, S.N., Xu, F., Sun, R.C. & Zhang, X.L. 2011. Comparative study of alkali-soluble hemicelluloses isolated from bamboo (*Bambusa rigida*). *Carbohydrate Research 346*(1): 111–120.

Yang, R.Y., Tsou, S.C., Lee, T.C., Hanson, P.M. & Lai, P.Y. 2005, November. Antioxidant capacities and daily antioxidant intake from vegetables consumed in Taiwan. In *Symposium on Taiwan-America Agricultural Cooperative Projects*. Agricultural Research Institute, Council of Agriculture.

Yu, G., Ji, S., Yun, Y., Cheng, K., Zhang, H., Jia, P., Wang, T. & Zhang, L. 2022. Effects of bamboo leaf extract intervention on the growth performance, antioxidant capacity, and hepatic apoptosis in suckling piglets. *Journal of Animal Science 100*(7): 1–11.

Zhang, J.J., Ji, R., Hu, Y.Q., Chen, J.C. & Ye, X.Q. 2011. Effect of three cooking methods on nutrient components and antioxidant capacities of bamboo shoot (*Phyllostachys praecox* CD Chu et CS Chao). *Journal of Zhejiang University Science B 12*(9): 752–759.

Zhao, P., Zhou, H., Zhao, C., Li, X., Wang, Y., Huang, L. & Gao, W. 2019. Purification, characterization and immunomodulatory activity of fructans from *Polygonatum odoratum* and *P. cyrtonema*. *Carbohydrate Polymers 214*: 44–52.

Zheng, Y., Zhang, S., Wang, Q., Lu, X., Lin, L., Tian, Y., Xiao, J. & Zheng, B. 2016. Characterization and hypoglycemic activity of a β-pyran polysaccharides from bamboo shoot (Leleba oldhami Nakal) shells. *Carbohydrate Polymers 144*: 438–446.

6 The Potential Application of Bamboo Leaves in Medicine and Therapy

Anitha Rani Elavarasan and Uma Maheshwari Srinivasa

6.1 INTRODUCTION

A highly valued component of traditional medicine, bamboo leaves (BLs) are known to have various potential health benefits, a topic that contemporary scientific research is progressively exploring. BLs, which are well known for their profusion of anti-inflammatory and antioxidant compounds and vital nutrients, have been used for centuries in Asia to treat various illnesses, including skin and digestive disorders. BLs have drawn attention for their potential benefits to weight management, immune system support, and cardiovascular health as interest in natural remedies is growing worldwide. BLs, which have become known for their resilience and adaptability, have been utilized in traditional medicine for thousands of years, particularly in Asia. Numerous bioactive substances, including flavonoids, lactones, phenolic acids, and polysaccharides, are found in BLs and have been associated with several medicinal benefits, such as antibacterial, anti-inflammatory, and antioxidant qualities (Cheng et al., 2023). BLs are positioned as a promising natural resource for the creation of pharmaceutical applications and health-promoting supplements because of these bioactive components (Sangeetha et al., 2015). Research on the potential medical benefits of BLs has grown steadily, revealing promising uses in the treatment of diabetes, heart disease, and some inflammatory disorders. The distinct chemical makeup of BLs may promote cellular health, lower oxidative stress, and boost immune function, according to studies like that conducted by Shang et al. (2014). This could open up new possibilities for natural, plant-based medicine. Furthermore, as interest in plant-based and sustainable healthcare options are growing worldwide, BL extracts offer a compelling substitute for synthetic medications.

This chapter investigates the state of science regarding the therapeutic qualities of BLs, considering their pharmacological effects, bioactive constituents, and possible uses in both therapeutic and preventive medicine. The objective of this review is to draw attention to the medicinal potential of BLs and stimulate more research into their use in medicine by combining recent findings and developments.

6.2 BAMBOO LEAVES AND THEIR CHARACTERISTIC FEATURES

The overall well-being and efficiency of bamboo plants depend on their leaves. Both the ecological benefits of the species and its historic medical uses are supported by their special characteristics. Since BLs are typically thin and long, their surface area is maximized for effective photosynthesis (Figure 6.1). The typical components of a leaf are a sheath that surrounds the stem, a blade that protrudes to capture light, and a ligule at the junction of the sheath and blade. Bamboo is renowned for growing quickly, a trait made possible by the effective use of nutrients and energy stored in the leaf's structure. The ecological flexibility and resilience of bamboo leaves are well known. These characteristics are crucial for the plant's resistance to environmental challenges like intense rain and wind (Kuehl, 2015; Liese et al., 2015).

DOI: 10.1201/9781003485933-6

FIGURE 6.1 Pictorial representation of bamboo leaves.

Numerous health benefits are provided by the quantity of bioactive substances, including flavonoids, phenolic acids, amino acids, and minerals, that are present in BLs. These compounds provide the leaves with their potent antioxidant properties, which reduce oxidative stress and the risk of chronic diseases. The presence of chlorophyll, polyphenols, and other phytochemicals in BLs is also responsible for their antibacterial and anti-inflammatory qualities, making them beneficial in traditional medical procedures (Singhal et al., 2013).

BLs have long been used for their medicinal properties in many different cultures. In Chinese medicine, they are used to rid the body of heat, treat infections, and improve mental clarity. BL extract is well known for its diuretic properties, which can help to treat kidney and urinary tract issues. Additionally, the high silica content of BL is very useful to bones because it supports joint health and bone density. It is for this reason that BL tea and extracts are increasingly being used in dietary supplements and pharmaceutical products (Farrelly, 1984). BL extract is widely used in skin care cosmetics due to its anti-aging function. BLs are a natural by-product of bamboo farming and thus, is sustainable giving way to its use for eco-friendly products such as biodegradable packaging (Kleinhenz and Midmore, 2001). Besides, BLs are of ecological and economic importance partly because they provide substantial contributions to carbon sequestration capacity in bamboo forests due to the high rate of carbon dioxide absorption through BLs.

6.3 BIOACTIVE COMPOUNDS IN BAMBOO LEAVES

BLs are rich in many bioactive substances that significantly increase their ability to support health. These substances are vital to the plant's defense systems and have a variety of clinical uses in both conventional and alternative medicine. BLs are currently mostly used to treat conditions like atherosclerosis, diabetes, liver diseases, and symptoms related to the nervous system because of their bioactive ingredients (Cheng et al., 2023). Several bioactive substances can be found in BLs; their chemical composition and potential health advantages are explained.

6.3.1 Flavonoids

Flavonoids are a broad class of phytonutrients which are found in a wide variety of plants, and BLs have an abundance of them. Flavonoids have strong antioxidant properties which scavenge free radicals from the body, reduce oxidative stress, and prevent cellular damage, making them the most sought after for treatment of many ailments, like cancer, inflammation, viral and bacterial diseases, etc. The anti-inflammatory properties of these flavonoids are essential for preventing chronic inflammation, which is the root cause of many diseases, including arthritis, heart disease, and several neurological conditions. Research has also been conducted on how flavonoids can improve mental clarity and reduce the risk of neurodegenerative diseases because they can change brain signaling pathways (Ni et al., 2012).

Flavonoids like isoorientin, orientin, and vitexin are present in BLs (Goyal et al., 2008). Orientin and isoorientin are two closely related flavonoids that have strong antioxidant qualities. Studies have demonstrated their ability to reduce inflammation, prevent heart disease, and strengthen cellular antioxidant defenses (Wang et al., 2010). Vitexin has been the subject of research due to its possible cardioprotective, anti-inflammatory and anticancer effects. Flavonoids also play a key role in protecting plants from ultraviolet (UV) radiation, infections, and herbivores.

6.3.2 Phenolic Acids

Ferulic acid, *p*-coumaric acid, and caffeic acid belong to a class of phenolic acids with strong antioxidant potential and are present in BLs. These acids have been associated with improved blood circulation, enhanced skin health, and lower chances of developing chronic conditions such as diabetes and heart disease. Their presence in BLs encourages the use of the plant in treatment processes that include improvement of cardiovascular and metabolic conditions (Park et al., 2010).

6.3.3 Amino Acids and Proteins

BLs are high in some essential and non-essential amino acids that are beneficial to the human body. These amino acids are required for practically every biological activity, including hormone production, tissue reconstruction, nutrient digestion, and protein building. These amino acids enhance the nutritional value of bamboo leaves, making them a suitable supplement to dietary supplements and health foods (Singhal et al., 2013).

6.3.4 Silica

Silica is a trace mineral located in BLs which is particularly beneficial for bones, skin, and hair. It plays a vital role in collagen production and maintaining connective tissue structure. BLs are used in traditional therapies due to their high silica concentration for improving hair, skin, and joint health (Kleinhenz and Midmore, 2001).

6.3.5 Polysaccharides

BLs contain polysaccharides which are complex carbohydrates that serve as energy storage and contribute to the tensile strength of plant cell walls. These are well known to have immune-modulatory properties in the human body. They have been used to treat autoimmune disorders and as an adjunct therapy to enhance an individual's immune system and decrease inflammation (Xie et al., 2015).

6.4 METHODS OF EXTRACTION OF BIOACTIVES FROM BAMBOO LEAVES

To get the benefits of BLs, researchers have employed different techniques for the extraction of phytochemicals.

6.4.1 Solvent Extraction

This traditional approach uses organic solvents like methanol, ethanol, or water to extract the bioactive compounds from dried or fresh BLs. The process usually necessitates maceration or refluxing to extract the desired compounds from the BLs. These methods are effective in removing a range of substances from the BLs, also including phenolic acids, flavonoids, and amino acids (Chung et al., 2011).

6.4.2 Ultrasound-Assisted Extraction (UAE)

This method utilizes ultrasound to create microcavities in solid plant material to enhance the extraction of bioactive components and the penetration of solvents into cellular materials. It facilitates the extraction of antioxidant compounds with a smaller amount of solvent and in a shorter duration (Li et al., 2006).

6.4.3 Supercritical CO_2 Extraction

This method improves the extraction of bioactive compounds and solvent improves in targeting the penetration into cellular materials by using ultrasonic waves to create the microcavities in plant material. It is particularly useful for efficiently removing antioxidants while using less solvent and reducing extraction time (Chen et al., 2024).

6.4.4 Other Techniques

Since BLs contain more than 100 different chemical compounds, including amino acids, phenolic acids, polysaccharides, coenzyme Q10, flavonoids and flavonoid glycosides, volatile components, and phenolic acids, methods like column chromatography, and membrane separation technologies are commonly used to extract these compounds. Advanced techniques for the separation such as preparative high-performance liquid chromatography (PHPLC), high-speed counter-current chromatography (HSCCC), simulated moving bed chromatography (SMB), and dynamic axial compression chromatography (DAC) are used to extract C-glycosides from the BL flavonoid.

6.5 ANALYSIS OF BAMBOO LEAVES EXTRACTS

Different methods have been used for the analysis of BL extracts, some of which are discussed as under.

6.5.1 High-Performance Liquid Chromatography (HPLC)

Using this method, the components of a mixture are identified, measured, and isolated. BL extracts are commonly used to measure specific flavonoids and phenolic acids, and it is particularly effective at analyzing substances that are soluble in water (Park and Kim, 2014b).

6.5.2 Gas Chromatography-Mass Spectrometry (GC-MS)

This technique is particularly helpful for volatile organic compounds because it uses gas chromatography and mass spectrometry for the thorough analysis of the compounds present in BL extracts to identify and quantify the smaller volatile compounds and essential oils (Dong et al., 2010).

6.5.3 Fourier Transform Infrared Spectroscopy (FTIR)

By measuring the infrared intensity against the wavelength (of infrared light) absorption, Fourier transform infrared (FTIR) spectroscopy creates an infrared spectrum that is useful for determining the general molecular structure and functional groups in complex extracts of BLs (Zhang et al., 2011).

6.6 PHARMACOLOGICAL PROPERTIES OF BAMBOO LEAVES

Discussed below are some of the medicinal properties that have been associated with BLs.

6.6.1 Antioxidant Effects

The abundance of flavonoids, polyphenols, and other antioxidants in BLs aids in their ability to scavenge free radicals. The primary way that these bioactive compounds function is by providing free radicals with an electron, neutralizing them, and preventing them from damaging cells (Kumar and Pandey, 2013). Isoorientin, vitexin, and catechins are among the phenolic acids and flavonoids present in bamboo leaves that directly scavenge free radicals. According to Park et al. (2010), they stabilize free radical species and stop the start of chain reactions that result in oxidative stress and cell damage. Iron and copper ions, which are known to catalyze the Fenton reaction and produce extremely reactive hydroxyl radicals, can be chelated by the antioxidants present in BLs (Chung et al., 2011). Endogenous antioxidant enzyme activities of glutathione peroxidase, catalase, and superoxide dismutase (SOD) are also encouraged by BLs (Hao et al., 2015). The antioxidant properties of BLs have significant beneficial health effects on the tissues and their biological mechanisms against oxidative stress which is one of the causes of many chronic diseases, particularly in humans. Cardiovascular systems, especially atherosclerosis progression, are highly aggravated by oxidative stress. BLs possess numerous antioxidants which prevent Low Density Lipoprotein (LDL) cholesterol oxidation, a major risk factor for heart difficulties (Li et al., 2010). More so, the cell and nuclear DNA of biological mechanisms are easily vulnerable to oxidative stress causing mutation and damage which may translate into cancer. Hence, Park et al. (2014a) suggested that the presence of antioxidant molecules in BLs may restrain oxidative DNA damage and therefore lower one's risk of developing cancer. The free radical scavengers in BLs work to maintain the brain homeostasis that may be compromised during pathologic processes. They recruit defensive mechanisms to combat oxidative stress in neuronal cells, which are important for the inhibition and treatment of neurodegenerative disorders such as Parkinson's and Alzheimer's diseases (Hao et al., 2015).

6.6.2 Anti-inflammatory Activity

Various bioactive components with anti-inflammatory activity, including flavonoids and phenolic acids, are present in these leaves. These compounds employ a diverse set of mechanisms to mitigate inflammatory responses. It has been shown that BL extracts restrict both the synthesis and activity of pro-inflammatory cytokines such as tumor necrosis factor-alpha (TNF-α) and interleukins (IL-6, IL-1β). Apart from that, they inhibit the activity of COX-2 and iNOS, both of which are important enzymes in the pathogenesis of inflammation (Park et al., 2010). Cytokines such as NF-κB, a major pathway related to inflammation and mediating transcriptional expression for the inflammatory genes, are one of the pathways that these extracts can alter. Hence, by targeting this pathway, BL contribute to the downregulation of transcriptional expression of various pro-inflammatory cytokine genes (Li et al., 2010). BL extracts also counteract ROS which might activate inflammatory pathways, thereby reducing inflammation at the cellular level (Chung et al., 2011). BL extracts have been demonstrated to have an anti-inflammatory effect on animal models of inflammation induced by carrageenan or lipopolysaccharides (LPS).

6.6.3 Antimicrobial Properties

The traditional use of BLs as an effective herbal drug is now supported by more scientific evidences. The bioactive compounds namely flavonoids, phenolic acids and alkaloids in BL exhibit antibacterial activity. These substances engage in various mechanisms of interaction with microbial cells. The phytoactives cause the rupture of bacterial and fungal cell membranes causing cell death. They also inhibit some enzymes essential for microbial metabolism and replication, which slows the growth and spread of these infections. A few substances found in BLs can either bind directly to genetic material or interfere with the synthesis of nucleic acids, which prevents the transcription and replication processes necessary for microbial survival.

Both gram-positive and gram-negative bacteria can be effectively combated by the strong antibacterial properties of BLs. Studies have demonstrated the efficacy of BL extracts against two common infection-causing bacteria: *Escherichia coli* and *Staphylococcus aureus* (Park et al., 2010). BLs also possess antiviral qualities that can combat enveloped viruses. Viruses, like Herpes simplex virus, are inhibited by BL extracts from replicating, lessening the intensity and length of outbreaks (Li et al., 2010). They also exhibit antifungal activity exhibiting action against a range of fungal pathogens, including those that cause crop spoilage and skin diseases. For instance, it has been demonstrated that BL extracts have the potential to inhibit the growth of *Candida* species, which are opportunistic yeasts that can infect humans with low immunity such as sick and old persons (Hao et al., 2015). Patients suffering from bacterial or fungal skin infections could also be treated with creams or ointments popularly prepared from BL extracts. BLs are also utilized to prevent bacterial spoilage of food, thereby extending food items' viability. The extracts can also be blended with other natural substances to form surface cleaners and personal sanitizers.

6.6.4 Anticancer Potential

In some pre-clinical studies, the BLs have been shown to exhibit convincing anticancer potential owing to their bioactive constituents. The modulation of estrogen receptors and associated signaling pathways by flavonoids and phenolic compounds is attributed by researchers to this effect (Park et al., 2010). The BL components can trigger programmed cell death, also known as apoptosis in cancer cells. This occurs by activating pro-apoptotic proteins or downregulating anti-apoptotic factors and by modifying several cell signaling pathways that cause the death of cancer cells (Chung et al., 2011). Some angiogenesis inhibitors are also known to be present in BLs. This is important because cancerous tumors are highly vascularized structures and therefore their growth and dissemination need to be controlled in terms of blood and oxygen supply. The ability of BLs to starve tumors by preventing the formation of new blood vessels has already been demonstrated (Park et al., 2014a). While many people believe that the compounds in BLs are only useful for their antioxygen properties, it has been confirmed that plant compounds can target cancerous cells to oxidative stress and thus induce damage and death to cancerous cells without affecting normal cells (Hao et al., 2015). In animal models, extracts from BLs have demonstrated effectiveness in preventing the growth of colorectal cancer cells. The mechanisms include the suppression of inflammation, which frequently occurs as a prelude to colorectal cancer, and the induction of apoptosis (Li et al., 2010). Activating the enzyme caspase, which is essential for the execution stage of cell apoptosis, has allowed extracts from BLs to successfully induce apoptosis in leukemia cells (Chung et al., 2011). Overall, BLs show significant nutraceutical properties when extracted using standardized techniques (Figure 6.2).

6.7 ETHNOMEDICINAL APPLICATIONS OF BAMBOO LEAVES

The entire bamboo plant is known to be a potential source of amino acids, vitamins, minerals, polysaccharides, phenolic, and volatile compounds (Shen et al., 2024). Reports suggest that the tribes of Adi belonging to Arunachal Pradesh, India, were familiar with the medicinal uses of *Bambusa tulda* Roxb., *Dendrocalamus giganteus* Wall. ex Munro, *Dendrocalamus strictus* (Roxb.) Nees., and *Schizostachyum capitatum* (Trin.) Rupr. They mainly utilize either leaf extract or its infusion to cure tetanus, stomach pain, and as steroids (Sharma and Borthakur, 2008). A similar observation with respect to the ethnomedicinal uses of bamboo was reported among the tribes of Gujarat, India (Patel, 2004). Indian Kani tribes were certain that seeds and leaf juice of *Bambusa arundinacea* (Retz.) would strengthen reproductive health and cartilage, thereby modulating osteoarthritis and osteoporosis. Further tribes of Odisha, India, were known to incorporate

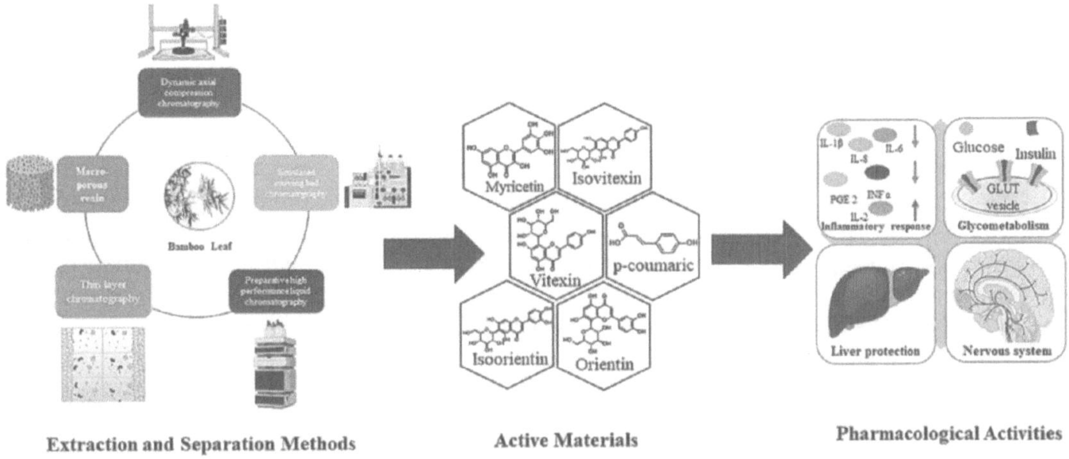

FIGURE 6.2 Overview of extraction and separation of bioactive compounds from bamboo leaves and their pharmacological properties.

bamboo in their daily diet to meet their nutritional demand and they fed *D. strictus* (Roxb.) Nees to the cattle suffering from diarrhea (Mishra and Chaudhury, 2012). The inhabitants of Eastern Asia and Japan are known to have embraced the ethnomedicinal properties of *Sasa senanensis* (Sangeetha et al., 2015). A traditional medicine system, Dai medicine, which is popular among the people of Xishuangbanna, the southeastern region of China, utilize the stem of *D. hamiltonii* Nees and Arn. ex Munro, *Thyrsostachys siamensis* Gamble, *Pseudostachyum polymorphum* Munro, and *D.* spp. to energize the functioning of kidneys (Yang et al., 2008). In the Yunnan province in the Chinese-Himalayan region, the leaves and young culms of *Phyllostachys* spp. are used to treat lung and throat infections, lung inflammations, and coughs. They also use the shoots of *Indosasaping bianensis* McClure to get relief from colds and headaches.

In addition to other bamboo species, the leaves of *Oxytenanthera abyssinica,* a native bamboo species in southeastern Benin, have been traditionally used as an herbal remedy (Honfo et al., 2015). A recent study conducted in the Central Siwalik region of Nepal explored the ethnobotanical uses of wide ranges of bamboo species where *B. bambos*, *B. tulda*, *B. nutans* subsp. *cupulata*, *D. strictus*, *B. nepalensis,* and *B. balcooa* were observed to be used solely for medicinal purposes. Their mode of administration was culm sheath hair powder, sap, and decoction of leaves and young shoots. These bamboo species were used to treat different diseases, from common colds to snake bites. Further, it has been found that nearly 69.66% of this ethnic group was dependent on bamboo for its ethnomedicinal purpose while 52.74% were dependent on bamboo for their nutritional needs (KC et al., 2024).

6.8 CASE STUDIES AND ANECDOTAL EVIDENCE OF MEDICINAL PROPERTIES OF BAMBOO LEAVES

A study conducted among 21–23 years of age of healthy women elucidated that the fiber-rich bamboo shoot intake was found to decrease the elevation of cholesterol levels effectively and showed improved bowel function. It concluded that the properties exerted by bamboo consumption in the human system would be eminent pliers in the regulation of chronic disease and colon cancer (Park and Jhon, 2009). Using randomized controlled trials, moxibustion external therapy using bamboo was carried out among 312 individuals suffering from lower back pain. This study suggested that the bamboo moxibustion treatment can modulate the suffering from chronic disease, especially facilitating ethnic indigenous groups with improved healthcare services in their grassroots (Wang et al., 2022). A similar clinical trial study conducted by Xue et al. (2024) also concluded that bamboo moxibustion therapy could be an alternative treatment to improve the quality of life in performing day-to-day activities among individuals suffering from lower back pain. Further, bamboo medicinal therapy was found to regulate serum levels of β-endorphin and substance-P, thereby exerting an analgesic effect against lower back pain. Chilicka et al. (2022) demonstrated a single-blinded placebo study among women of 19–23 years of age, where it was claimed that the sonophoresis with a combined composition of green tea, bamboo extract, and lactic acid relieved sufferings from acne vulgaris. This finding underlined that the method of administration was found to be effective and safe. Further, it was uncovered that the anti-inflammatory, wound healing, and antimicrobial properties of bamboo efficiently reduced acne vulgaris and its contributing factors such as skin eruptions.

Studies have also investigated the modulation of cariogenic oral bacteria and improvement of oral health through herbal toothpaste formulated using bamboo salt. A double-blinded parallel randomized controlled clinical trial study in 60 dental students between the age ranges of 18–30 years was carried out to confirm the potential of bamboo salt as an alternative source in place of conventional toothpaste (Biria et al., 2022). Similarly, oral health clinical trials conducted by Choi et al. (2017) used aqueous and oil-based mouth rinses containing bamboo salt, magnolia bark, and *Centella asiatica* extracts. The 34 healthy adults who participated in this study showed that the incorporation of bamboo salt was found to exert antigingivitis properties through its antimicrobial and inflammatory properties. Simultaneously, the dietary supplementary study using bamboo extract given to 27 women with deforming arthritis confirmed that the anti-inflammatory and antioxidant properties played a crucial role in reducing their microcirculatory disorders (Tokhiriyon et al., 2019). A similar response was documented in clinical trials conducted among 130 osteoarthritis individuals using dietary supplements containing bamboo extract (50 mg). After follow-up of medication of dietary supplements for 6 months, it was shown that the locomotory abilities were improved and pain was reduced among the individuals (Puigdellivol et al., 2019). Preliminary clinical trials were conducted in nine patients suffering from mild to severe atopic dermatitis using derma gels where constituents included total bamboo leaves extract (50%) and other herbal oils such as olive, coconut, curcumin, and sweet gourd. This scientific work has elucidated that the BL extract would strengthen skin cell layers and prevent microbial infection in the skin lesion, augmenting the pharmaceutical usage of BLs (Len et al., 2023). A case study carried out by Chen et al. (2020) among fifty adolescents shed light on the protective mechanism of water-soluble flavone C-glycoside antioxidant from BLs against the internal toxicity built upon exposure to preprocessed food containing acrylamide and glycidamide. C-glycoside antioxidant was found to alter the distribution kinetics of acrylamide and glycidamide and facilitate its excretion rate.

6.9 AN OVERVIEW OF EXISTING CLINICAL TRIALS

A few clinical studies have highlighted bamboo's benefits in various health domains, including cardiovascular health, pain management, skin care, oral health, and detoxification (Table 6.1). It has been found that most clinical trials conducted using bamboo were intended for external applications such as herbal medicinal toothpaste formulations, mouth rinses, moxibustion, and sonophoresis using bamboo extract (Chilicka et al., 2022; Choi et al., 2017; Len et al., 2023; Wang et al., 2022). Further, the study design was found to be either single-blinded or double-blinded or randomized controlled trials. Certain study designs were unable to maintain placebo groups either due to lower sample size or other unavoidable reasons. These are the essential interventions to remove the bias or background noises in understanding the pharma dynamics of therapeutic interventions (Iflaifel et al., 2022; Monaghan et al., 2021). It has been observed that the clinical trials of bamboo were carried out in combination with other bioactive components, such as coconut oil, olive oil, green tea extract, lactic acid, etc. The external application of bamboo therapeutic works via transdermal penetration; hence, intended to escape the complexity of the digestion process and its dose/toxicity mediated responses. However, few studies have been found to overcome the above limitations by

TABLE 6.1
Clinical Trials Intended to Explore the Curative Effect of Bamboo

Sl. No	Target Disease	Participant Nature	Study Design	Mode of Therapeutics Administration	Study Outcome	References
1	Osteoarthritis	Total of 130 osteoarthritis patients were included	Six months observational study of single treatment group	Ingestion of Artipotect®, a dietary supplement containing 50 mg/tablet of bamboo extract	Significant reduction in pain and improved locomotor function was observed	Puigdellivol et al., 2019
2	Osteoarthritis	The study was carried on outpatient basis where 27 osteoarthritis cases were considered (Women gender of age 44-49)	The subjects were divided into control and treatment groups	Oral dietary supplement of 3.6 mg bamboo combined other antioxidant was prescribed for treatment groups for 2 months	Decrease in pain indexes was recorded	Tokhiriyon et al., 2019
3	Acrylamide and glycidamide detoxification	A sample size of 50 Chinese adolescents were enrolled	Participant and investigators were blinded and randomization techniques used to categorize the participant into control and treatment group were the study was intended to carry for 10 days	Capsulated oral dietary supplement containing water-soluble flavone-C-glycoside-rich antioxidant extracted from bamboo leave	Lessen the toxicity exposure of acrylamide and glycidamide	Chen et al., 2020
4	Acne vulgaris	Women of total 60 with mild to moderate facial acne was chosen and their age werein the range of 19-23 years	The study was designated to a single blind placebo controlled trials carried for 14 days	Incorporation of ultrasound gel combined with the extract if green tea, bamboo, mimose and lactic acid via sonophoresis procedure	Emphasized the role of anti-inflammatory of bamboo in alleviating skin eruptions	Chilicka et al., 2022
5	Nasal allergy, dyslipidemia, hyperglycemia, and cognitive functions			Oral administration of chyawanprash containing bamboo	Reduced the chief complaints of the disease condition mentioned	Sanger et al., 2021; Sharma et al., 2022
	Chronic bronchitis	Individuals with the history of chronic bronchitis for past two who were between the age group of 18-70 were included and the samples size were 75 who were	It was an open-label prospective study carried for 84 days	Talisadichurna powder containing bamboo was mixed honey and orally fed at a dosage of 3 gm twice a day	No adverse effect were recorded upon talisadichurna administration. Chronic bronchitis's cardinal symptoms were found to be improved	Tomar et al., 2022

(Continued)

TABLE 6.1 *(Continued)*
Clinical Trials Intended to Explore the Curative Effect of Bamboo

Sl. No	Target Disease	Participant Nature	Study Design	Mode of Therapeutics Administration	Study Outcome	References
6	Lumbar disc herniation	Nearly 312 individuals with lumbar disc herniation complication were enrolled	Here, in order to detect any adverse effect, data collectors and statisticians alone were blinded. The subjects were randomly segregated into two groups (control and treatment) for 14 days	Transdermal absorption of Ma's bamboo-based medicinal moxibustion	Suggested that this therapy can improve health care system in ethnic areas	Wang et al., 2022
7	Atropic dermatitis	Atropic dermatitis cases of 9 subjects (mild to severe atropic dermatitis were investigated	Neither randomization or placebo controls were used. Treatment was given for 3 months followed by observational investigation after 3 months	External application of derma gel containing total bamboo leaf extract containing mixed herbal oils	Bamboo leaf extract had positively suppressed atropicdermatitis symptoms	Len et al., 2023
8	Diabetes mellitus	Hyperglycemic individuals of total 212 (30-70 age) where 108 male and 104 female took part in this study	Randomized open-label drug coded trials compared with control drug metformin hydrochloride of 500 mg at a dosage of twice per day	Administrated with two tablet (twice a day) of Dolabi containing *Bambusabambos* as one of major component	Reduced fasting blood glucose and post prandial blood glucose levels in hyperglycemic patients	Sheikh et al., 2023
9	Lumbar disc herniation	312 individuals with lumber disc herniation syndrome of wide age range 18-65 years were included	Clinical trials were characterised by two-center, single-blind, and randomized controlled trial	Transdermal absorption of Ma's bamboo-based medicinal moxibustion	This therapeutic design was found to be effective to that of acupuncture	Xue et al., 2024

designing dietary supplements. These studies have comprehensively validated the safety of those formulations, which is a commendable achievement in the clinical trials for elucidating bamboo ethnomedicinal properties (Len et al., 2023; Wang et al., 2022). Further, most of the studies obtained qualitative data sets for validation while the studies conducted by (Park and Jhon, 2009) and Chen et al. (2020) were able to assess the biochemical changes driven via bamboo medication.

6.10 INTEGRATION OF BAMBOO LEAF EXTRACTS IN MODERN THERAPIES

The amalgamation of data collected from *in vitro* and *in vivo* studies cohesively suggests that BL extracts or derivative phytochemicals act against various clinical disorders. Prior administration of bamboo extract was effective enough to suppress tumor inoculation, growth, incidence, and its survival as BLs efficiently modulate cytokine and free radicals production, thereby controlling the progression of carcinogenesis (Seki and Maeda, 2010). Goyal et al. (2011) confirmed high availability of phenolic content and flavonoids in *Dendrocalamus strictus,* which exhibits greater antioxidant activity. Wedler et al. (2014) exemplified the dual activity of leaf extracts from *Phyllostachys edulis* in targeting angiogenesis at lower concentrations and inflammation at higher doses. A comparative study conducted on ethyl acetate and aqueous extracts of *B. vulgaris* leaves illustrate the wound-healing properties of bamboo via collagen proliferation, with better results of ethyl acetate–based extract. This implied that flavonoid responses as healing promotors are rapid than that of amino acid, vitamins and ascorbic fractions of bamboo leaf extract (Lodhi et al., 2016).

S. coreana Nakai leaf extract was reported to exhibit anti-adipogenic properties via inhibiting pro-inflammatory response, adipocyte differentiation, and cell migration. This research finding helps to assess that the flavonoids present in the *S. coreana* species, which includes orientin, isoorientin, and vitexin, mediated mitigation strategies over the downregulation of nitric oxide production and inducible nitric oxide synthase (iNOS) expression, thereby substantiating anti-inflammatory and anti-adipogenic properties (Yang et al., 2017). Another study claimed the chemoprotective effect of BL extract and found that it could be a promising, safe, antimalarial candidate without suppressing the liver and renal function. It has been demonstrated that the presence of significant flavonoids and phenol content in *B. vulgaris* is effective against malarial infection (Anigboro, 2018). Flavonoids present in BLs were reported to modulate the cascade of the Nrf2/HO-1/NQO1 pathway in protecting $HepG_2$ cells from oxidative injury where alleviated lipid accumulation, biochemical indexes, reactive oxygen species, and mitochondrial membrane potential increase the susceptibility of damage. The study also shows that BL flavonoids enriched with cinnamon essential oil exert positive effects over intestinal microflora, and undoubtedly improve the immune reflexes (Yu et al., 2019).

6.11 DEVELOPMENT OF PHARMACEUTICALS AND SUPPLEMENTS

Industrial pharmaceutical formulation largely utilizes the oxidant quenching properties of antioxidants to the rapid deterioration of the medicine. Since the absence of understanding the usage of natural antioxidants, maintaining quality of medicine has led to both underutilization of natural antioxidants and elevated use of synthetic antioxidants (Schillaci et al., 2014). Furthermore, incorporation of the synthetic antioxidants might increase the risk of cancer and exert toxic effects into the physiological system (Nirmala et al., 2018). A plethora of data suggest that the antioxidants from culm, shoots, and leaves of bamboo species such as *Bambusa vulgaris, B. arundinaceae, D. asper, Sasa borealis, S. senanensis,* and *S. quelpaertensis* have proven to be a potential candidate against free radicals (Fan et al., 2015; Felisberto et al., 2017; Seki et al., 2010; Senthilkumar et al., 2011). Studies demonstrated by Lu et al. (2006), Zhang et al. (2007), and Fan et al. (2015) assure that BL antioxidants could effectively slow lipid peroxidation and bio-amine formation in food commodities. In China, BL-based antioxidants have been recognized as novel food additives to preserve edible oil, fish, and meat products. Though bamboo's antioxidant property is widely acknowledged as a promising candidate in preventing the release of free radicals, its potential as an antioxidant in the pharmaceutical industrial sector remains unknown. These findings pave the way to promote the utilization bamboo as a natural antioxidant in therapeutic drugs which should be additionally substantiated by scientific evidences.

Chyawanprash, an Ayurvedic medication prescribed as an immune booster, uses BLs as a major constituent (Alias et al., 2023). Another Ayurvedic preparation, Sitopaladichurna, is known to be very effective against coughs, respiratory disorders, numbness, and digestive illness (Sawarkar et al., 2023). Several studies have highlighted the presence of phytosterols, bacteriocins, and exopolysaccharides in fermented bamboo shoots, demonstrating their use as food additives and supplements (Chavhan et al., 2015; Das et al., 2013; Zeng et al., 2016). Many pharmaceutical drug formulations are largely dependent on phytosterols for active sterols production which are present in greater proportion in *B. tulda* and *D. giganteus* (Choudhury et al., 2012; Santosh et al., 2021; Sarangthem and Singh, 2003).

Numerous dietary supplements of bamboo or its derived products such as extract are also available in the market. Large populations are dependent on these products for their health benefits which include antioxidant, anti-inflammatory, immune booster, and wound-healing properties. Ancient purity, swanson, wish, Nutrics® (contains bamboo's silica), mother nature's, ostrovit, and trexgenics are some of dietary supplements containing bamboo as major components (Sawarkar et al., 2023). Food fortification is also another food processing technique to promote bamboo's consumption for its nutraceutical value. Cookies, nuggets, crackers, biscuits, and candy have been fortified with fermented bamboo shoots or non-fermented shoots or fiber of various bamboo species such as *B. auriculata, B. bambos, B. tulda, B. polymorpha, B. balcooa, B. vulgaris, D. asper,* and *D. strictus* not only to improve the organoleptic properties of the above-mentioned food products but also acts as a preserving agent. These value-added food products were also reported to enhance nutritional composition (Choudhury et al., 2015; Das et al., 2013; Gazi et al., 2017; Mustafa et al., 2016; Thomas et al., 2016).

Bamboo shoots added to deep-fried battered and breaded fish balls have been shown to decrease fat absorption (Zheng et al., 2016). The dietary fiber of bamboo shoots is reported to modulate total cholesterol levels and improve bowel movement which has a significant role in enhancing the nutraceutical value of food (Park and Jhon, 2009). Properties associated with antioxidant, anti-inflammatory, anticancerous, antidiabetic, antibiotic, and anti-angiogenesis of bamboo are widely accepted as sustainable resources for the pharmaceutical and dietary supplement industry (Verma et al., 2022).

6.12 SAFETY, DOSAGE, AND REGULATORY CONSIDERATIONS

As aforementioned, bamboo as a pharmaceutical agent has undoubtedly improved the healthcare system for centuries. Yet the gap in understanding the safety regulation needs to be approached such that it could provide essential insights to support the evolution of bamboo from ethnomedicine to evident pharmaceutical drug. Several studies demonstrated the safety in administration of BL extract. Particularly the acute and sub-chronic toxicity experimental procedure designed by Lu et al. (2005) and Lu et al. (2006) employing mice and rats manifested even at the dosage of 10 g/kg/day and 4.3 g/kg/day upon 90 days administration did not show any adverse effect. Similarly, there was negligible effect on gestation index, fertility, sex ratio, and embryo-fetal number observed in the toxicity study conducted in pregnant rats. On the contrary, bamboo shoots ethanolic extract dosing (300 mg/kg/day) has compromised the fertility index in male rats via altering the epididymis structural and functional integrity. Though this dosage did not exhibit any systemic toxicity, it was potent enough to decrease the mating activity within four days of treatment. Further, it is worthy to note that upon withdrawing the treatment, the fertility rate was restored. These findings of bamboo interference on fertility was further supported by studies that claimed the BL extract dosage of 250 to 500 mg/kg/day is potentially enough to induce abortion and to increase mortality rate in rabbits under gestation. It has been concluded that at this dosage there are significant changes in modulation of serum progesterone, follicle stimulating hormones, luteinizing hormones, alkaline phosphatase activity, and glucose concentration (Yakubu and Bukoye, 2009; Yakubu et al.,2009). Panee (2015) suggests the consumption of bamboo during pregnancy might hinder the thyroid function, thereby promoting the birth defects rate. While looking into the toxicity assessment of bamboo stem products, no abnormalities in clinical symptoms, hematology parameters, and in sperm test in rats and mice were recorded when fed at the dosage of 10 g/kg/day (acute toxicity dose) and 830 mg/kg/day (sub-chronic toxicity dose) (Zhang et al., 2004). Currently, many studies have been focused on integrating novel technologies in formulating therapeutic drugs.

Bamboo-based nanoparticles have also been evaluated to combat diseases. Research findings using diabetes-induced Wister rats' histological assessment of the liver indicated the presence of normal cell architecture depicting the possibility of safety in using bamboo nanoparticles that mediated diabetes amelioration (Jayarambabu et al., 2021). The prevalence of antinutritional components of bamboo, such as cyanogenic glucoside, phytate, glucosinolate, oxalate, saponin, and tannin, should also be taken into consideration while assessing the safety of bamboo-incorporated therapeutic drugs which may adversely affect human health, toxicate, and hinder the absorption metabolism of other essential nutrients.

Bamboo nanofibers impart better emulsification of oil, shape retention, viscosity, water retention, and dispersion stability of the food matrix. A safety assessment of bamboo formulated as cellulose nanofibers among 22 Japanese men and women by Shimokawa et al. (2024) for 12 weeks showed that although there were significant changes in parameters, such as anthropometry, urinary, physical, and hematology, no adverse effects were recorded. Further, an increase in bowel movement, fecal excretion frequency, and fecal volume was found. A similar observation was earlier documented in the study conducted by Park and Jhon et al. (2009). Further, bamboo nanofibers are also reported to enhance the targeted delivery of curcumin without exerting any side effects (Chang et al., 2023). The addition of BL extract of certain species like *Phyllostachys Sieb.*in oil, meat, and seafood products has been certified (Lu et al., 2006).

Importantly, some studies have claimed the presence of heavy metals, such as lead, arsenic, chromium, cadmium, copper, and magnesium, in bamboo species. The metal toxicity levels were even reported to exceed food safety limits for human effect. Thus, precautions are required for the proper handling of bamboo products to reduce the risk of heavy metal toxicity (Brodziak-Dopierała et al., 2018; Tian et al., 2020).

As discussed, the quality and safety of the ethnomedicine directly depends on the quality of the raw commodities. The commodities used in therapeutic drugs go through a complex supply chain operating among producers, middlemen, and consumers which demands the necessity to maintain quality at each checkpoint (Heinrich et al., 2011; Shahidullah and Haque, 2010). Consequently, the transparency of operation of the inter-link existing between this supply chain is poorly understood, affecting the authenticity of the end product and requires greater attention to ensure the safety of ethnomedicine practice (Alam and Belt, 2009; Kala et al., 2006). This implies bamboo-based health and food products requires efficient quality and safety measures. The regulatory authorities worldwide have formulated certain frameworks to keep such nutraceuticals and herbal health products under constant observation by emphasizing training, standard of practice, code of ethics, and peer review (Caldis et al., 2001). They assist in regulation and promotion of complementary healthcare systems. The European nations have set high bars on prioritizing the herbal medicinal product through implementing European Pharmacopoeia and Traditional Herbal Medicinal Products Directive (EMA, 2013). Although they have effectively regulated the herbal medicinal practice, manufacturers belonging to countries like India and China are known to suffer from large, sophisticated inputs that incur (Patel, 2006).

6.13 POTENTIAL FOR NEW THERAPEUTIC DISCOVERIES

Bamboo extracts have been recognized as opening new horizons in drug development and therapeutics. Several research findings illustrated the greater efficacy of flavonoid components against oxidative stress, inflammation, diabetes, cancer, and adipogenesis than the aqueous components of bamboo (Goyal et al., 2011; Lodhi et al., 2016; Wedler et al., 2014; Yang et al., 2017), helping to understand the curative effects of bamboo in regulating disease.

Synchronizing the nanotechnology in effectively combining two different bioactives could boost the stability, steady, and targeted delivery of those bioactives. For instance, the antidiabetic potential of leaves of *B. arundinacea* were shown to have antidiabetic potential (Goyal et al., 2017; Indira et al., 2023; Menaria, 2016; Rani et al., 2023; Senthilkumar et al., 2011); the zinc oxide nanoparticles of *B. arundinacea* synthesis were also found to show an antihyperglycemia effect where biochemical parameters of blood, such as blood glucose level, triglycerides, and cholesterol, were reduced significantly in the bamboo treatment group when compared to the control diabetes group (Jayarambabu et al., 2021). Likewise, curcumin encapsulated with bamboo shoot cellulose fiber was found to achieve the above merits, and suggested that it could synergistically benefit gut micro-biota. It is evident that bamboo could function as a carrier medium for the effective targeted deliverance of other bioactives (Chang et al., 2023). Similarly, Yang et al. (2019) integrated BL extract with cinnamon essential oil and concluded that it synergistically boosts the immune system and improves the intestinal micro-biota.

These studies indicate wide potential usage of bamboo and its extracts in novel therapies.

6.14 CHALLENGES IN RESEARCH AND COMMERCIALIZATION

Despite having greater anecdotal evidence and traditional knowledge of the ethnomedicinal properties of bamboo, they suffer from a huge gap in obtaining scientific confirmation to claim these health benefits. The absence of rigorous clinical trials established empirical and scientific research models creates the biggest challenge in the implementation and commercialization of bamboo or its related products as a medicinal commodity (Lyu et al., 2019; Patel and Mehta, 2021; Zeng et al., 2022). The chemical composition of bamboo and its potency varies by its growth conditions, geographical location, and species. Further availability of lesser trained professionals and equipment has led to improper handling of bioactives of bamboo for its therapeutic uses (Eswarappa and Bhatia, 2013; Kaur et al., 2022; Sawarkar et al., 2023). Further, certain bamboo species are highly susceptible to fungal attacks which might hinder the post-harvest handling of bamboo. Hence, it is evident that there is a prevalence of improper post-harvest management and preservation techniques which is ultimately known to affect the lifespan of bamboo bioactives. These variabilities not only cause a delay in the standardization of bamboo-based medicinal formulations but also fail to maintain consistency in those product quality parameters (Kaur et al., 2016; Van derLugt et al., 2006).

There are few studies to claim that bamboo could contain cyanides which impose acrid flavor due to the delay in harvesting. The formation of cyanides is due to the action of hydrolytic enzymes on antinutritional compounds triggered by the action of plant cell disruption (Lu et al., 2006; Nirmala et al., 2007; Zhang et al., 2024). Approximately 50–60 mg of free cyanide is available in bamboo plants and could effectively act as a lethal dose for adult humans. Although processing methods such as soaking, boiling, fermentation, and canning could reduce cyanide levels, incomplete processing of bamboo can still pose adverse health effects (Choudhury et al., 2010; Suresh Kumar et al., 2015). Pertaining to the toxins, intact research and development (R&D)is necessary to explore their biopesticide activity, which paves the way for effective use of these toxins (Satya et al., 2010). While considering these findings, the prevalence of the antinutritional entity might pose some threat towards the commercialization of bamboo in pharmaceutical and nutraceutical formulations (Zhang et al., 2024). Although bamboo is a fast-growing and fast-regenerating plant, its extensive exploitation to meet global medicinal demand by employing vigorous harvesting techniques could affect the ecosystem and local environmental conditions, thereby necessitating the implementation of fair and ethical policies for channeling bamboo ethnomedicinal trading processes (Sawarkar et al., 2023). Except for in-built natural tolerance, there are no other strong breeding practices to develop stress-tolerant bamboo variants to overcome these adverse effects when also accounting for the underutilization and immobilization of bamboo. Knowledge sharing, forums, certification programs (species awareness, cultivation, harvesting, and post-harvest handling), research capacity building, and proper networking channels are required to conquer the above-mentioned drawbacks (Palombini and Nogueira, 2023; Ramakrishnan et al., 2020; Scheba et al., 2017).

Although bamboo is highly known for its medicinal potential, only 15% and 5% of its genera and total bamboo species were extensively studied by the researchers. Further, the main concern is that the therapeutic properties of bamboo species originated from Asian regions. The herbaceous species of bamboo belonging to Central and South American regions remain unexplored and their chemical composition in comparison to highly explored Asian regions can enhance the knowledge to bridge the gaps between bamboo species. Bamboo is known for its rich source of alkaloids and coumarins, including C-glycosylated ones; however, due to a lack of knowledge on hyphenated techniques, the other classes of chemical constituents remain unexplored. A plethora of data have been accumulated on the elucidation of medicinal properties of bamboo using their extract and these extracts hold complex combinations of chemical components, making it not only difficult to understand their molecular mechanism of action but also the dose-dependent responses. It is worthy to note that several research works highlight the potency of single bamboo species alone. For example, alkaloid components such as galantamine are known to be the better chemical constituents in Alzheimer's, though its presence in bamboo species was identified, only anticholinesterase properties was studied. Hence, it is necessary to alleviate the preliminary research to *in vivo* stages which uncover the hidden potential of entire bamboo species (Gagliano et al., 2022). Only limited data is available on bamboo's supply and demand trend which might potentially affect the marketing strategy of bamboo.

Ethnobotanical knowledge of bamboo is largely known to deprive communities which is not sufficient to address the effective utilization of bamboo. There also exists a lack of machinery to carry out unit operations, and high labor demands, aided by a lack of sufficient knowledge on the optimum duration of cultivation, propagation, and harvest, determine the quality of bamboo products (Kaur et al., 2022). Hence, local governance, technological intervention, and policy-related assistance are a must to enhance the livelihood and propagation of high-priority bamboo species (Darabant et al., 2016; KC et al., 2024).

6.15 ETHICAL AND ECOLOGICAL CONSIDERATIONS IN HARVESTING BAMBOO

Due to its diverse attribution towards micro- and macro-level services such as medicine, food, and construction, the bamboo sector drove a market share of USD$59.3 billion in 2021, with a growth rate of 4.5% annually (INBAR, 2021). A bamboo study conducted in Tanzania concluded that bamboo has performed better in carbon sequestration (Masisi et al., 2022). A typically grown bamboo could comprehensively sequester up to 119.5 Mg C ha^{-1} (Xu et al., 2018). A plethora of data suggests the capability of bamboo in balancing the dynamics of aboveground and below-ground environments (Cortinovis et al., 2022; Yang et al., 2019; Yen and Wang, 2013). Some studies elucidate the crucial involvement of bamboo in the conservation and purification of water resources, thereby maintaining the hydrological cycle (Nath and Das, 2012; Peprah et al., 2014). A few studies suggest that bamboo not only connects with the ecosystem but also holds a synergistic relationship with the community in social, traditional, and cultural practices (Alemayehu and Hido, 2023; Awazi et al., 2023; Chantarat et al., 2023; Torres et al., 2022). It has also been suggested that inclusion and encouragement of bamboo cultivation could be a resilient strategy to protect climate changes by improving community livelihood. Bamboo cultivation has been known to create an ecological balance which plays an essential role in wildlife habitat management, land restoration, soil, and biodiversity conservation (Alamerew et al., 2024; Lathwal et al., 2023; Yu et al., 2023). These findings suggest that bamboo cultivation and harvest should implement sustainable management practices with the traditionally managed bamboo species to be prioritized over intensive ones due to low-input demands (Pérez et al., 2014; Ruiz-Pérez et al., 2001). However, several management and bioactive extraction practices have caused huge disturbances to the bamboo plantations (Pérez et al., 2014; Xu, 2011). Further, practices like monoculture plantations can significantly increase the loss of biodiversity, soil erosion, and structural alteration, causing an imbalance in the ecosystem. Prompting the natural regeneration of bamboo and emphasizing proper selective cutting methods could ease the establishment of ethical bamboo harvesting practices (Liese, Welling and Tang, 2015). Consequently, these contexts should be considered while drafting the ethical bamboo harvesting and extract collection policies which vindicate the fair involvement of indigenous groups (Bystriakova et al., 2004; Hoare, 2015; Rathour et al., 2022; Scheba et al., 2017).

6.16 CONCLUSION

BLs have enormous potential in medicine due to a wide variety of bioactives and phytochemicals that include antioxidants such as phenolic acids, flavonoids, etc. These compounds have been shown to be a promising natural remedy for inflammation, microbials, and cancers. Bamboo extracts can also help in immune function, work against oxidative stress, and improve heart health. It also has shown anti-aging effects that come from leaves. However, we need more extensive clinical studies to completely understand and evaluate its safety as well as efficacy. Thus, exploring BLs in medicine and therapy has a greater promise, warranting future scientific investigation.

REFERENCES

Alam, G., & Belt, J. 2009. Developing a medicinal plant value chain: Lessons from an initiative to cultivate Kutki (*Picrorhizakurrooa*) in Northern India. *The Journal of Infection in Developing Countries 1–14.*

Alamerew, A. N., Kozak, R., Shrestha, A. K., Zhu, Z., & Wang, G. 2024. A way for livelihood improvement: Systematic review on bamboo forest research. *Trees, Forests and People* 100571.

Alemayehu, A., & Hido, A. 2023. Determinates and governance of bamboo production and marketing in Ethiopia: A critical review. *Advances in Bamboo Science* 100047.

Alias, A. H. D., Razak, N. Q. A., Yusoff, M. H. M., Chin, K. H., Kamal, M. L., Uyup, N. H., & Shafie, M. H. 2023. Antioxidant and antimicrobial activities of essential oils extracted from bamboo (*Bambusavulgaris*) leaves and its application in skincare products: A review. *Biocatalysis and Agricultural Biotechnology 54*: 102930.

Anigboro, A. A. 2018. Antimalarial efficacy and chemopreventive capacity of bamboo leaf (*Bambusavulgaris*) in malaria parasitized mice. *Journal of Applied Sciences and Environmental Management 22*: 1141–1145.

Awazi, N. P., Quandt, A., & Kimengsi, J. N. 2023. Endogenous livelihood assets and climate change resilience in the Mezam Highlands of Cameroon. *GeoJournal 88*(3): 2491–2508.

Biria, M., Rezvani, Y., Roodgarian, R., Rabbani, A., & Iranparvar, P. 2022. Antibacterial effect of an herbal toothpaste containing bamboo salt: A randomized double-blinded controlled clinical trial. *BMC Oral Health* 22(1): 193.

Brodziak-Dopierała, B., Fischer, A., Szczelina, W., & Stojko, J. 2018. The content of mercury in herbal dietary supplements. *Biological Trace Element Research 185*: 236–243.

Bystriakova, N., Kapos, V., & Lysenko, I. 2004. *Bamboo biodiversity: Africa, Madagascar and the Americas (No. 19)*. UNEP/Earthprint.

Caldis, K. S., McLeod, H. D., & Smith, P. R. 2001. The fall of the bamboo curtain: A review of complementary medicine in South Africa. *South African Actuarial Journal 1*: 63–93.

Chang, Y., Wang, Q., Huang, J., Luo, X., Huang, Y., Wu, Y., & Zheng, Y. 2023. Curcumin-loaded bamboo shoot cellulose nanofibers: Characterization and in vitro studies. *Foods 12*: 3512.

Chantarat, P., Poolsiri, R., Wannalangka, I., Kaitpraneet, S., Puangchit, L., & Jenke, M. 2023. Aboveground biomass productivity and nutrient use dynamics of clumping tropical bamboos in Northern Thailand. *Forests 14*(7): 1450.

Chavhan, D. M., Hazarika, M., Brahma, M. L., Hazarika, R. A., & Rahman, Z. 2015. Effect of incorporation of fermented bamboo shoot on physicochemical and microbial quality of pork pickle. *Journal of Food Science and Technology 52*: 1223–1227.

Chen, X., Jia, W., Wang, Q., Han, J., Cheng, J., Zeng, W., & Zhang, Y. 2020. Protective effect of a dietary flavonoid-rich antioxidant from bamboo leaves against internal exposure to acrylamide and glycidamide in humans. *Food & Function 11*: 7000–7011.

Chen, Y., Wang, Y., He, L., Wang, L., Zhao, J., Yang, Z., Li, Q., & Shi, R. 2024. Supercritical CO2 extraction of terpenoids from *Indocalamuslatifolius* leaves: Optimization, purification, and antioxidant activity. *Foods 13*(11): 1719.

Cheng, Y., Wan, S., Yao, L., Lin, D., Wu, T., Chen, Y., Zhang, A., & Lu, C. 2023. Bamboo leaf: A review of traditional medicinal property, phytochemistry, pharmacology, and purification technology. *Journal of Ethnopharmacology 306*: 116166.

Chilicka, K., Rogowska, A. M., Rusztowicz, M., Szyguła, R., Yanakieva, A., Asanova, B., & Wilczyński, S.2022. The effects of green tea (*Camellia sinensis*), bamboo extract (*Bambusavulgaris*) and lactic acid on sebum production in young women with acne vulgaris using sonophoresis treatment. *Healthcare 10*: 684.

Choi, S., Paik, D. I., & Jin, B. H. 2017. Comparative effects of aqueous single-phase and oil-water two-phase mouthrinses containing bamboo salt, magnolia bark and *Centella asiatica* extracts on reducing gingivitis: A randomized clinical trial. *Journal of Korean Academy of Oral Health 41*(3): 181–187.

Choudhury, D., Sahu, J. K., & Sharma, G. D. 2010. Biochemistry of bitterness in bamboo shoots. *Assam University Journal of Science and Technology 6*: 105–111.

Choudhury, D., Sahu, J. K., & Sharma, G. D. 2012. Value addition to bamboo shoots: A review. *Journal of Food Science and Technology 49*: 407–414.

Choudhury, M., Badwaik, L. S., Borah, P. K., Sit, N., & Deka, S. C. 2015. Influence of bamboo shoot powder fortification on physico-chemical, textural and organoleptic characteristics of biscuits. *Journal of Food Science and Technology 52*: 6742–6748.

Chung, H. S., Moon, H. S., Ju, J. C., Cho, Y. S., Kim, M. K., & An, N. J. 2011. Phenolic-rich extract from bamboo leaves (*Phyllostachyspubescens* Mazel) inhibits obesity in rats. *Journal of Medicinal Plants Research 5*: 3140–3145.

Cortinovis, C., Olsson, P., Boke-Olén, N., & Hedlund, K. 2022. Scaling up nature-based solutions for climate-change adaptation: Potential and benefits in three European cities. *Urban Forestry & Urban Greening 67*: 127450.

Darabant, A., Rai, P. B., Staudhammer, C. L., & Dorji, T. 2016. Designing and evaluating bamboo harvesting methods for local needs: Integrating local ecological knowledge and science. *Environmental Management 58*: 312–322.

Das, A., Nath, D. R., Kumari, S., & Saha, R. 2013. Effect of fermented bamboo shoot on the quality and shelf life of nuggets prepared from desi spent hen. *Veterinary World 6*(7): 419–423.

Dong, Y., Wan, X., & Liang, Z.2010. Chemical composition of bamboo leaves (*Bambusa spp.*) determined by gas chromatography–mass spectrometry (GC-MS). *Journal of Agricultural and Food Chemistry 58*(12): 7354–7359.

Eswarappa, V., & Bhatia, S. K. 2013. Implications, challenges, and recommendations. *Naturally Based Biomaterials and Therapeutics: The Case of India* 1:83–90

Fan, W., Yi, Y., Zhang, Y., & Diao, P. 2015. Effect of an antioxidant from bamboo leaves combined with tea polyphenol on biogenic amine accumulation and lipid oxidation in pork sausages. *Food Science and Biotechnology 24*: 421–426.

Farrelly, D., 1984. *The Book of Bamboo, Sierra Club Books*. San Francisco, California.

Felisberto, M. H. F., Beraldo, A. L., & Clerici, M. T. P. S. 2017. Young bamboo culm flour of *Dendrocalamusasper*: Technological properties for food applications. *LWT-Food Science and Technology 76*: 230–235.

Gagliano, J., Anselmo-Moreira, F., Sala-Carvalho, W. R., & Furlan, C. M. 2022. What is known about the medicinal potential of bamboo? *Advances in Traditional Medicine 22*: 467–495.

Gazi, S. K. N., Hosseini, S. E., Yakhchali, S. H., & Borajee, M. 2017. The Application of the statistical methods to measure the effect of adding a bamboo fiber and combine it with animal collagen on the physical, chemical and sensory characteristics of Nags tilapia fish. In *International Conference on Literature, History, Humanities and Social Sciences* 28–31.

Goyal, A. K., Middha, S. K., & Sen, A. 2008. Evaluation of the DPPH radical scavenging activity, total phenols and antioxidant activities in Indian wild *Bambusavulgaris* 'Vittata' methanolic leaf extract. *Journal of Natural Medicines 62*(4): 497–502.

Goyal, A. K., Middha, S. K., & Sen, A. 2011. In vitro antioxidative profiling of different fractions of *Dendrocalamusstrictus*(Roxb.) Nees leaf extracts. *Free Radicals and Antioxidants 1*: 42–48.

Goyal, A. K., Middha, S. K., Usha, T., & Sen, A. 2017. Analysis of toxic, antidiabetic and antioxidant potential of *Bambusabalcooa* (Roxb.) leaf extracts in alloxan-induced diabetic rats. *3 Biotech 7*: 1–11.

Hao, J., Han, W., Huang, S., Xue, B., & Deng, X. 2015. Bamboo leaf flavonoids extracts alleviate oxidative stress in HepG2 cells. *BioMed Research International 2015*: 303876.

Heinrich, M., Dhanji, T., & Casselman, I. 2011. Açai (*Euterpeoleracea* Mart.)—A phytochemical and pharmacological assessment of the species' health claims. *Phytochemistry Letters 4*: 10–21.

Hoare, A. 2015. *Tackling illegal logging and the related trade. What progress and where next.* 79.

Honfo, H., Tovissodé, F. C., Gnanglè, C., Mensah, S., Salako, V. K., Assogbadjo, A. E., & Kakaï, R. G. 2015. Traditional knowledge and use value of bamboo in southeastern Benin: Implications for sustainable management. *Ethnobotany Research and Applications 14*: 139–153.

Iflaifel, M., Partlett, C., Bell, J., Cook, A., Gamble, C., Julious, S., & Sprange, K. 2022. Blinding of study statisticians in clinical trials: A qualitative study in UK clinical trials units. *Trials 23*: 535.

INBAR. 2021. *Bamboos market size, share and trends analysis report by application and segment forecasts, 2022–2030.* https://www.grandviewresearch.com/industry-analysis/bamboos-market

Indira, A., Joshi, B., Oinam, S., Koul, A. & Chongtham, N. 2023. Potential of bamboo in the prevention of diabetes-related disorders: Possible mechanisms for prevention. In: *Bamboo Science and Technology.* Springer Nature Singapore, Singapore, 89–124.

Jayarambabu, N., Rao, T. V., Kumar, R. R., Akshaykranth, A., Shanker, K., & Suresh, V. 2021. Anti-hyperglycemic, pathogenic and anticancer activities of *Bambusaarundinacea* mediated Zinc Oxide nanoparticles. *Materials Today Communications 26*: 101688.

Kala, C. P., Dhyani, P. P., & Sajwan, B. S. 2006. Developing the medicinal plants sector in northern India: Challenges and opportunities. *Journal of Ethnobiology and Ethnomedicine 2*(1): 15.

Kaur, P. J., Kardam, V., Pant, K. K., Naik, S. N., & Satya, S. 2016. Characterization of commercially important Asian bamboo species. *European Journal of Wood and Wood Products 74*: 137–139.

Kaur, P. J., Yadav, P., Gupta, M., Khandegar, V., & Jain, A. 2022. Bamboo as a source for value added products: Paving way to global circular economy. *Bio Resources 17*(3): 1.

KC, B. M., Lamichhane, J., Khanal, S. N., & Gauchan, D. P. 2024. Traditional utilization of bamboo in the Central Siwalik region, Nepal. *PLosOne 19*(1): e0296886.

Kleinhenz, V., & Midmore, D. J. 2001. Aspects of bamboo agronomy. *Advances in Agronomy 74*: 99–153.

Kuehl, Y. 2015. Resources, yield, and volume of bamboos. *Bamboo: The Plant and Its Uses 10*: 91–111.

Kumar, S., & Pandey, A. K. 2013. Chemistry and biological activities of flavonoids: An overview. *The Scientific World Journal 2013*: 162750.

Lathwal, M., Rani, M., Indira, A., & Chongtham, N. 2023. Bamboo: A sustainable alternative for biochar production. In *Bamboo Science and Technology*, pp. 265–295. Singapore: Springer Nature Singapore.

Len, T. T., Nhung, H. N., & Thanh, T. C. 2023. A case series evaluating the efficacy of herbal oil-based *Bambusavulgaris* (Bamboo) leaf extracts as a topical treatment for atopic dermatitis. *Phytomedicine Plus 3*: 100480.

Li, X., Chen, Y., Lai, Y., Yang, Q., Hu, H., & Wang, Y. 2010. Analysis of the nutritional components in bamboo leaves of *Phyllostachyspubescens. Biomass Chemistry and Engineering 44*: 15–18.

Li, X., Zhang, Z., & Zhang, Y. 2006. Ultrasound-assisted extraction of bioactive compounds from medicinal plants. *Ultrasound in Medicine & Biology 32*(10): 1495–1500.

Liese, W., Welling, J., & Tang, T. K. H. 2015. Utilization of bamboo. *Bamboo: the Plant and Its Uses 1*: 299–346.

Lodhi, S., Jain, A. P., Rai, G., & Yadav, A. K. 2016. Preliminary investigation for wound healing and anti-inflammatory effects of *Bambusavulgaris* leaves in rats. *Journal of Ayurveda and Integrative Medicine 7*: 14–22.

Lu, B., Wu, X., & Shi, J. 2006. Toxicology and safety of antioxidant of bamboo leaves. Part 2: Developmental toxicity test in rats with antioxidant of bamboo leaves. *Food Chemistry Toxicology 44*: 1739–1743.

Lu, B., Wu, X., Tie, X., Zhang, Y., & Zhang, Y. 2005. Toxicology and safety of anti-oxidant of bamboo leaves. Part 1: Acute and subchronic toxicity studies on anti-oxidant of bamboo leaves. *Food and Chemical Toxicology 43*: 783–792.

Lyu, B., Zeng, C., Deng, S., Liu, S., Jiang, M., Li, N., & Chen, Q. 2019. Bamboo forest therapy contributes to the regulation of psychological responses. *Journal of Forest Research 24*: 61–70.

Masisi, B., Zabel, A., Blaser, J., & Augustino, S. 2022. Fighting climate change with bamboo in Africa: The case of Kyela, Rungwe and Mufindi districts – Tanzania. *Advances in Bamboo Science 1*: 100009.

Menaria, J. 2016. Antidiabetic activity of leaves extract of *Bambusa arundinacea. Pharm Chem J* 3(2): 197–200.

Mishra, S., & Chaudhury, S. S. 2012. Ethnobotanical flora used by four major tribes of Koraput, Odisha, India. *Genetic Resources and Crop Evolution 59*: 793–804.

Monaghan, T. F., Agudelo, C. W., Rahman, S. N., Wein, A. J., Lazar, J. M., Everaert, K., & Dmochowski, R. R. 2021. Blinding in clinical trials: Seeing the big picture. *Medicina 57*: 647.

Mustafa, U., Naeem, N., Masood, S., & Farooq, Z. 2016. Effect of bamboo powder supplementation on physicochemical and organoleptic characteristics of fortified cookies. *Food Science and Technology 4*: 7–13.

Nath, A. J., & Das, A. K. 2012. Ecological implications of village bamboo as global climate change mitigation strategy: A case study in Barak Valley, Assam, North East India. *International Journal of Climate Change Strategies and Management 4*(2): 201–215.

Ni, Q., Xu, G., Wang, Z., Gao, Q., Wang, S., & Zhang, Y. 2012. Seasonal variations of the antioxidant composition in ground bamboo *Sasaargenteastriatus* leaves. *International Journal of Molecular Sciences 13*(2): 2249–2262.

Nirmala, C., Bajwa, H. K., & Oinam, S. 2024. Bamboo mediated green synthesis of silver nanoparticles-a new approach towards utilization of an underutilized plant. *Advances in Bamboo Science 6*: 100061.

Nirmala, C., Bisht, M. S., Bajwa, H. K., & Santosh, O.2018. Bamboo: A rich source of natural antioxidants and its applications in the food and pharmaceutical industry. *Trends in Food Science & Technology 77*: 91–99.

Nirmala, C., David, E., & Sharma, M. L. 2007. Changes in nutrient components during ageing of emerging juvenile bamboo shoots. *International Journal of Food Sciences and Nutrition 58*: 612–618.

Palombini, F. L., & Nogueira, F. M. 2023. *Bamboo Science and Technology*. Springer, Singapore.

Panee, J. 2015. Potential medicinal application and toxicity evaluation of extracts from bamboo plants. *Journal of Medicinal Plant Research 9*: 681.

Park, C. S., & Kim, T. O. 2014a. Quantitative analysis of flavonoid content in leaves of bamboo (*Phyllostachyspubescens*) using high-performance liquid chromatography. *Journal of Agricultural Chemistry and Environment 3*: 24–29.

Park, E. J., & Jhon, D. Y. 2009. Effects of bamboo shoot consumption on lipid profiles and bowel function in healthy young women. *Nutrition 25*(7–8): 723–728.

Park, J. H., & Kim, Y. H. 2014b. Methods of analysis for bamboo leaves extracts. *Journal of Pharmaceutical Analysis 4*(6): 303–308.

Park, S. Y., Lee, K. H., Lee, H. S., & Kim, S. H. 2010. *Antioxidant and anti-inflammatory activities of bamboo (Bambusavulgaris) leaves. Journal of Ethnopharmacology 128*(1): 172–177.

Patel, A. B. 2004. Traditional bamboo uses by the tribes of Gujarat. *Indian Journal of Traditional Knowledge 4*: 179–184.

Patel, M., & Mehta, P. 2021. Bamboo a supplement to human health: A comprehensive review on its ethnopharmacology, phytochemistry, and pharmacological activity. *The Natural Products Journal 11*: 140–168.

Patel, P. 2006. Impact assessment of THMPD on trade of Ayurveda products. In *Ayurveda and its scientific aspects: Opportunities for globalisation*, Council of Scientific and Industrial Research, New Delhi, India.

Peprah, T., Essien, C., Owusu-Afriyie, K., Foli, E. G., Govina, J., & Oteng-Amoako, A. A. 2014. Exploring the use of bamboo for accelerated reclamation of degraded mined sites in Ghana.

Perez, M. R., Rodriguez, L. G., Yang, X., Xie, J., & Fu, M. 2014. From basic raw material goods to cultural and environmental services: The Chinese bamboo sophistication path. *Ecology and Society 19*(4): 1–10.

Puigdellivol, J., ComellasBerenger, C., PérezFernández, M., CowalinskyMillán, J. M., Carreras Vidal, C., Gil, I., & Aguilar Hernández, M. E. 2019. Effectiveness of a dietary supplement containing hydrolyzed collagen, chondroitin sulfate, and glucosamine in pain reduction and functional capacity in osteoarthritis patients. *Journal of Dietary Supplements 16*: 379–389.

Ramakrishnan, M., Yrjälä, K., Vinod, K. K., Sharma, A., Cho, J., Satheesh, V., & Zhou, M. 2020. Genetics and genomics of Mosobamboo (*Phyllostachysedulis*): Current status, future challenges, and biotechnological opportunities toward a sustainable bamboo industry. *Food and Energy Security 9*: e229.

Rani, E. A., Janani, R., Chonche, M. J., & Vallikannan, B. 2023. Lactucaxanthin regulates the cascade of retinal oxidative stress, endoplasmic reticulum stress and inflammatory signaling in diabetic rats. *Ocular Immunology and Inflammation 31*: 320–328.

Rathour, R., Kumar, H., Prasad, K., Anerao, P., Kumar, M., Kapley, A., ... & Singh, L. 2022. Multifunctional applications of bamboo crop beyond environmental management: An Indian prospective. *Bioengineered 13*(4): 8893–8914.

Ruiz-Pérez, M., Maoyi, F., Xiaosheng, Y., & Belcher, B. 2001. Bamboo forestry in China: Toward environmentally friendly expansion. *Journal of Forestry 99*(7): 14–20.

Sangeetha, R., Diea, Y. K. T., Chaitra, C., Malvi, P. G., & Shinomol, G. K. 2015. The amazing bamboo: A review on its medicinal and pharmacological potential. *Indian Journal of Nutrition 2*: 1–7.

Sanger, N. S., Kankakee, A., Singh, R., Gupta, A., Shrikhande, B., & Ganu, G. 2021. A randomized controlled clinical trial to evaluate the potential of ayurvedic medicines in patients with mild symptoms of COVID-19. *European Journal of Molecular & Clinical Medicine 8*: 3017–3027.

Santosh, O., Bajwa, H. K., Bisht, M. S. & Chongtham, N. 2021. Application of bamboo in the food and pharmaceutical industry. In: *Biotechnological Advances in Bamboo*. Springer, Singapore, 401–429.

Sarangthem, K., & Singh, T. N. 2003. Microbial bioconversion of metabolites from fermented succulent bamboo shoots into phytosterols. *Current Science 84(12):* 1544–1547.

Satya, S., Bal, L. M., Singhal, P., & Naik, S. N. 2010. Bamboo shoot processing: Food quality and safety aspect (a review). *Trends in Food Science & Technology* 21(4): 181–189.

Sawarkar, A. D., Shrimankar, D. D., Kumar, M., Kumar, P., & Singh, L. 2023. Bamboos as a cultivated medicinal grass for industries: A systematic review. *Industrial Crops and Products 203*: 117210.

Scheba, A., Blanchard, R., & Mayeki, S. 2017. Bamboo For green development? The opportunities and challenges of commercialising bamboo in South Africa *South Africa. Human Sciences Research Council (HSRC).* 1–44.

Schillaci, C., Nepravishta, R., & Bellomaria, A. 2014. Antioxidants in food and pharmaceutical research. *Albanian Journal of Pharmaceutical Sciences 1*: 9–15.

Seki, T., Kida, K., & Maeda, H.2010. Immunostimulation-Mediated anti-tumor activity of bamboo (*Sasasenanensis*) leaf extracts obtained under 'Vigorous' condition. *Evidence-Based Complementary and Alternative Medicine 7*: 447–457.

Seki, T., & Maeda, H. 2010. Cancer preventive effect of Kumaizasa bamboo leaf extracts administered prior to carcinogenesis or cancer inoculation. *Anticancer Research 30*: 111–118.

Senthilkumar, M. K., Sivakumar, P., Changanakkattil, F., Rajesh, V., & Perumal, P. 2011. Evaluation of anti-diabetic activity of *Bambusavulgaris* leaves in streptozotocin induced diabetic rats. *International Journal of Pharmaceutical Sciences and Drug Research 3*: 208–210.

Shahidullah, A. K. M., & Haque, C. E. 2010. Linking medicinal plant production with livelihood enhancement in Bangladesh: Implications of a vertically integrated value chain. *The Journal of Transdisciplinary Environmental Studies 9*: 1.

Shang, Y. F., Kim, S. M., & Um, B. H. 2014. Optimisation of pressurised liquid extraction of antioxidants from black bamboo leaves. *Food Chemistry 154*: 164–170.

Sharma, R., Kakodkar, P., Kabra, A., & Prajapati, P. K. 2022. Golden ager Chyawanprash with meager evidential base from human clinical trials. *Evidence-Based Complementary and Alternative Medicine 2022*(1): 9106415.

Sharma, T. P., & Borthakur, S. K. 2008. Ethnobotanical observations on bamboos among Adi tribes in Arunachal Pradesh. *Indian Journal of Traditional Knowledge 7*: 594–597.

Sheikh, H., Alam, M. K., & Samad, M. A. 2023. Comprehensive evaluation of the clinical efficacy of an antidiabetic polyherbal formulation. *Turkish Journal of Agriculture-Food Science and Technology 11*(10): 1905–1916.

Shen, H., Wang, Y., Shi, P., Li, H., Chen, Y., Hu, T., & Yu, L. 2024. Effects of the species and growth stage on the antioxidant ad antifungal capacities, polyphenol contents, and volatile profiles of Bamboo leaves. *Foods 13*(3): 480.

Shimokawa, T., Hayashi, N., & Takao, T. 2024. Safety assessment of bamboo-derived cellulose nanofibers in food for healthy adults: A long-term, open-label ingestion trial. *Japan tappi journal* 78(5): 441–451.

Singhal, P., Bal, L. M., Satya, S., Sudhakar, P., & Naik, S. N. 2013. Bamboo Shoots: A novel source of nutrition and medicine. *Critical Reviews in Food Science and Nutrition 53*(5): 517–534.

Suresh Kumar, P., Devi, P., Usha Kumari, K., Sangeetha, A., & Choudhary, V. K. 2015. Bamboo shoots, food of future-nutritional, antioxidants, medicinal and economic importance under changing climatic conditions. *III International Symposium on Underutilized Plant Species 1241*: 647–658.

Thomas, R., Jebin, N., Saha, R., & Sarma, D. K. 2016. Antioxidant and antimicrobial effects of Kordoi(*Averrhoacarambola*) fruit juice and bamboo (*Bambusapolymorpha*) shoot extract in pork nuggets. *Food Chemistry 190*: 41–49.

Tian, Z., Liu, X., Sun, W., Ashraf, A., Zhang, Y., Jin, X., & He, B. 2020. Characteristics of heavy metal concentrations and risk assessment for giant pandas and their habitat in the Qinling Mountains, China. *Environmental Science and Pollution Research 27*: 1569–1584.

Tokhiriyon, B., Poznyakovsky, V. & Beliaev, N. 2019. Biologically active complex for the functional support of the connective tissues: Scientific rationale, clinical evidence. *International Journal of Pharmaceutical Research & Allied Sciences 8(1)*: 115–122.

Tomar, R., Yadav, B., Singh, H., Hetalben, A., Khanduri, S., Sharma, B. S., & Dhiman, K. S. 2022. A clinical study to evaluate the efficacy of ayurvedic intervention in management of chronic bronchitis. *Journal of Research in Ayurvedic Sciences* 6(3): 101–108.

Torres, B., Cayambe, J., Paz, S., Ayerve, K., Heredia-R, M., Torres, E., ... & García, A. 2022. Livelihood capitals, income inequality, and the perception of climate change: A case study of small-scale cattle farmers in the Ecuadorian Andes. *Sustainability 14*(9): 5028.

Van derLugt, P., Van den Dobbelsteen, A. A. J. F., & Janssen, J. J. A. 2006. An environmental, economic and practical assessment of bamboo as a building material for supporting structures. *Construction and Building Materials 20*: 648–656.

Verma, S., Mili, M., Hada, V., Kalita, P., Hashmi, S. A. R., Rathore, S. K. S., & Srivastava, A. K. 2022. Bamboo leaves: Sustainable feedstock in pharmaceuticals, food, and agricultural sector. *Environmental Technology Reviews 11*: 266–281.

Wang, J., Tang, F., Yue, Y., Guo, X., & Yao, X. 2010. Development and validation of an HPTLC method for simultaneous quantitation of isoorientin, isovitexin, orientin, and vitexin in bamboo-leaf flavonoids. *Journal of AOAC International 93*(5): 1376–1383.

Wang, X. Z., Xue, K. Y., Chen, P. N., Xiao, C. H., Cui, J., & Fu, J. 2022. Ma's bamboo-based medicinal moxibustion therapy of low back pain in lumbar disc herniation: Study protocol for a randomized controlled trial. *Trials 23*(1): 446.

Wedler, J., Daubitz, T., Schlotterbeck, G., & Butterweck, V. 2014. In vitro anti-inflammatory and wound-healing potential of a *Phyllostachysedulis* leaf extract—Identification of isoorientin as an active compound. *PlantaMedica 80*: 1678–1684.

Xie, L., Zhang, Y., & Liu, H. 2015. Phytochemical properties and health benefits of bamboo leaves. *Journal of Ethnopharmacology 169*: 89–96.

Xu, J. 2011. China's new forests aren't as green as they seem. *Nature* 477(7365): 371.

Xu, M., Ji, H., & Zhuang, S. 2018. Carbon stock of Moso bamboo (*Phyllostachys pubescens*) forests along a latitude gradient in the subtropical region of China. *PLOS ONE* 13(2): e0193024.

Xue, K., Wang, X., Xiao, C., Zhang, N., Liu, M., Fu, J., & Cui, J. 2024. Clinical efficacy and safety of a modified moxibustion therapy for low back pain in lumbar disc herniation: A two-center, randomized, controlled, non-inferiority trial. *Journal of Pain Research 17*: 1853–1865.

Yakubu, M. T., & Bukoye, B. B. 2009. Abortifacient potentials of the aqueous extract of *Bambusavulgaris* leaves in pregnant Dutch rabbits. *Contraception 80*: 308–313.

Yakubu, M. T., Bukoye, B. B., Oladiji, A. T., & Akanji, M. A. 2009. Toxicological implications of aqueous extract of *Bambusavulgaris* leaves in pregnant Dutch rabbits. *Human & Experimental Toxicology 28*: 591–598.

Yang, C., Ni, H., Zhong, Z., Zhang, X., & Bian, F. 2019. Changes in soil carbon pools and components induced by replacing secondary evergreen broadleaf forest with Moso bamboo plantations in subtropical China. *Catena 180*: 309–319.

Yang, J. H., Choi, M. H., Yang, S. H., Cho, S. S., Park, S. J., Shin, H. J., & Ki, S. H. 2017. Potent anti-inflammatory and antiadipogenic properties of bamboo (*SasacoreanaNakai*) leaves extract and its major constituent flavonoids. *Journal of Agricultural and Food Chemistry 65*: 6665–6673.

Yang, Q., Duan, Z. B., Wang, Z. L., He, K. H., Sun, Q. X., & Peng, Z. H. 2008. Bamboo resources, utilization and ex-situ conservation in Xishuangbanna, South-eastern China. *Journal of Forestry Research 19*: 79–83.

Yen, T. M., & Wang, C. T. 2013. Assessing carbon storage and carbon sequestration for natural forests, man-made forests, and bamboo forests in Taiwan. *International Journal of Sustainable Development & World Ecology 20*(5): 455–460.

Yu, L., Wei, J., Li, D., Zhong, Y., & Zhang, Z. 2023. Explaining landscape levels and drivers of Chinese moso bamboo forests based on the plus model. *Forests 14*(2): 397.

Yu, Y., Li, Z., Cao, G., Huang, S., & Yang, H. 2019. Bamboo leaf flavonoids extracts alleviate oxidative stress in HepG2 cells via naturally modulating reactive oxygen species production and Nrf2-mediated antioxidant defense responses. *Journal of Food Science 84*: 1609–1620.

Zeng, H., Chen, J., Zhai, J., Wang, H., Xia, W., & Xiong, Y. L. 2016. Reduction of the fat content of battered and breaded fish balls during deep-fat frying using fermented bamboo shoot dietary fiber. *LWT 73*: 425–431.

Zhang, Y., Wu, X., Ren, Y., Fu, J., & Zhang, Y. 2004. Safety evaluation of a triterpenoid-rich extract from bamboo shavings. *Food and Chemical Toxicology 42*(11): 1867–1875.

Zhang, X., Liu, J., & Wang, Z. 2011. Analysis of chemical composition in bamboo leaves by Fourier transform infrared spectroscopy. *Journal of Plant Research 124*(4): 459–464.

Zhang, Y., Chen, J., Zhang, X., Wu, X., & Zhang, Y. 2007. Addition of antioxidant of bamboo leaves (AOB) effectively reduces acrylamide formation in potato crisps and French fries. *Journal of Agricultural and Food Chemistry 55*: 523–528.

Zhang, Y., Wu, L., Li, Y., Yang, J., Yang, H., Zhao, Y., & Chen, G. 2024. Bamboo shoot and its food applications in last decade: An undervalued edible resource from forest to feed future people. *Trends in Food Science & Technology 146*: 104399.

7 Secondary Metabolites of Bamboo and Their Medical Applications

Varshini Ramesh, Sadari Manusha, Varsha Nagabhushan, Yuvaraj Sivamani, and Sumitha Elayaperumal

ABBREVIATIONS

AE-BCT Aqueous extract of *Bambusae caulis*
BCT *Bambusae caulis*
BSCP Bamboo shoot crude polysaccharide
BSP Bamboo shoot angiotensin converting enzyme inhibitory peptide
CPS *Chimonobambusa quadrangularis* polysaccharide
HMC Human mast cell
IL Interleukin
WBP Water-soluble polysaccharide from bamboo shoots of *Phyllostachys praecox*
XOS Xylo-oligosaccharide

7.1 INTRODUCTION

Bamboo, one of the notable genera of woody grasses, falls under the subfamily *Bambusoideae* and the family Poaceae. This subfamily comprises over 70 genera and 1,450 species and can be found in various climates (Gratani et al., 2008). According to Singhal et al. (2013), bamboo can also be found in a dry dipterocarp forest, a mixed deciduous forest, and a tropical evergreen forest. The plant, in addition to its industrial and medical uses, has become one of the most bearing plants across the globe since all of its parts contain high levels of nutrients and antioxidants. This includes its use in medicine or as food for culms, leaves, shoots, and culm sheaths (Scurlock et al., 2000).

Its nutritional provision is getting advanced focus as driven by the increasing knowledge of its low-fat content, high fibrous content which is digestible, has little toxicity residues, and high mineral content, including potassium, calcium, manganese, zinc, chromium, copper, and iron, as well as phosphorus and selenium in smaller proportions; bamboo vegetation is gaining traction across the globe (Bystriakova et al., 2003). Especially ancient and medieval histories of China and India have centuries of reference to the use of bamboo, particularly in their systems of medicine, boiling down to almost prehistoric times (Bamboo Shoot a Neglected Natural Resource: A Source of Food and Prosperity for North-East India, n.d.). Using this age-old tradition, contemporary medicinal preparations made from bamboo are being created to address a variety of health issues. Some of them include vinegar, bamboo extract, and salt (Effects of Bamboo Shoot Consumption on Lipid Profiles and Bowel Function in Healthy Young Women). Analytical compound BMP-x has several biological functions, including antioxidative, anticancer, antihypertensive, antimicrobial, and anti-inflammatory, which have been shown by the secondary metabolites obtained from bamboo stalks and leaves (Hu et al., 2000; Yang et al., 2017) (Figure 7.1). Table 7.1 shows some of the commonly known therapeutic applications of medicinal bamboo species.

 DOI: 10.1201/9781003485933-7

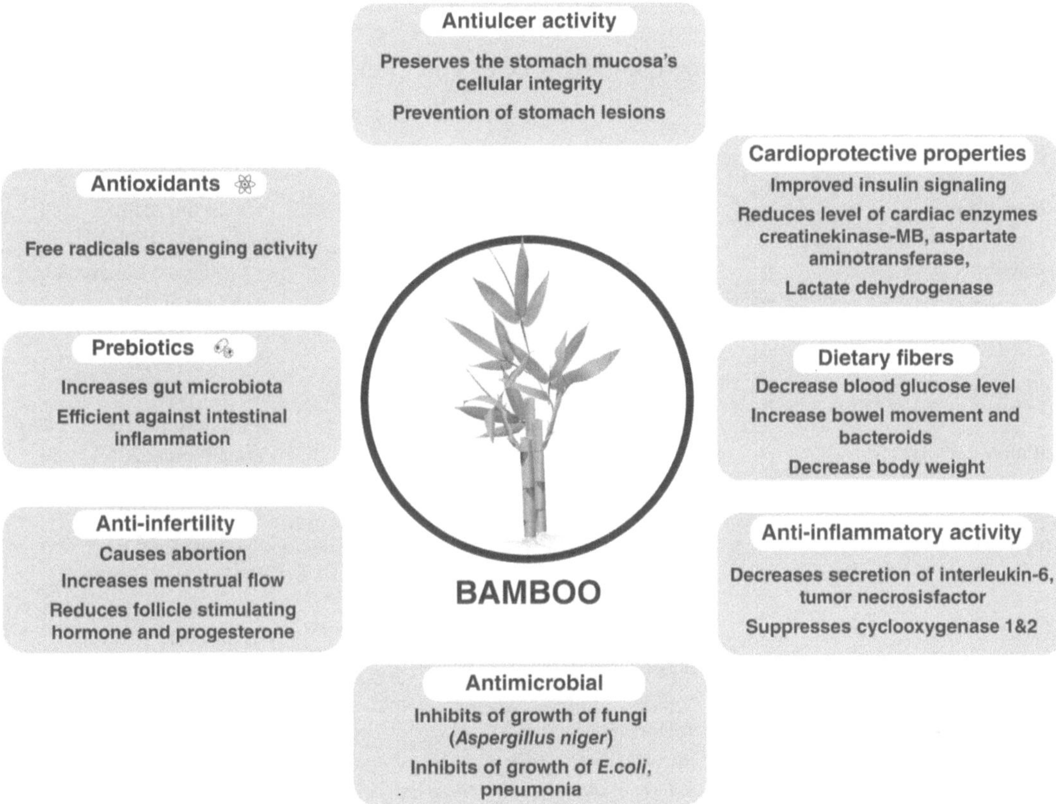

FIGURE 7.1 Schematic representation of antimicrobial properties of bamboo.

TABLE 7.1
Therapeutic Applications of Some Bamboo Species and Their Application

Application	Bamboo Species	Reference
Antioxidants	*Phyllostachys nigra* var. henonis	(Hu et al., 2000)
	Phyllostachys pubescens	(Mu et al., 2004)
	Phyllostachys bambusoides	
	Sasa argenteastriatus	(Ni et al., 2012)
Anticancer	*Bambusa arundinacea*	(Kalaiarasi et al., 2015)
	Bambusa nutans	
	Bambusae caulis	(Kim et al., 2013)
	Kumaizasa bamboo	(Seki et al., 2010)
	Phyllostachys heterocycla var. Pubescens	(Abdelhameed et al., 2020)
	Bamboo salt	(Zhao et al., 2013)
	Pseudosasa japonica	(Panee, 2009)
Antimicrobial activity	*Phyllostachys pubescens*	(Afrin et al., 2012)
	Moso bamboo	(Shen et al., 2014)
	Bamboo seeds	(Soumya et al., 2014)
	Bambusa vulgaris	(Owolabi and Lajide, 2015)
	Bamboo leaf hydrosol	(Yuan et al., 2019)
	Black bamboo leaves	(Arifani et al., 2021)

(Continued)

TABLE 7.1 *(Continued)*
Therapeutic Applications of Some Bamboo Species and Their Application

Application	Bamboo Species	Reference
Antihypertensive effect	Bamboo shavings	(Jiao et al., 2007)
	Tali bamboo	(Liu et al., 2012)
Dietary fibers	*Fargesia spathacea*	(Wu et al., 2020)
	Bamboo shoot shell	(Luo et al., 2017)
Prebiotics	*Chimonobambusa quadrangularis*	(Yakubu and Bukoye, 2009)
	Gigantochloa levis	(Azmi et al., 2012)
	Phyllostachys praecox	(He et al., 2016)
	Bamboo shoots	(Deng et al., 2023)
Anti-infertility	*Bambusa vulgaris*	(Gill, 1992)
	Bambusa arundinacea	(Vanithakumari et al., 1989)
Antiulcer activity	*Bambusa balcooa*	(Upreti et al., 2016)
	Sasa albomarginata	(Otani et al., 1990)
	Bambusa arundinacea	(Jivani, 2011)
Anti-inflammatory	Purple bamboo salt	(Shin et al., 2003)
	Phyllostacys edulis J. Houz	(Tundis et al., 2023)
	Bamboo tali leaves	(Hidayah and Hafshah, 2023)
	Sasa coreana Nakai	(Yang et al., 2017)
	Bamboo vinegar	(Ho et al., 2013)
	Phyllostacys edulis	(Wedler et al., 2014)
Cardioprotective properties	*Bambusa blumeana*	(Rosales et al., 2013)
	Bamboo leaf extract	(Zhang et al., 2017)
	Bambusa arundinacea, Bambusa balcooa, Bambusa tulda, Bambusa vulgaris, Phyllostachys bambusoides, Phyllostachys nigra, Sasa borealis, Sasa quelpaertensis	(Indira et al., 2023)
	Phyllostachys pubescens	(Fu et al., 2005)
	Ginkgo biloba	(Zhang Ying et al., 2000)
	Bamboo leaves	(Fu et al., 2006)

7.2 BIOACTIVE COMPOUNDS IN BAMBOO

Secondary metabolites, often called bioactive compounds, are abundant in plants and are used for the treatment of various human diseases (Santosh et al., 2016). Bioactive chemicals found in bamboo shoots include phytosterols, which are the building blocks of many pharmaceutical steroids, phenols, and free radical terminators (Srivastava, 1990). Bioactive substances, primarily including phenols, phytosterols, and dietary fibers, that function as organic antioxidants, fortify along with shield vital organ systems of the body, such as the central nervous system, and improve cardiovascular health.

The majority of the plants have phenolic compounds, which have attracted both the scientific community and the general population due to their possible application as bactericidal and antioxidant substances, among other health advantages they provide (Nirmala et al., 2014). Their anti-inflammatory, antiallergic, antibacterial, cardioprotective, and vasodilating qualities are a few health benefits of phenols. One of the many ways that phenols function as antioxidants is by their interaction with biomembranes, oxidative enzyme inhibition, transition metal chelation, quenching of reactive oxygen species (ROS), and scavenging of free radicals (Brewer, 2011).

The part of food generated from plants that human digestive enzymes are unable to fully break down is known as dietary fiber. It increases the bulk of the stool and is regarded as a bulking agent and mordant. The bulking agent is further subdivided into dissoluble and non-dissoluble based on their solubility in water. Bamboo shoots have two types of fiber content: the bulk of acid-detergent fiber is made of cellulose and lignin, in contrast to neutral-detergent fiber, which refers to the ligno-cellulosic material that cannot be digested and is made up of cellulose, hemicelluloses, and lignin (Chongtham et al., 2011).

The bioactive elements found in the non-saponifiable portion of plant oils are called phytosterols. Many pharmaceutical significant steroids like glucocorticoids, oral birth control pills, anti-inflammatories, artificial anabolic steroids, and estrogenic hormones are derived from phytosterols (Chongtham et al., 2011).

7.3 THE HEALTH ADVANTAGES OF BAMBOO EXTRACTS

7.3.1 Antioxidants

Antioxidants are compounds in food that help in free radical scavenging. Antioxidants primarily benefit the body's health by lowering inflammation (Srivastava, 1990). Antioxidants help the food and pharmaceutical sectors avoid rancidity, degradation, and discoloration that occur during processing and storage (Schillaci et al., 2014). In this study, some methanol and hot water extracts from powdered bamboo, much utilized in Japan, as opposed to the extracts made from control, α-tocopherol, and ascorbic acid, exhibited a more potent scavenging effect toward superoxide anion and hydroxyl radical (Nagai et al., 2009).

In China, bamboo leaves have been authorized as organic antioxidants, besides being considered medicinal. Hence, they may be incorporated as an alternative food additive in seafoods, meats, and edible oils (Lu et al., 2006). The leaves of several bamboo species have yielded a few compounds with antioxidative properties that have been identified. From the extracts obtained from the leaf of *Phyllostachys nigra* var. *henonis*, phenolic compounds including luteolin-7-glucoside, chlorogenic acid, along with caffeic acid were discovered and their antioxidant and radical scavenging properties were assessed (Hu et al., 2000). Bamboo leaf flavonoids have demonstrated the ability to regulate blood cholesterol levels by increasing HDL levels and decreasing LDL levels, which in turn helps in reducing the risk of arteriosclerosis, and exhibit antioxidant, antifatigue, and anti-aging characteristics.

The antioxidative effect of the methanol obtained from the culm and leaves of *Phyllostachys pubescens* (Moso bamboo) and *Phyllostachys bambusoides* (madake bamboo) were examined, employing the DPPH (1,1-diphenyl-2-picrylhydrazyl) scavenging activity assay, reduction power, and linoleic acid peroxidation inhibition activity. As per the findings, DPPH radical scavenging activity is lower than the activities of the antioxidants from ethanol extracts of madake bamboo leaves and Moso bamboo culms (Mu et al., 2004). *Sasa argenteastriatus* is a newly discovered bamboo species that has a leafy structure and possesses a great deal of active antioxidants and other compounds in leaves. Further, details were also provided about the changes over seasons, if any, on the *S. argenteastriatus* leaf antioxidant profile. In particular, chlorogenic acid, isoorientin, and vitexin were the main components of *S. argenteastriatus* leaves; however, isovitexin and chlorogenic acid both demonstrated similar seasonal variations (Ni et al., 2012). Small cytokines called chemokines are part of a very large family of molecules that regulate leukocyte trafficking. The most widely accepted classification of those small proteins distinguishes four main divisions of chemokines: C, CXC, CC, and CX3C. Another commonly used traditional East Asian medicine, *Bambusae caulis* in Liquamen, is known for its additional immune-regulating and antioxidant effects. One other study investigates *B. caulis* in Liquamen for information about potential benefits and concludes that the substance may suppress thymus expression. In the hypothesis of the authors, antioxidant action may be also conferred by *B. caulis* in Liquamen to human keratinocytes (Qi et al., 2012).

7.3.2 Anticancer

According to research, there are promising prospects for the use of bamboo stems in preventing cancer and other diseases. Bamboo stalks contain substantial amounts of compounds – phytosterols, phenols, and dietary fibers, for instance, which have strong anticancer properties. To assess the immunostimulating efficacy of bamboo extracts and their anticancer properties, a focused study was conducted on the antitumor efficacy of leaf extracts that were prepared under harsh conditions following active immunostimulation (Sharma and Nirmala, 2018). The extracts induced natural killer curcumin cells and macrophages, showing that these might be the primary immuno-enhancers that have a cancer aversion effect (Kimura et al., 2002). In addition, it was demonstrated how synthesized AgNPs with compounds derived from the leaf of *Bambusa arundinacea* and *Bambusa nutans* are biosynthesized, characterized, and have cytotoxic effect on PC3 cell lines. Out of the two, the NPs showed greater cytotoxic effects than that of the leaf extract, so it can be suggested that the NPs can be used in therapeutic management of cancer (Kalaiarasi et al., 2015). According to Traditional Chinese medicine, the fragments of bamboo of *Bambusae caulis* in Taenia (BCT) are used in the treatment and management of high blood pressure and heart ailments. In the study, the aqueous extract of BCT (AE-BCT) inhibited the proliferation of metastatic B16F10 lung cancer cells in C57BL/6j mice and also inhibited the spread of HT1080 cancer cells. AE-BCT significantly inhibited factors such as the formation of colonies, cellular movement, invasion of surrounding tissues, and even intercellular adhesion while also inhibiting NF-kB activity and ERK phosphorylation. Therefore, AE-BCT is posited as a potentially non-toxic phytomedicine for the management of metastatic disease (Kim et al., 2013). A new extraction method for Kumazasa bamboo leaves has been prepared which resulted in an enhanced immunostimulating activity of the obtained extract. The given extract encourages the recruitment of NK cells and macrophages and the synthesis of IL-2, IL-12, and IFN-γ in mice bearing a tumor. It also possessed a tumor growth suppression effect in the models of tumors grown in mice besides enhancing survival duration (Seki et al., 2010).

In addition, a new sterol–glucoside–fatty acid conjugate was synthesized from the methanol extract of the bamboo stem peel, *Phyllostachys heterocycla* var. Pubescens, was obtained and the cytotoxicity and apoptosis induced by it have been evaluated. The sterol–glucoside–fatty acid derivative was proven to be have more cellular toxicity in the case of MCF-7 cells and promote breast cancer cell death through apoptosis. It also showed an affinity for binding to the receptor for vascular endothelial growth factor and tyrosine-specific protein kinase (Abdelhameed et al., 2020). Bamboo salt, a traditional Korean food, has been found to have strong anticancerous results in HCT-116 human's large intestine cancerous cells. Cell proliferation was 53% reduced by 1% bamboo salt that had been roasted nine times, significantly higher than other salts. Bamboo salt also promoted programmed cell death in cancer cells by increasing the levels of caspase-9, caspase-3, downregulating inflammation genes, and exerting high antimetastatic effects on cancer-causing cells compared to various salts (Zhao et al., 2013).

It has been shown that the soy sauce called sarijang, which contains bamboo salt, has anti-inflammatory along with anticancer properties. The pro-apoptotic properties in sarijang were investigated *in vitro* using a human leukemia cell type, U937. Research was held to identify the use of sarijang to activate caspases and cause apoptosis in U937 human leukemia cells. A pan-caspases play in the process. Finding the active components in sarijang is necessary since these results imply that it might be used as antineoplastic agents to command U937 leukemia cells (Choi et al., 2013). In a different study, researchers looked at how *Pseudosasa japonica* leaves affected the way DMBA (7,12-Dimethylbenz[a] Anthracene) developed to cause mammary gland cancer in Sprague-Dawley rats. In contrast to the command group, it was found that oral bamboo extract administration given 3 weeks before an injection of DMBA postponed the beginning of mammary gland cancer by almost 7 days. Additionally, following 11 weeks of DMBA treatment, the extract demonstrated the ability to limit the tumor's growth rate by 67% and lower the incidence of tumor development by 44% (Panee, 2009).

7.3.3 Antimicrobial Activity

Bamboo is a naturally adaptable and fast-growing plant that is resistant to extreme climates due to the high antibacterial quality it possesses. Many studies have been done in the quest to determine the bioactive components that are responsible for the antibacterial action of bamboo. *Phyllostachys pubescens*, an Australian bamboo variety, is prevalent in antibacterial activity also (Afrin et al., 2012). Their antibacterial activities were tested on a gram-negative bacterium known as *Escherichia coli*. However, it was shown that the antimicrobial substance in *P. pubescens* was mainly from lignin. Antimicrobial properties of lignin are attributed to its aromatic and phenolic functional groups. It has been proposed in several works that lignin, for instance, contained toxic substances found in the stems of bamboo (Ramful et al., 2022). Concomitantly, a few other traditional medicines related to bamboo have also been employed to treat fever and detoxication, respectively, in Asian and Indian traditional Ayurvedic medicine for many ages. One such species is Moso bamboo, a common vegetative, monopodial bamboo species belonging to the family of *Bambusoideae*, which is prevalent in the subtropics of China. Ethnomedicine has used holistic approaches for centuries and the therapeutic uses include the use of bamboo against inflammation and enhancing immunity. Yogurt was cultured using Moso bamboo seeds and the study focused on the diversity and antibacterial potential of endophytic fungi which were cultured from the seeds. The unpurified extracts of the isolates revealed varying levels of activity against pathogenic fungi associated with bacteria, which was observed during submerged culture fermentation. The results of this investigation, which at first centered on the antibacterial effect of entophytic fungi associated with Moso bamboo seeds, indicated that these fungi could potentially act as a source of plant defense activators and bioactive compounds (Shen et al., 2014).

In the age-old traditions of culinary practices in the south Indian states of Western Ghats, traditional medicine has used the seeds of bamboo for their special blooming and seeding characteristics. Twenty-eight phytoconstituents were investigated which had potential antibacterial and antioxidative activity, including steroid compounds, aromatics, and long chain aliphatic carboxylic acids, and were derived from the oil of the mixture of bamboo's watery and ethanol-based fraction of seeds (Soumya et al., 2014). Extracts of *Bambusa vulgaris*, for example, were found to be effective against *Aspergillus niger* and *Verticillium alboatrum*, suppressing their growth. They also showed the ability to kill gram-positive and gram-negative bacteria that include *Staphylococcus aureus*, *Bacillus cereus*, *E. coli*, and *Klebsiella pneumoniae*, respectively. This accounts for the traditional practices where *B. vulgaris* has been used in the management of sexually transmitted diseases and to treat injuries (Owolabi and Lajide, 2015). One of the studies looked at the main volatiles of bamboo leaf hydrosols as well as other chemical and antibacterial properties. Of the 45 compounds isolated, the most active were found to be β-ionone, geranylacetone, damasone, and ionone. Hydrosols, on the other hand, were most effective against gram-positive microorganisms and were moderately resistant to heat, pH, and UV radiation (Yuan et al., 2019). In a different study concerning the active phytochemicals in black bamboo leaves, attention was concentrated on the antioxidant and antibacterial activities of its phenolic compounds and flavonoids. As a result, the total phenolic content determination expressed as gallic acid equivalents indicated that 157.426 mg was present in the boiled bamboo leaves ethanolic extract which was tested. In the case of total flavonoid content, a target of 19.406 mg was achieved and was tested using a quercetin solution. The DPPH assay value showed an antioxidant activity of 46.8672 μg/mL and an IC_{50} BHT value of 9.5454 μg/mL which was derived from the extract of the leaves under investigation. These results attest to the possible antibacterial and antioxidative activities of the bamboo leaves (Arifani et al., 2021).

7.3.4 Antihypertensive Effect

The positive effect of bamboo extracts in treating hypertension has been confirmed by different studies. A study focused on assessing the antihyperlipidemic and antihypertensive effect of extracts made from powdered bamboo shavings on hyperlipidemic rats. The results confirmed that the bamboo shaving extracts reduced systolic pressure in spontaneously hypertensive animals and also lowered total serum cholesterol and triglycerides levels (Jiao et al., 2007). Another related study confirmed the cardiovascular protection properties of bamboo extracts on hypertensive rats as well as in rats fed with a high-fat diet, as it could increase NO levels in serum and kidneys; improved the status of oxidant stress; markedly inhibited angiotensin-converting enzyme in the lungs and systolic blood pressure; and reduced MDA levels, LDL, overall cholesterol, and triglycerides. The findings further confirmed that the effect of crude extracts was more pronounced than the synthesized peptide, suggesting a synergistic interaction of phenolic components (*p*-coumaric acid, ferulic acid, caffeic acid, homoorientin, and orientin) found in the bamboo crude extract (Liu et al., 2012). A more recent study has shown the blood pressure–controlling effects of bamboo extracts. Sunarti and Octaviani (2020), in their study on ethanol extracts of tali bamboo (*Gigantochloa apus)*, showed that the blood pressure in hypertensive male white rats could be effectively lowered with Tali bamboo extracts when used at a concentration of 40 mg/kg.

7.3.5 Dietary Fibers

Plant parts that can be eaten include dietary fiber and related carbohydrate polymers. Dietary fiber does not seem to be absorbed or chemically altered in a person's small intestines but is modified in the large intestines, which is, however, only partially or entirely carried out by the person's gut's microbial flora (Dong et al., 2019). The shoots, a source of dietary fiber, improve bowel movement and cholesterol profiles. Researchers have verified the advantages of the impact of bamboo stems on levels of cholesterolemia and bowel regularity in young and fit females on a diet with adequate amount of fibers (Chongtham et al., 2011). The gut flora ferments dietary fiber to create short-chain fatty acids, which are essential for controlling the immune system and host metabolism (de Lange et al., 1989). Based on their water solubility, they can be separated into two groups: dissoluble and non-dissoluble dietary fibers. The physiological and physicochemical characteristics of dissoluble and non-dissoluble dietary fibers derived from the stems of bamboo were demonstrated in research. The outcome demonstrated that insoluble dietary fiber has greater and superior adsorption capabilities for cholesterol. Potential prebiotic effects were demonstrated, whereas soluble dietary fibers from *Fargesia spathacea* may be employed as bioactive components for making functional foods (Wu et al., 2020). Another study examined the effects of bamboo, a premium supplier of non-dissoluble dietary fiber, on the synthesis of gut microflora and short-chain fatty acid formation. Alkaline hydrogen peroxide, complex enzymatic hydrolysis, and commercial bamboo fiber BF90 were used in the extraction of insoluble dietary fibers. When alkaline hydrogen peroxide, complex enzymatic hydrolysis, and BF90's *in vitro* fecal fermentation properties were examined, it was discovered that they increased the synthesis of total short-chain fatty acids after 24 hours and changed the microbial diversity and composition of the microflora of the gut, particularly by reducing the proportion of firmicutes to bacteroidetes and raising the number of bacteroids. The research indicates that certain bacteria in the human colon may be able to partially consume insoluble dietary fibers derived from bamboo (Ge et al., 2022). Furthermore, several enzymes can be used to break down bamboo shoot shells, which are frequently thrown away in the processing industries, into nutritional fibers. The final fibers obtained have the ability to bind to bile acids, fats, cholesterol, and nitrites and also tend to have a significant swelling and water retention capacity. It lowers body weight gain by 2.48% and reduces total cholesterol, triglycerides, and low-density lipoprotein (LDL) levels. Hence, it is plausible that the fiber present in the bamboo shoot shell may have the potential to be incorporated as a food ingredient in the functional food industry (Luo et al., 2017). Processing the bamboo stems shell found three types of dietary fibers known as overall fiber, dissoluble fiber, and non-dissoluble fiber. Studies have shown the hypoglycemic effect of these fibers by increasing the inhibition of α-amylase activity, inhibition against α-glucosidase activity, and delaying glucose diffusion. In diabetic-induced rats, bamboo shoot shell fibers improved oral glucose tolerance and reduced blood sugar levels while soluble dietary fiber dramatically raised blood insulin levels, indicating that bamboo shoot shell fibers can be used as functional foods (Zheng et al., 2019).

7.3.6 Prebiotics

Plant-based polysaccharides consist of indigestible saccharides or roughage that have been demonstrated to have prebiotic properties. The connection between human health and gut bacteria is receiving a lot of attention. As a result, there is a growing need for functional food products like prebiotics, which are believed to contribute to and improve the health of an individual by regulating the gut microbiota. Prebiotics are undigested substances, like carbohydrate fibers, that improve the health of the individual by favorably promoting the development along with the metabolic function of a particular gut microbial colony. Although α-amylase and other hydrolases in the upper digestive tract are unable to fully break down indigestible carbohydrate fibers, gut bacteria can use them as carbon sources during bacterial anaerobic fermentation (Chen et al., 2020a).

A study was conducted by using three polysaccharide fragments (CPS70, CPS75, and CPS80) from the *Chimonobambusa quadrangularis* bamboo shoot. The prebiotic activity and *in vitro* digestibility of the fractions were assessed. When subjected to human digestive juices along with α-amylase hydrolysis, all three exhibited high non-digestibility. When used as a substitute carbon source for glucose during the fermentation of probiotics within the lab, they markedly increased the growth of probiotic bacteria and improved the synthesis of lactic acid, acetic acid, propionic acid, and butyric acid. The prebiotic potential of CPS75 was the highest, as evidenced by its largest impact on the generation of volatile fatty acids along with probiotic bacterial growth. *Chimonobambusa quadrangularis* polysaccharide (CPS) prebiotic activity was greatly affected by the amount of ethanol utilized in the precipitation process (Chen, et al., 2020b). The β-glucan components found in bamboo shoot crude polysaccharides (BSCPs), which were isolated from *Gigantochloa levis*, have a potential use in the culinary and medicinal industries. When compared to FOS, BSCP showed significantly higher growth rates for *Bifidobacterium animalis, Bifidobacterium longum,* and *Lactobacillus acidophilus*, while both BSCP and FOS showed slower *Salmonella choleraesuis* survival. Accordingly, BSCP appears to be a promising prebiotic (Azmi et al., 2012).

According to research, hydrophilic polysaccharides were isolated from the stems of the bamboo *Phyllostachys praecox* (WBP), and their characteristics and prebiotic properties were examined. Following purification, two parts, WBP-1 and WBP-2, were discovered to be heteropolysaccharide-protein complexes. The existence of α and β glycosidic linkages between the sugar units was further confirmed by FTIR and NMR spectra, respectively. Moreover, a notable increase in *Bifidobacteria adolescentis* and *Bifidobacteria bifidum* populations was recorded with WBP-1 and WBP-2 which also stimulated the production of organic acids, suggesting potential prebiotic activity (He et al., 2016). In order to regulate the gut microbiota, the use of xylo-oligosaccharides (XOSs) with predominant proportions of longer X2-X3 units as a prebiotic were investigated. The initial XOS extract to increase the proportions of X2-X3 units was prepared using fresh bamboo shoots and thermally treated, followed by enzymatic hydrolysis. The XOS extracts enriched with the greater proportion of X2-X3 units were more effective in stimulating the synthesis of short-chain fatty acids and probiotic growth. XOS extracts which contain more fractions with X2-X3 parts have shown better effectiveness for alleviating intestinal inflammation *in vivo* in a colitis mouse model. Here, the basic XOS extract (S2) produced from bamboo stems by endoxylanase hydrolysis and employing hydrothermal treatment was enzymatically treated to obtain XOS containing X2-X3 regions (S1). *In vitro* experiments were performed to compare the different prebiotic effects of S1 and S2 on the proliferation of beneficial gut bacteria in the standard gut. In the intestinal microbiota, the S1 group considerably elevated *Bacteroides* and *Streptomyces*, while the S2 group elevated *Bacteroides*. In addition, the XOS solutions decreased the quality of *Pasteurella* and *Alcaligenes*, demonstrating its potential as an anticolitis agent and gut microbiota regulator (Deng et al., 2023).

7.3.7 Anti-infertility

Regulating fertility is not a novel idea. In traditional medicine, ancient medicinal practices have used a variety of plants or their derivatives as contraceptive medications throughout various nations worldwide. Women have attempted to regulate their fertility with varying degrees of social pressure throughout history. Traditional uses for several herbal medicines include contraceptives, abortifacients, and oxytocics (Ritchie, 2001).

Tropical and subtropical regions are home to the plants *Bambusa vulgaris*, particularly in the rainy lowlands and rainy season. *B. vulgaris* leaf extract consists of tannins, phenolics, glycosides, saponins, flavonoids, and anthraquinones. It has been claimed that bamboo leaves can be utilized as a febrifuge, styptic, and eye solution. Bamboo is utilized in Nigerian traditional medicine as a menstrual promoter, abortion-causing agent, an appetizer, and a treatment for gonorrhea and other respiratory conditions (Gill, 1992). Extracts at 250 mg/kg body weight significantly reduced the number of viable fetuses, while the amount of 500 mg/kg body weight led to zero production of live fetuses. While fetal survival was limited to 29% and 0%, abortion at the same doses occurred at 60% and 100%, respectively. The extract also lowered the levels of luteinizing hormone, follicle-stimulating hormone, and progesterone in the blood. This proves that *B. vulgaris* has the ability to cause abortions (Yakubu and Bukoye, 2009).

According to a study, *Bambusa arundinacea* shoots showed contraceptive impacts on both sexes when given at a reasonable dosage for a predetermined amount of time. The placenta would descend within the next 2 hours if 1 kg of tender stems of *B. arundinacea* was taken shortly after parturition, indicating that it possessed sufficient antifertility potential. In a study, a 70% ethanol extract of *B. arundinacea* young shoots was prepared and given to male and female rats for about a week at a dose of approximately 300 mg/kg per day. The results demonstrated that the fertility of the male rats decreased by 15% after 1 week and that of the female rats decreased by 23% after just 4 days of treatment. However, after 8 days of not eating *B. arundinacea* tender shoots, full recovery was visible. Similarly, spermatozoa produced by the cauda epididymis showed decreased motility in conjunction with a notable drop in spermatozoa quantity in cauda epididymis and caput. Nonetheless, the extract's complete nontoxicity was demonstrated by the pyruvic/oxaloacetic transaminase activity and serum protein profile. Additionally, the weight of the prostate, vas deferens, epididymis, and testis was reduced (Vanithakumari et al., 1989).

7.3.8 Antiulcer Activity

Around the world, bamboo has been utilized in traditional medicines in numerous nations. A multi-author study's findings indicate that substances including phenols, terpenoids, flavonoids, and tannins have positive and therapeutic effects on health (Singh et al., 2012). The antiulcer effects of *Bambusa balcooa*'s flavonoids and tannins is noteworthy since they preserve the stomach mucosa's cellular integrity. The bamboo species *B. balcooa* was collected to analyze the phytochemical study comparing the extracts of the stem and leaves in six different solvents. The highest yield of *B. balcooa* leaves with methanol as the extracting solvent was achieved, at 34.55%. The presence of flavonoids, cardiac glycosides, reducing sugar, coumarins, sterols, and other polyphenolic substances was revealed by phytochemical analysis. *B. balcooa* leaf extract used in a methanol pharmacological investigation showed an antiulcer efficacy of 14.66% (Upreti et al., 2016). Another study revealed that giving a hot water extract orally (folin) of *Sasa albomarginata* greatly decreased the occurrence of stomach ulcers caused by indomethacin, ethanol, and water immersion and restrained stress (Otani et al., 1990). The histological analysis of the rats' gastric mucosa after folin treatment allowed the researchers to assess the antiulcer effect of bamboo. It was established that the small thrombi played a role in safeguarding the outermost epithelium as well as maintaining the structural integrity of the stomach mucosa, particularly in the prevention of stress demonstrative ulceration.

Studies have shown that Folin can protect red blood cells from destruction, hasten their clumping in acidic conditions, as well as inhibit histamine release from mast cells in rats. There was also a decrease in hyperemia and reduction of acid mucopolysaccharides in the ethanol-induced ulcer. On the basis of the observations, it can be suggested that the protective action of Folin against gastric lesions is due to its microscopically hemostatic properties complemented with a membrane-stabilizing activity (Muniappan and Sundararaj, 2003).

B. arundinacea leaves, when applied as a methanolic extract, have been shown in another study to have antiulcer properties against paw edema caused by carrageen. The antiulcer activity in albino rats was studied by administering this methanolic extract to the rats during the course of the experiment. Following completion of the entire experiment, it was determined that the best anti-inflammatory medication for treating inflammatory disorders such as rheumatoid arthritis and peptic ulcers was produced by combining *B. arundinacea* methanolic extract with non-steroidal anti-inflammatory medications (Jivani, 2011).

7.3.9 Anti-inflammatory Activity

The body uses inflammation as a physiological defense mechanism against pathogens, external substances, and irritants. A drug or treatment that lowers swelling or inflammation is said to be anti-inflammatory. Bamboo, among the many plants, may have anti-inflammatory properties.

Purple bamboo salt is a uniquely prepared salt made from regular salt and bamboo, following a traditional formula. In Korea, it has been used for the management and prevention of certain forms of sickness. Using human mast cell line (HMC-1) as a model, an experiment was performed to examine the anti-inflammatory properties of purple bamboo salt. When the 1 mg/mL purple bamboo salt was administered together with calcium ionophore A23187, the secretion of interleukin-6, tumor necrosis factor-alpha, and interleukin-1 beta was inhibited. Also, in HMC-1 cells, tumoral necrosis factor alpha mRNA expression was inhibited by purple bamboo salt at a similar experimental setup (Shin et al., 2003).

Bamboo plant *Phyllostacys edulis* J. Houz is one of the forest resources which restructures forest ecosystems to our advantage economically among other benefits such as health. The extracts of the bamboo leaf and sheath have been evaluated for their radical scavenging activity, anti-inflammation, and antioxidant activities. The leaves had a total phenol and flavinoid worth 73.92 mg. The fractions also demonstrated remarkable radical scavenging activity. Bamboo sheath extract was found to cause less ROS in HepG2 liver cells without loss of feasibility, while another HepG2 treated with bamboo leaf extract was found to be cytotoxic. The study claimed that bamboo sheath and leaf extract had anti-inflammatory and antioxidant activity, indicating that it can be utilized in nutraceuticals, skin care, and medications (Tundis et al., 2023). This study aimed to evaluate the content of secondary metabolites and further determine the anti-inflammatory efficiency of the bamboo tali leaves' crude extract: the inhibitory concentration fifty (IC_{50}) or the liquid volume of the crude extract at half maximal inhibition. Bamboo tali leaves were screened for active phytochemicals after soaking the leaves in 96% ethanol for a certain period. Subsequently, the anti-inflammatory activity of the extract was investigated *in vitro* by using the bovine serum albumin protein denaturation inhibition method. The ethanol leaf extract of bamboo tali revealed the presence of phenols, alkaloids, flavonoids, and saponins. These results indicate that bamboo tali leaf ethanol extract has anti-inflammatory activity with an IC_{50} value less than 100 ppm (Hidayah and Hafshah, 2023). The anti-inflammatory and antiadipogenic activities of bamboo leaf extracts, especially the ethylacetate fractions from the Korean species *Sasa coreana* Nakai, have been the focus of attention. The extracts were effective in reducing inflammation by attenuating nitric oxide synthase expression in lipopolysaccharide-induced macrophages. The ethylacetate fraction of the methanol extracts of *S. corneana* Nakai leaves also inhibited adipocyte differentiation and migration, suggesting its potential as an anti-inflammatory and anti-obesity medication. The research results suggest that *S. coreana* Nakai has a potential value in the preservation of healthy adipose tissue (Yang et al., 2017).

An additional investigation looked at the anti-inflammatory effects of bamboo species *Sasa veitchii* and *Phyllostachys nigra* in cell cultures. The extracts were tested for their ability to inhibit cyclooxygenase-1, -2, and nuclear factor-κB–dependent gene expression. The extracts were juxtaposed with those of bamboos that were not known to possess any anti-inflammatory effect. Tandem mass spectrometry and HPLC analyses were performed. In comparison to solid phase extraction fractions, the bioactive strength of the crude extracts was higher. It is interesting to note that both species possess anti-inflammatory activities in spite of their acting on different molecular pathways (Van Hoyweghen et al., 2014).

In a study, it was discovered that bamboo vinegar, the by-product liquid created as a result of the bamboo charcoal process, has anti-inflammatory properties in both outside the living and within the living. Without altering the emergence of the tumor necrosis factor-α or cyclooxygenase-2, it decreased the emergence of nitric oxide synthase along with interleukin-6 release in macrophages. The discovery of cresol, a significant anti-inflammatory molecule in bamboo vinegar, suggested that it would be possible to use it to anti-inflammatory drugs for illnesses linked to inflammation (Ho et al., 2013). In immortalized human keratinocytes, a study was conducted into the anti-inflammatory properties of a bamboo leaf extract of *Phyllostachys* edulis in relation to the oversynthesis of interleukin-8 and vascular endothelial growth factor caused by tumor necrosis factor-α. Additionally, it assessed the ability of albino mouse fibroblasts to repair wounds. Interleukin-8 and vascular endothelial growth factor were not released in response to tumor necrosis factor-α, while interleukin-6 levels in treated cells were decreased by isoorientin and bamboo leaf extract. Isomerin and bamboo leaf extract may have complementary anti-inflammatory properties, according to the study (Wedler et al., 2014).

7.3.10 Cardioprotective Properties

As early as 10,000 BCE, pharmacopeias all across the world have recorded the known medicinal benefits of bamboo. In addition to lowering blood pressure, they have been shown in recent research to have anticancer, antidiabetic, and cardiovascular disease prevention qualities. In the modern world, diabetes is the most prevalent and long-lasting health issue. Pancreatic-impaired insulin sensitivity and β-cell failure are the hallmarks of this metabolic condition, which can lead to numerous related problems.

A study looked at the cardioprotective effects of crude extract from the *Bambusa blumeana* leaf against isoproterenol-induced myocardial infarction in Sprague-Dawley rats. The rats were dopamine exposed after almost 3 weeks of tocopherol and leaf extract administration. The findings showed that the extract decreased the levels of the cardiac markers, namely, creatine kinase-MB, aspartate aminotransferase, and lactate dehydrogenase, and also lessened the inflammatory processes. The histopathological findings supported the protective effect of the extract, stating that moderate to pronounced myocardial degeneration and necrosis were observed (Rosales et al., 2013). The rats that were predisposed to diabetes-encouraged cardiac fibrosis were included in the study so that the potential protective effects of the bamboo leaf extract could be judged for addressing cardiac problems. After the administration of streptozotocin which induces diabetic conditions in the rats, cardiac function was evaluated. The effect of bamboo leaf extract treatment on cardiac function and oxidative stress was proved by the reduction in cardiac inflammation and changes in the morphology of myocardial tissue. It was also observed that the treatment reduced the nuclear factor-κB, interleukin-6, and secreted transforming growth factor-β1 protein content of the heart. This suggests that bamboo leaf extract may be utilized as a novel treatment option for diabetic patients (Zhang et al., 2017). Bamboo extracts procured from bamboo species such as *B. arundinacea, B. balcooa, Bambusa tulda, B. vulgaris, P. bambusoides, P. nigra, Sasa borealis, Sasa quelpaertensis,* and many others have been studied and found as an active substance capable of reducing blood glucose levels. Preclinical studies demonstrated that bamboo extracts lower blood glucose levels while enhancing insulin signaling and concurrently activating 5′ adenosine monophosphate-activated protein kinase.

The antioxidant potential of bamboo – largely due to its polyphenols, flavonoids, and phytosterols and polysaccharides – helps scavenge ROS to reduce oxidative stress and thereby avert diabetes-related complications. This natural resource can be used as a preventive nutraceutical along with the production of functional food and medicinal products (Indira et al., 2023).

The serum cholesterol levels of the high-cholesterol mice were given various dosages of bamboo leaf extract. It has been noted that the leaves of *P. pubescens* provide protection against lipoapoptosis produced by palmitic acid (Fu et al., 2005). Studies has proven that the biological outcomes of bamboo leaf extract of *Ginkgo biloba* are comparable to dramatically lower blood triglycerides and cholesterol, increase HDL levels, and decrease LDL levels in a study conducted on high blood-lipid rats (Zhang Ying et al., 2000). A study result indicated that orientin, a chemical derived from bamboo leaves, has a strong cardioprotective impact on cardiomyocytes and the rat heart following treatments with hypoxia/reoxygenation and ischemia/reperfusion. It decreased cardiomyocyte apoptosis and inhibited myocardial apoptosis. Moreover, orientin downregulated Bax expression while upregulating bcl-2 expression, which raised the bcl-2/Bax ratio. In myocardial and cardiomyocytes damaged by ischemia/reperfusion and hypoxia/reoxygenation, it also decreased the expression of cytochrome-c and caspases-3 (Fu et al., 2006).

7.4 TRADITIONAL MEDICINAL USES

Bamboo has been used for therapeutic reasons since time began and has been incorporated in the pharmacopoeias of traditional medicines of many Asian countries (Tripathi, 2011). For instance, bamboo is used to treat gastrointestinal issues; mental disorders linked to aging in humans; and respiratory infections in ancient Chinese, Indian, and Tibetan pharmacy (Laws, 2013). Researchers collected data on the ethnobotanical applications of bamboo among the Adi tribes in the Indian state of Arunachal Pradesh. The residents mentioned using *Dendrocalamus strictus* leaves as an abortifacient, using *B. tulsa* for tetanus treatment, *Dendrocalamus giganteus* for the manufacture of steroid drugs, and using *Schizostachyum capitatum* leaves as an infusion for stomach pains (Sharma and Borthakur, 2008). For hundreds of years, eastern Asians have utilized *Sasa senanensis* alkaline leaf extract, as a possible natural medication source. In addition, traditional Japanese medicine uses the leaves' extract as an anti-inflammatory agent (Sangeetha et al., 2015).

People in the southeastern Chinese region of Xishuangbanna reported that *Dendrocalamus hamilttonii, Thyrsostachys siamensis, Pseudostachyum polymorphum,* and other bamboo species' stems are used in Dai medicine, a conventional drug from the area that is said to be an energizing treatment for kidney disease (Yang et al., 2008). Research on the ancient works of bamboo in the Chinese Himalayan region of the Yunnan province revealed that people used the leaves of *Phyllostachys glauca* to cure lung inflammation and coughs, the sap and young culms of *Phyllostachys heterocycla* to treat tracheal inflammation and coughs, and the shoots of *Indosasa pingbianensis* to cure common colds and headaches (Yang et al., 2014). Indian Kani tribes believe that *B. arundinacea* wild seeds increase human fertility. *Bambusa* leaf juice has also been utilized to improve cartilage in cases of osteoporosis and osteoarthritis (Sangeetha et al., 2015).

Tall bamboo is used to alleviate fever in China because of its strong taste, fidgeting, and slightly bitter flavor (Kiruba et al., 2007). It has been used to treat psoriasis and eczema and plays a significant part in maintaining the unification of bone, skin, teethes, gums, hair, nails, and artery walls (Vanithakumari et al., 1989). Bamboo plants were utilized by the Chinese in the past to treat a variety of illnesses. For example, bamboo shavings, also known as Chinesezhuru, are made by shaving the outside of a bamboo rod and then cutting the center strip of the rod into longitudinally and thin cut segments which are utilized. It looks white-green and is used to treat acute fevers, convulsions, heat-related bleeding, sweet vomiting, mild coldness, clearing heat, and clearing phlegm, among other conditions. Bamboos tabasheer, also known as tabashir, are widely utilized as cures for fever in youngsters as well as epilepsy. Additionally, tabasheer is used in convulsions, fevers, loss of consciousness linked to phlegm-heat, and has anti-convulsive properties (Das et al., 2021).

In folk medicine, *B. arundinacea* plant extract has been employed to cure various inflammatory disorders. The leaves, stems, and roots are employed as laxatives, diuretics, and astringents in Ayurveda. A root ointment is reputed to be a traditional treatment for hard tumors and cirrhosis. *B. arundinacea* shoot is used to remove worms from ulcers. The decoction of *B. arundinacea* leaf buds is used to promote menstrual flow. In leprosy, hematenesis, fever, and hemoptysis, leaves are helpful. Additionally, leaves are utilized for snake bites and cough-related paralysis issues. An ointment derived from the roots of *B. arundinacea* is reported to be used as a folk treatment for tumors, particularly those of the stomach, liver, spleen, abdomen, and cirrhosis. Asthma and coughs are treated with tabasheer, a siliceous secretion of *B. arundinacea.* It is said to be an aphrodisiac, cooling agent, and tonic. Leaves are administered to horses with colds and coughs (Soni et al., 2013).

To induce abortion of an early conception, tribal women in Salem, Tamil Nadu, chew *B. arundinacea* leaves in the morning and evening for 1 to 3 days (Bhaduri, 1968). Indian traditional medicine has used *B. arundinacea* preparations to treat a variety of inflammatory illnesses. The herb also possesses antiulcer properties. It is believed that treating inflammatory disorders might greatly benefit from these two characteristics present in the same extract. It is commonly recognized that the majority of anti-inflammatory medications on the market cause ulcers (Muniappan and Sundararaj, 2003). A decoction of leaves is used for dysmenorrhea and amenorrhea to stimulate menses and reduce menorrhagia symptoms. After the delivery, the woman is expected to drink a bamboo nodes decoction to enhance the flow of lochia (Kirtikar, 1918). *B. arundinacea* seeds wild in the southern Western Ghats by the Kani tribes of the Kanyakumari district. There has been a description of how indigenous people gather, store, and consume seeds (Soni et al., 2013).

7.5 CONCLUSION

The potential medical benefits of bamboo, a plant utilized in traditional medicine, are now being researched. The majority of species found in Asian nations have had extensive research done on their chemical makeup and biological characteristics. Certain species have demonstrated antioxidant capacity, which may help treat conditions including anticancer and anti-inflammatory disorders. Bamboo extracts may also contain bioactive peptides and polysaccharides that can exhibit a wide range of biological activities, including but not limited to, anti-free radical, anti-aging, antifatigue, antibacterial, and antiviral activities. These materials can be incorporated in food and cosmetic preparations as well as functional dietary supplements. Moreover, most of the studies have focused on the extracts which are a mixture of components and therefore are not easy to determine the dose-response relationships or even explain the mechanism of action of the particular bioactive compound. In addition, specific *in vivo*–tested species have not been published with regards to bioavailability or pharmacokinetics. It may be safe to say that based on the findings of this research, there do exist commercially important bioactive compounds in bamboo that can be used for therapeutic purposes. However, it remains to be established whether the amount of these materials or the extracts which contain them are sufficient for the production of herbal medicines; this is the case in a few studies which consider or present yields for purified substances. To conclude, further *in vitro* tests and photochemistry studies are still needed in order to evaluate the full scope of the medicinal potential.

REFERENCES

Abdelhameed, R. F. A., Nafie, M. S., Ibrahim, A. K., Yamada, K., Abdel-Kader, M. S., Ibrahim, A. K., Ahmed, S. A., Badr, J. M., & Habib, E. S. 2020. Cytotoxic, apoptosis-inducing activities, and molecular docking of a new sterol from bamboo shoot skin *Phyllostachys* heterocycla var. pubescens. *Molecules 25*(23): 5650.

Afrin, T., Tsuzuki, T., Kanwar, R. K., & Wang, X. 2012. The origin of the antibacterial property of bamboo. *Journal of the Textile Institute 103*(8): 844–849.

Arifani, E. N., Santoso, U., & Supriyadi, S. 2021. The correlation of the total phenolic and flavonoid content on its antioxidant and antimicrobial activity of bamboo leaf extract. *Key Engineering Materials 884*: 256–263.

Azmi, A. F. M. N., Mustafa, S., Md. Hashim, D., & Manap, Y. A. 2012. Prebiotic activity of polysaccharides extracted from *Gigantochloa levis* (Buluh beting) shoots. *Molecules 17*(2): 1635–1651.

Bhaduri, B. E. A. 1968. Antifertility activity of some medicinal plants. *Indian Journal of Experimental Biology 6*: 252–253.

Brewer, M. S. 2011. Natural antioxidants: Sources, compounds, mechanisms of action, and potential applications. *Comprehensive Reviews in Food Science and Food Safety 10*(4): 221–247.

Bystriakova, N., Kapos, V., Lysenko, I., & Stapleton, C. M. A. 2003. Distribution and conservation status of forest bamboo biodiversity in the Asia-Pacific Region. *Biodiversity and Conservation 12*(9): 1833–1841.

Chen, G., Hong, Q., Ji, N., Wu, W., & Ma, L. 2020a. Influences of different drying methods on the structural characteristics and prebiotic activity of polysaccharides from bamboo shoot (*Chimonobambusa quadrangularis*) residues. *International Journal of Biological Macromolecules 155*: 674–684.

Chen, G., Ran, C., Li, C., Xiong, Z., & Ma, L. 2020b. Comparisons of prebiotic activity of polysaccharides from shoot residues of bamboo (*Chimonobambusa quadrangularis*) via different ethanol concentrations. *Journal of Food Biochemistry 44*(5): e13171.

Choi, E. A., Park, C., Han, M. H., Lee, J. H., Kim, G. Y., Choi, B. T., & Choi, Y. H. 2013. Induction of apoptosis by sarijang, a bamboo salt sauce, in U937 human leukemia cells through the activation of caspases. *Experimental and Therapeutic Medicine 6*(2): 381–387.

Chongtham, N., Bisht, M. S., & Haorongbam, S. 2011. Nutritional properties of bamboo shoots: Potential and prospects for utilization as a health food. *Comprehensive Reviews in Food Science and Food Safety 10*(3): 153–168.

Das, B., Kumar Singh, P., & Munna, J. 2021. Miraculous medicinal properties of bamboo. *Journal of Medicinal Plants 9*(5): 28–30.

De Lange, C. F. M., Sauer, W. C., Mosenthin, R., & Souffrant, W. B. 1989. The effect of feeding different protein-free diets on the recovery and amino acid composition of endogenous protein collected from the distal ileum and feces in pigs. *Journal of Animal Science 67*(3): 746.

Deng, J., Yun, J., Gu, Y., Yan, B., Yin, B., & Huang, C. 2023. Evaluating the in vitro and in vivo prebiotic effects of different xylo-oligosaccharides obtained from bamboo shoots by hydrothermal pretreatment combined with endo-xylanase hydrolysis. *International Journal of Molecular Sciences 24*(17): 13422.

Dong, J., Wang, L., LÜ, J., Zhu, Y., & Shen, R. 2019. Structural, antioxidant and adsorption properties of dietary fiber from foxtail millet (*Setaria italica*) bran. *Journal of the Science of Food and Agriculture 99*(8): 3886–3894.

Fu, X., Wang, M., Li, S., & Li Y. 2005. The effect of bamboo leaves extract on hemorheology of normal rats. *Journal of Chinese Medicinal Materials 28*(2): 130–132.

Fu, X.-C., Wang, M.-W., Li, S.-P., & Wang, H.-L. 2006. Anti-apoptotic effect and the mechanism of orientin on ischaemic/reperfused myocardium. *Journal of Asian Natural Products Research 8*(3): 265–272.

Ge, Q., Li, H., Zheng, Z., Yang, K., Li, P., Xiao, Z., Xiao, G., & Mao, J. 2022. In vitro fecal fermentation characteristics of bamboo insoluble dietary fiber and its impacts on human gut microbiota. *Food Research International 156*: 111173.

Gill, L. S. 1992. Ethnomedical uses of plants in Nigeria UNIBEN Press. *University of Benin, Benin City, Edo State, Nig*: 176–177.

Gratani, L., Crescente, M. F., Varone, L., Fabrini, G., & Digiulio, E. 2008. Growth pattern and photosynthetic activity of different bamboo species growing in the botanical garden of Rome. *Flora - Morphology, Distribution, Functional Ecology of Plants 203*(1): 77–84.

He, S., Wang, X., Zhang, Y., Wang, J., Sun, H., Wang, J., Cao, X., & Ye, Y. 2016. Isolation and prebiotic activity of water-soluble polysaccharides fractions from the bamboo shoots (*Phyllostachys praecox*). *Carbohydrate Polymers 151*: 295–304.

Hidayah, N., & Hafshah, M. 2023. In vitro anti-inflammatory activity of bamboo Tali leaf (*Gigantochloa apus*) ethanol extract. *Biology, Medicine, & Natural Product Chemistry 12*(2): 601–606.

Ho, C.-L., Lin, C.-Y., Ka, S.-M., Chen, A., Tasi, Y.-L., Liu, M.-L., Chiu, Y.-C., & Hua, K.-F. 2013. Bamboo vinegar decreases inflammatory mediator expression and NLRP3 inflammasome activation by inhibiting reactive oxygen species generation and protein kinase C-α/δ activation. *PLoS One 8*(10): e75738.

Hu, C., Zhang, Y., & Kitts, D. D. 2000. Evaluation of antioxidant and prooxidant activities of bamboo *Phyllostachys nigra* Var. *Henonis* leaf extract in vitro. *Journal of Agricultural and Food Chemistry 48*(8): 3170–3176.

Indira, A., Joshi, B., Oinam, S., Koul, A., & Chongtham, N. 2023. Potential of bamboo in the prevention of diabetes-related disorders: Possible mechanisms for prevention. In *Bamboo Science and Technology* (pp. 89–124). Singapore: Springer Nature Singapore.

Jiao, J., Zhang, Y., Lou, D., Wu, X., & Zhang, Y. 2007. Antihyperlipidemic and antihypertensive effect of a triterpenoid-rich extract from bamboo shavings and vasodilator effect of friedelin on phenylephrine-induced vasoconstriction in thoracic aortas of rats. *Phytotherapy Research 21*(12): 1135–1141.

Jivani, N. P. 2011. Phytopharmacological properties of *Bambusa* arundinacea as a potential medicinal tree: An overview. *Journal of Applied Pharmaceutical Science* 1: 27–31.

Kalaiarasi, K., Prasannaraj, G., Sahi, S. V., & Venkatachalam, P. 2015. Phytofabrication of biomolecule-coated metallic silver nanoparticles using leaf extracts of in vitro-raised bamboo species and its anticancer activity against human PC3 cell lines. *Turkish Journal of Biology 39*(2): 223–232.

Kim, A., Im, M., Yim, N.-H., Jung, Y. P., & Ma, J. Y. 2013. Aqueous extract of *Bambusae caulis* in Taeniam inhibits PMA-induced tumor cell invasion and pulmonary metastasis: Suppression of NF-κB activation through ROS signaling. *PLoS One 8*(10): e78061.

Kimura, Y., Suto, S., & Tatsuka, M. 2002. Evaluation of carcinogenic/co-carcinogenic activity of Chikusaku-eki, a bamboo charcoal by-product used as a folk remedy, in BALB/c 3T3 cells. *Biological and Pharmaceutical Bulletin 25*(8): 1026–1029.

Kirtikar, K. R., & Basu, B. D. 1918. *Indian Medicinal Plants* (Vol. 2). Publisher not identified Basu, Bhuwaneśwari Âśrama.

Kiruba, S., Jeeva, S., Das, S., & Kannan, D. 2007. Bamboo seeds as a means to sustenance of the indigenous community. *Indian Journal of Traditional Knowledge 6*(1): 199–203.

Lachance, P. A., & He, Y. H. 1998. Hypocholesterolemic composition from bamboo shoots. PCT International Patent PCT/US98/12556.

Laws, B. 2013. *Plants That Changed the Course of History* (pp. 98–103). Rio de Janeiro: Sextant.

Liu, L., Liu, L., Lu, B., Xia, D., & Zhang, Y. 2012. Evaluation of antihypertensive and antihyperlipidemic effects of Bamboo shoot angiotensin converting enzyme inhibitory peptide in vivo. *Journal of Agricultural and Food Chemistry 60*(45): 11351–11358.

Lu, B., Wu, X., Shi, J., Dong, Y., & Zhang, Y. 2006. Toxicology and safety of antioxidant of bamboo leaves. Part 2: Developmental toxicity test in rats with antioxidant of bamboo leaves. *Food and Chemical Toxicology 44*(10): 1739–1743.

Luo, X., Wang, Q., Zheng, B., Lin, L., Chen, B., Zheng, Y., & Xiao, J. 2017. Hydration properties and binding capacities of dietary fibers from bamboo shoot shell and its hypolipidemic effects in mice. *Food and Chemical Toxicology 109*: 1003–1009.

Mbah, C. J., & Onah, C. M. (2019). Physicochemical properties and bioanalytical methods of therapeutic agents utilized in the management of Alzheimer's disease. *Acta Scientific Pharmaceutical Sciences*, 3(10), 35–41

Mu, J., Uehara, T., Li, J., & Furuno, T. 2004. Identification and evaluation of antioxidant activities of bamboo extracts. *Forestry Studies in China 6*(2): 1.

Muniappan, M., & Sundararaj, T. 2003. Antiinflammatory and antiulcer activities of *Bambusa* arundinacea. *Journal of Ethnopharmacology 88*(2–3): 161–167.

Nagai, T., Suzuki, N., Tanoue, Y., Kai, N., & Nagashima, T. 2009. Antioxidant activities of the extracts from bamboo powder as underutilized resource. *Journal of Food Agriculture & Environment 7*(2): 228–232.

Ni, Q., Xu, G., Wang, Z., Gao, Q., Wang, S., & Zhang, Y. 2012. Seasonal variations of the antioxidant composition in ground bamboo Sasa argenteastriatus leaves. *International Journal of Molecular Sciences 13*(2): 2249–2262.

Nirmala, C., Bisht, M. S., & Laishram, M. 2014. Bioactive compounds in bamboo shoots: Health benefits and prospects for developing functional foods. *International Journal of Food Science & Technology 49*(6): 1425–1431.

Nirmala, C., Sheena, H., & David, E. 2009. Bamboo shoots: A rich source of dietary fibres. In *Dietary Fibres, Fruit and Vegetable Consumption and Health* (pp. 15–30). Nova Science Publishers: USA.

Otani, K., Yanaura, S., Yuda, Y., Kawaoto, H., Kajita, T., Hirano, F., & Inouye, S. 1990. Histo-chemical studies on the anti-ulcer effect of bamboo grass in rats. *International Journal of Tissue Reactions 12*(6): 319–332.

Owolabi, M. S., & Lajide, L. 2015. Preliminary phytochemical screening and antimicrobial activity of crude extracts of *Bambusa vulgaris* Schrad. Ex JC Wendl. (Poaceae) from southwestern Nigeria. *American Journal of Essential Oils and Natural Products 3*(1): 42–45.

Panee, J. 2009. Bamboo extract in the prevention of diabetes and breast cancer. In: *Complementary and Alternative Therapies and the Aging Population* (pp. 159–177). Elsevier.

Pisoschi, A. M., & Negulescu, G. P. 2011. Methods for total antioxidant activity determination: A review. *Biochem Anal Biochem 1*(1): 106.

Qi, X. F., Kim, D. H., Yoon, Y. S., Song, S. B., Teng, Y. C., Cai, D. Q., & Lee, K. J. 2012. *Bambusae* caulis in liquamen suppresses the expression of thymus and activation-regulated chemokine and macrophage-derived chemokine in human keratinocytes due to antioxidant effect. *Evidence-Based Complementary and Alternative Medicine 2012*(1): 617494.

Ramful, R., Sunthar, T. P. M., Kamei, K., & Pezzotti, G. 2022. Investigating the antibacterial characteristics of Japanese Bamboo. *Antibiotics 11*(5): 569.

Ritchie, H. E. 2001. The safety of herbal medicine use during pregnancy. *Frontiers in Fetal Health 3*(10): 259–266.

Rosales, S. L. O., Anog, B. F. A., Blanza, R. M., Cuartero, M. A. D., Landicho, M. C., Talledo, E. G. D., & Magbojos, C. R. 2013. Cardioprotective activity of *Bambusa blumeana* Schultes leaf crude extract against isoproterenol induced myocardial infarction in Sprague-Dawley rats. *The Steth 7*: 41–57.

Sangeetha, R., Diea, Y. K., Chaitra, C., Malvi, P. G., & Shinomol, G. K. 2015. The amazing bamboo: A review on its medicinal and pharmacological potential. *Indian Journal of Nutrition 2*(1): 1–7.

Santosh, O., Bajwa, H. K., Bisht, M. S., & Nirmala, C. 2016. Bamboo shoot for food fortification: Analysis of nutrients and bioactive compounds in juvenile shoots of two bamboo species. In Development of Economies Through Innovation & Sustainable Growth: Vision 2020 53–62.

Schillaci, C., Nepravishta, R., & Bellomaria, A. 2014. Antioxidants in food and pharmaceutical research. *Albanian Journal of Pharmaceutical Sciences 1*: 9–15.

Scurlock, J. M. O., Dayton, D. C., & Hames, B. 2000. Bamboo: An overlooked biomass resource? *Biomass and Bioenergy 19*(4): 229–244.

Seki, T., Kida, K., & Maeda, H. 2010. Immunostimulation-Mediated anti-tumor activity of bamboo (*Sasa senanensis*) leaf extracts obtained under 'Vigorous' condition. *Evidence-Based Complementary and Alternative Medicine 7*(4): 447–457.

Sharma, T. P., & Borthakur, S. K. 2008. Ethnobotanical observations on Bamboos among Adi tribes in Arunachal Pradesh *Indian Journal of Traditional Knowledge 7*: 594–597.

Sharma, V., & Nirmala, C. 2018, August. Therapeutic potential of bamboo shoots against cancer: An overview. In *Proceedings of 11th World Bamboo Congress.*

Shen, X.-Y., Cheng, Y.-L., Cai, C.-J., Fan, L., Gao, J., & Hou, C.-L. 2014. Diversity and antimicrobial activity of culturable endophytic fungi isolated from Moso bamboo seeds. *PLoS One 9*(4): e95838.

Shin, H., Lee, E., Kim, C., Shin, T., Kim, S., Song, Y., Lee, K., Hong, S., & Kim, H. 2003. Anti-inflammatory activity of Korean folk medicine purple bamboo salt. *Immunopharmacology and Immunotoxicology 25*(3): 377–384.

Singh, S. A., Bora, T. C., & Singh, N. R. 2012. Preliminary phytochemical analysis and antimicrobial potential of fermented *Bambusa balcooa* shoots. *The Bioscan 7*(3): 391–394.

Singhal, P., Bal, L. M., Satya, S., Sudhakar, P., & Naik, S. N. 2013. Bamboo Shoots: A novel source of nutrition and medicine. *Critical Reviews in Food Science and Nutrition 53*(5): 517–534.

Soni, V., Jha, A. K., Dwivedi, J., & Soni, P. 2013. Traditional uses, phytochemistry and pharmacological profile of *Bambusa arudinacea* Retz. *Tang [Humanitas Medicine] 3*(3): 20.1–20.6.

Soumya, V., Muzib, Y. I., & Venkatesh, P. 2014. GC-MS characterization, *in vitro* antioxidant and antimicrobial activity of newly isolated oil from edible wild Bamboo rice (*Bambusa bambos*). *Journal of Biologically Active Products from Nature 4*(3): 209–215.

Srivastava, R. C. 1990. Bamboo, new raw material for phytosterol. *Current Science 59*(24): 1333–1334.

Sunarti, & Octaviani, P. 2020. The effect of antihypertensive of tali's bamboo shoot ethanol extract (*Gigantochloa apus* (Schult.&Schult.F.)) to male white rats Sprague Dawley. In *Proceedings of the 1st International Conference on Community Health (ICCH 2019).*

Tripathi, Y. C. 2011. Bamboo on the touchstone of modern phytotherapy. In *Productivity Enhancement and Value Addition of Bamboos* (pp. 315–326). Excel India: New Delhi.

Tundis, R., Augimeri, G., Vivacqua, A., Romeo, R., Sicari, V., Bonofiglio, D., & Loizzo, M. R. 2023. Anti-inflammatory and antioxidant effects of leaves and sheath from Bamboo (*Phyllostacys edulis* J. Houz). *Antioxidants 12*(6): 1239.

Upreti, A., Byanju, B., Pokhrel, B., Kuinkel, S., Bhusal, A., Thapa, N., Suwal M. M., Lamichhane J, & Gauchan, D. P. 2016. *Bambusa balcooa* Roxb.: A novel remedy for peptic ulcer. *Kathmandu University Journal of Science, Engineering and Technology 12*(2): 88–95.

Van Hoyweghen, L., De Bosscher, K., Haegeman, G., Deforce, D., & Heyerick, A. 2014. *In vitro* inhibition of the transcription factor nf-κb and cyclooxygenase by Bamboo extracts. *Phytotherapy Research 28*(2): 224–230.

Vanithakumari, G., Manonayagi, S., Padma, S., & Malini, T. 1989. Antifertility effect of bambusa arundinacea shoot extracts in male rats. *Journal of Ethnopharmacology 25*(2): 173–180.

Wedler, J., Daubitz, T., Schlotterbeck, G., & Butterweck, V. 2014. In vitro anti-inflammatory and wound-healing potential of a *Phyllostachys edulis* leaf extract – Identification of isoorientin as an active compound. *Planta Medica 80*(18): 1678–1684.

Wu, W., Hu, J., Gao, H., Chen, H., Fang, X., Mu, H., Han, Y., & Liu, R. 2020. The potential cholesterol-lowering and prebiotic effects of bamboo shoot dietary fibers and their structural characteristics. *Food Chemistry 332*: 127372.

Yakubu, M. T., & Bukoye, B. B. 2009. Abortifacient potentials of the aqueous extract of *Bambusa vulgaris* leaves in pregnant Dutch rabbits. *Contraception 80*(3): 308–313.

Yang, J. H., Choi, M.-H., Yang, S. H., Cho, S. S., Park, S. J., Shin, H.-J., & Ki, S. H. 2017. Potent anti-inflammatory and antiadipogenic properties of bamboo (*Sasa coreana* Nakai) leaves extract and its major constituent flavonoids. *Journal of Agricultural and Food Chemistry 65*(31): 6665–6673.

Yang, J.-P., He, H., & Lu, Y.-H. 2014. Four flavonoid compounds from *Phyllostachys edulis* leaf extract retard the digestion of starch and its working mechanisms. *Journal of Agricultural and Food Chemistry 62*(31): 7760–7770.

Yang, Q., Duan, Z., Wang, Z., He, K., Sun, Q., & Peng, Z. 2008. Bamboo resources, utilization and ex-situ conservation in Xishuangbanna, South-eastern China. *Journal of Forestry Research 19*(1): 79–83.

Yang, R. Y., Tsou, S. C., Lin, T. C., Hsu, P. M., & Lee, P. Y. (2005). Antioxidant capacities and daily antioxidant intake from vegetables consumed in Taiwan. Agricultural Research Institute, Council of Agriculture, 69–77.

Yuan, W., Han, X., Wu, Y., Hu, J., & Lv, Z. 2019. Composition and antimicrobial activity of bamboo (*Phyllostachys Heterocycla cv. Pubescens*) leaf hydrosols. *Current Topics in Nutraceutical Research 18*(2): 207–219.

Zhang, L., Mao, Y., Pan, J., Wang, S., Chen, L., & Xiang, J. 2017. Bamboo leaf extract ameliorates cardiac fibrosis possibly via alleviating inflammation, oxidative stress and apoptosis. *Biomedicine & Pharmacotherapy 95*: 808–817.

Zhang Ying, Z. Y., Fen Lei, F. L., Chen Xia, C. X., Wu XiaoQing, W. X., & Sheng Jian, S. J. 2000. Bamboo beer-a new kind of healthy beer. *Journal of Bamboo Research 19*(1): 33–37.

Zhao, X., Kim, S.-Y., & Park, K.-Y. 2013. Bamboo salt has *in vitro* anticancer activity in HCT-116 cells and exerts anti-metastatic effects *in vivo*. *Journal of Medicinal Food 16*(1): 9–19.

Zheng, Y., Wang, Q., Huang, J., Fang, D., Zhuang, W., Luo, X., Zou, X., Zheng, B., & Cao, H. 2019. Hypoglycemic effect of dietary fibers from bamboo shoot shell: An in vitro and in vivo study. *Food and Chemical Toxicology 127*: 120–126.

8 Medicinally Important Bamboo of *Phyllostachys* Genus

Aditi Sharma, Manu Pant, Hema Pandey, Rita Bungla, Anjali Rawat, Anshika Chauhan, and Vaibhav Painuly

8.1 INTRODUCTION

According to the Food and Agricultural Organisation of the United Nations (FAO), almost 800 million people have been affected by food and nutritional security (Da Silva, 2014). FAO is estimated to enhance food production by 60% to meet rising global demands by 2050 (Tilman, 1999). This calls for seeking alternative nutritious food supplies capable of growing in stressful environmental conditions (Chongtham et al., 2011). Bamboo serves as a potential candidate to fulfil the requirement of overcoming food and nutrition insecurity, thereby promoting various health benefits (Chand et al., 2017; Goyal and Brahma, 2014). Recent studies have observed the presence of a variety of bioactive components in bamboo leaves, including polysaccharides, amino acids, essential oils, volatile compounds, flavonoids, phenolic acids, phytosterols, etc., which impart multiple pharmacological and medicinal properties (Shen et al., 2022). The presence of these active ingredients facilitates in improving blood circulation; regulating blood lipids; protecting the liver; and operating as an anti-oxidant, anti-osteoporosis, anti-inflammatory, anti-virus, and anti-fatigue (Kimura et al., 2022).

Globally, the distribution of bamboo species can be observed in almost all the continents except Antarctica and Europe; although the introduction of certain bamboo species was observed in some parts of Europe a few years ago. They inhabit a variety of ecological regions including warm and moist tropical or warm temperate climates; further extending to cool mountainous regions, highland cloud forests, grasslands, etc. (Akinlabi et al., 2017). The distribution of bamboo species also depends on altitudinal and latitudinal diversity, ranging from sea level to 4,000 m and 46°N 47°S, respectively (Liese and Tang, 2015).

Phyllostachys is a genus of bamboo with around 80 species, distinguished not only because of their structural uses but also for their medicinal properties. *Phyllostachys* species are temperate bamboo belonging to the sub-family Bambusoideae under the Poaceae family, primarily native to China along with being introduced in Korea. The species of *Phyllostachys* hold significant agricultural value, particularly in forestry, as their culms are utilised in the wood industry. Furthermore, when they develop young shoots, they are harvested for fresh markets or as processed vegetables. Additionally, due to their attractive appearance, these species are often cultivated in gardens and parks for ornamental purposes (Nemenyi et al., 2015). Among its versatile uses, it is consumed as food and medicine and used as fuel, wood substitute, and as an anti-erosion crop (Tundis et al., 2023).

Phyllostachys bamboos have been used as a significant part of Traditional Chinese Medicine (TCM), in Ayurveda, and in regions like South Asia, Africa, and South America. Each species in this genus covers a unique profile of phytochemicals which ultimately contribute to their anti-inflammatory, anti-oxidant, anti-diabetic, and anti-cancer properties (Figure 8.1, Table 8.1). This chapter reveals the extensive medicinal applications of *Phyllostachys* bamboo species, and as promising candidates as natural alternatives and supplements for numerous health conditions.

DOI: 10.1201/9781003485933-8

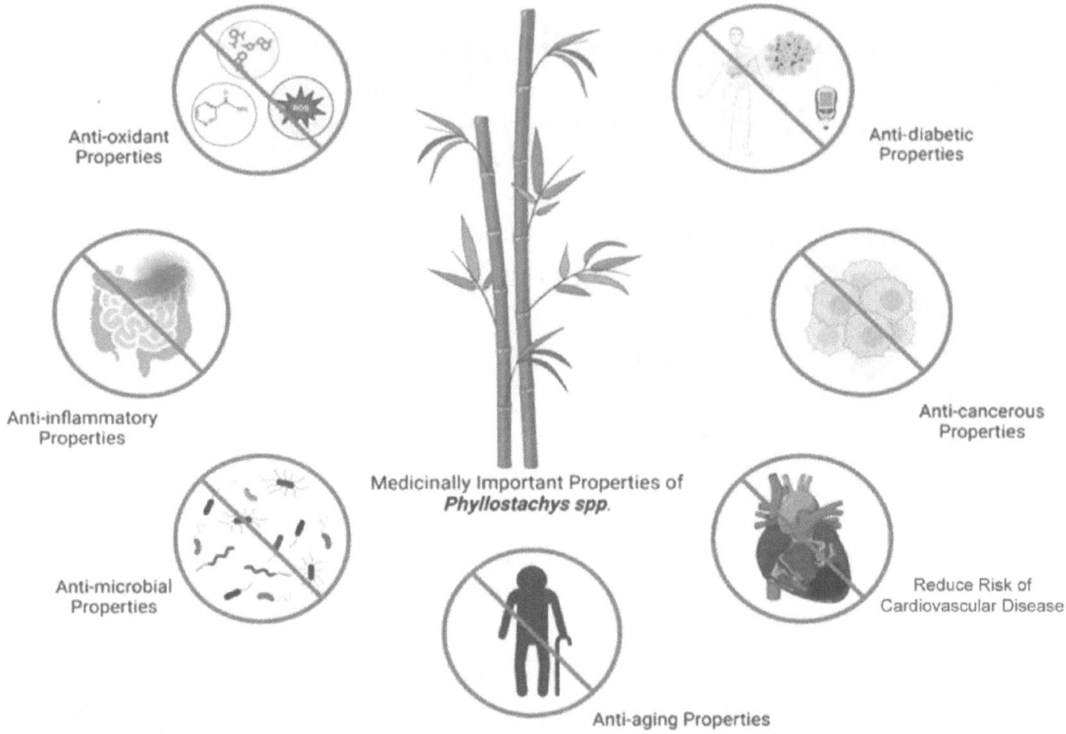

FIGURE 8.1 The illustration highlights the diverse medicinal properties of *Phyllostachys* spp., showcasing its potential health benefits.

8.2 MEDICINALLY IMPORTANT *PHYLLOSTACHYS* SPP.

This section deals with information on some of the medicinally active *Phyllostachys* bamboo species.

8.2.1 *PHYLLOSTACHYS EDULIS*

P. edulis also known as Moso bamboo, is one of the most important species of *Phyllostachys* genus. The word '*edulis*' refers to edible shoots which are large, woody, perennial, and can grow rapidly, reaching heights up to 50 meters. *P. edulis* thrives in subtropical to temperate climates, especially in regions with moderate to high rainfall. It prefers well-drained, slightly acidic to neutral soils, and can grow in both mountainous and lowland areas. It is native to China and is globally distributed including parts of East Asia such as Japan, Korea, and Taiwan. In India, *P. edulis* is cultivated in Assam, Meghalaya, and Sikkim, where climate and soil conditions are favourable (Santosh et al., 2021).

Leaf extracts of *P. edulis* and their constituents exhibit anti-inflammatory properties by blocking signalling pathways, while the flavonoids extracted from bamboo leaves stabilize the imbalance of pro-inflammatory and inflammatory cytokines. Dietary fibres present in the shoots of this bamboo species reduce fat, thus maintaining blood pressure and obesity. The shoots possess anti-cancerous properties and play an important role in the absorption of polyphenols which can easily oxidise *in vitro* and can be easily affected by gastric acids (Tang et al., 2021). Isoorientin is the most active compound present in *P. edulis* that contributes to anti-inflammation by suppressing Tumor Necrosis

TABLE 8.1
Species under *Phyllostachys* Genus with Bioactive Compounds Imparting Medicinal Properties

S. No.	Scientific Name	Common Name	Bioactive Compounds	Medicinal Properties	References
1.	*Phyllostachys acuta*	Sharp-pointed bamboo or acuta bamboo	–	High anti-oxidant activity, anti-inflammatory diseases, anti-bacterial and anti-fungal properties	Yu et al., 2024
2.	*Phyllostachys aurea*	Fish pole bamboo, golden bamboo or walking stick bamboo	Phytochemical; polyphenols; fatty acids; primary alcohols; alkyl esters; aldehydes; alkanes; tocopherols; triterpenols; triterpenyl palmitates; primary amides	Anti-oxidant, anti-cancer, anti-diabetic, anti-microbial and anti-angiogenic properties	Verma et al., 2022
3.	*Phyllostachys aureosulcata* 'Spectabilis'	Yellow groove bamboo	Scopoletin	Anti-microbial, anti-inflammatory, anti-cancerous, hepatoprotective, anti-oxidant	Wang et al., 2013
4.	*Phyllostachys bambusoides*	Giant timber bamboo, or Japanese timber bamboo	Phytosterols; lignophenol derivatives	Neuroprotective activity, Anti-inflammatory, Anti-ulcer, and Anti-oxidant properties	Akao et al., 2004; Mu et al., 2004;
5.	*Phyllostachys bissetii*	Green bamboo	Phenolic compounds	Anti-oxidant	Nemenyi et al., 2015
6	*Phyllostachys edulis*	Moso bamboo	Isoorientin; 3-O-(3′-methylcaffeoyl) quinic acid; 5-O-caffeoyl-4-methylquinic acid; 3-O-caffeoyl-1-methylquinic acid; Phyllostadimer A	Anti-oxidant and anti- inflammatory properties, inhibit liposomal lipid peroxidation	Kweon et al., 2001; Panee, 2015; Suga et al., 2003; Wedler et al., 2014
7.	*Phyllostachys flexuosa*	Sinuate bamboo	Phenolic compounds	Anti-oxidant	Nemenyi et al., 2015
8.	*Phyllostachys glauca*	Hedge bamboo or running bamboo	Isoorientin; orientin; vitexin; isovitexin; luteo-lin-6-C-arabinoside; luteolin; tricin.	Anti-oxidant	Guo et al., 2013b
9.	*Phyllostachys humilis*	Scottish bamboo	Isoorientin; farobin A; C-pentosyl-luteolin; cas siaoccidentalin B, O; C-dideoxyhexosyl-luteo lin; O-deoxyhexosyl-C-deoxyhexosylluteolin; tricin-5-O-glucoside; O-hexosyl-tricin; tricin; 5-feruloylquinic acid; sinapic acid derivative; caffeic acid; sinapoyl hexose.	Anti-oxidant	Nemenyi et al., 2015
10.	*Phyllostachys iridescens*	Evergreen Bamboo	–	Anti-oxidant	Nemenyi et al., 2015
11.	*Phyllostachys mannii*	Decora	–	Anti-oxidant	Nemenyi et al., 2015; Rana et al., 2022
12.	*Phyllostachys nidularia*	Big-node bamboo	Scopoletin	Anti-viral, Anti-tumor, Anti-osteoporosis, Anti-sepsis, and Anti-coagulation	Wang et al., 2013

(Continued)

TABLE 8.1 *(Continued)*
Species under *Phyllostachys* Genus with Bioactive Compounds Imparting Medicinal Properties

S. No.	Scientific Name	Common Name	Bioactive Compounds	Medicinal Properties	References
13.	*Phyllostachys nigra*	Black bamboo	Isoorientin; isovitexin; orientin; vitexin; trans-coniferyl alcohol; p-coumaric acid; n-feruloyl serotonin; caffeic acid ethyl ether; tricin; coumaryl alcohol; coumaric acid ethyl ether; ferulic acid ethyl ether	Cardioprotective, anti-oxidant, anti-aging, anti-bacterial, anti-free radical, DPPH radical scavenging	Fu et al., 2006; Sangeetha et al., 2015; Shang et al., 2014
14.	*Phyllostachys praecox*	Early bamboo	Polyphenols	Hypertension, arteriosclerosis, cardiovascular disease and anti-cancer properties	Wróblewska et al., 2018
15.	*Phyllostachys prominens*	Prominent bamboo	Amarusine A; 5-O-caffeoylquinic acid, 3,5,3′,5′-tetramethoxy-4-hydroxyl-(8-O-cinnamyl alcohol)-7-O-glucoside; 4, 4′, 9′-trihydroxyl-3, 5, 3′, 5′ tetramethoxy-7, 7′-monoepoxylignan-9-O glucoside; 3,5-dimethoxy-4,4′-dihydroxyl-9-O benzylacrylicester-phenylpropano-7-O glucopyranoside; luteolin-7-O-glucoside; luteolin-8-C-α-L-arabinose; isoorientin; tricin-7-O-β-D-glucoside; tricin-5-O-β-D-glucoside; isovitexin-2″-xylopyranoside; phenolic glycosides; α phenolic acid, α phenylpropanoid, and amarusine,	DPPH scavenging, removing active oxygen preventing Hemal sclerosis, improving nutrition for tissue, and preventing aging	Xu et al., 2016,
16.	*Phyllostachys pubescens*	Moso bamboo	Phenylpropanoid; coumarins; tricin; water-soluble neutral polysaccharides	Anti-fungal, anti-bacterial, anti-oxidant activity, anti-allergic, DPPH and ABTS free radical scavenging, induce apoptotic cell death in HepG2 and MCF-7 cancer, acts as a tyrosine kinase inhibitor	Fujimura et al., 2005; Ibrahim et al., 2021; Jin et al., 2011; Mu et al., 2004; Song et al., 2021; Tanaka et al., 2011; Tanaka et al., 2013; Tao et al., 2018; Tao et al., 2019; Xiao et al., 2020; Yong-Chun et al., 2012
17.	*Phyllostachys sulphurea*	Sulphur bamboo	Scopoletin	Anti-oxidant	Nemenyi et al., 2015
18.	*Phyllostachys violascens*	Evergreen bamboo	-	Anti-oxidant	Nemenyi et al., 2015
19.	*Phyllostachys viridiglaucescens*	Green-Glaucous bamboo	-	Anti-oxidant	Nemenyi et al., 2015
20.	*Phyllostachys vivax f. aureocaulis*	Giant Timber bamboo	-	Anti-oxidant	Nemenyi et al., 2015

Factor (TNF) alpha induced production of pro-inflammatory cytokines and chemokines along with acting as a therapeutic agent to treat skin diseases. *P. edulis* also plays an important role in wound healing by the formation of new collagen structures to support other cells responsible for effective wound healing (Wedler et al., 2014).

Phyllostchys heterocycla (a wild variant of *P. edulis*) is a unique ornamental bamboo with medicinal importance. The oils obtained from the leaves of three cultivars of *P. heterocycla* (viz., *Pubescens, Gracilis,* and *Heterocycla*), have proven to exhibit anti-microbial and anti-oxidant activities (Jin et al., 2011). The ethyl acetate extract of *P. heterocycla* var. *pubescens* has also been observed to present its compelling cytotoxic activity and confirmed vitality percentage of 2.14% of HepG2 cells. This extract yielded five compounds, including novel recognised compound, 7-hydroxy, 5-methoxy, methyl cinnamate. This compound was tested against HEpG2 and MCF-7 cells, depicting IC_{50} values of 7.43 and 10.65 μM, respectively. It induced a significant increase in apoptotic cell death by arresting the cell cycle at the G1 phase; up to 58.6% which was a 12.44-fold increase in comparison to 4.71% of the controlled group, making *P. heterocycla* var. *pubescens* a potential anti-cancer agent (Abdelhameed et al., 2021). The shoot extract of *P. heterocycla* has also been established as a hypoglycaemic agent, besides being a popular functional food ingredient. In a study by Liu et al. (2018), the shoots of *P. heterocycla* were observed to contain a type-II arabinogalactan-PBSS2 (a 1,3-linked β-D-galactan with branching degree of 61.1% at O-6 position), which can effectively inhibit glucose absorption in a time-dependent manner, suggesting its ability as a natural hypoglycaemic agent.

8.2.2 *Phyllostachys pubescens*

Phyllostchys pubscenes (P. pubscenes), also known as Moso bamboo or giant timber bamboo, is grown in tropical and subtropical region and is native to China (Isagi and Torii, 1997; Son et al., 2017). *P. pubescens* is rich in proteins with essential amino acids, carbohydrates, vitamins, minerals, phenolic acid, steroids, and fibres and is consumed as food in China and Southeast Asia. *P. pubescens* leaves show anti-fungal and anti-oxidant activity due to the presence of flavonoids (isoorientin, isovitexin, orientin, and vitexin), phenylpropanoids (*p*-coumaric acid and chlorogenic acid) and coumarins (skimin, scopolin, umbellifirone, psoralen, and xanthotoxin) (Milani et al., 2020; Seo and Song, 2021; Yong-Chun et al., 2012). It is also a potential source of phenols (syringic acid, *p*-hydroxybenzoic acid, and protocatechuic acid), sterols (stigmasterol, campesterol, and β-sitosterol), fibres (lignin, hemicellulose, and cellulose), fatty acids (oleinic acid, palmitic acid, linolenic acid, and linoleic acid), organic acids (malic acids, oxalic acid, and citric acid), and vitamins (ascorbic acid and tocopherol) (Lu et al., 2010; Nirmala et al., 2018; Park and Jhon, 2010). *P. pubescenes* leaves are known for their high content of flavonoids that contribute to their anti-oxidant properties as they neutralize free radicals, thus lowering oxidative stress which may further cause various chronic illnesses. *P. pubescens* serves as a valuable natural resource for boosting immune health, slowing down cellular aging, and supporting metabolic functions (Guo et al., 2013a). The extracts from the different parts of *P. pubescens* have different chemical compositions with anti-oxidant, anti-cancerous, anti-bacterial, and anti-allergic activities. The compounds present in the extract also show cytotoxic activity selectively towards cancer and apoptotic investigation (Abdelhameed et al., 2021). The oil extracted from the shoot of the *P. pubescens* shows hypolipidemic activity by reducing relative liver weight by 27.2%, atherogenic index and level of hepatic lipid, triglyceride by 6.7%, low density cholesterol by 41.0%, and total cholesterol by 26.8% that is due to the high content of unsaturated fatty acid and phytosterol which competes with exogenous cholesterol for absorption by brush border thus increasing the excretion of cholesterol (Lu et al., 2010). Additionally, *P. pubescens* shoot skins have shown anti-bacterial activity when the dichloromethane-soluble fraction of methanol extracts from these shoot skins were analysed, showing a strong effectiveness against *Staphylococcus aureus* and moderate effectiveness against *Escherichia coli*. Key active compounds contributing to antibacterial properties were recognized as beta-sitosterol, stigmasterol, and di-hydrobrassicasterol which can successfully inhibit bacterial proliferation. The anti-bacterial capabilities of these phytosterols were associated with their distinct side-chain structures (Tanaka et al., 2013). Another study showed a potent anti-hypertensive effect of *P. pubescens* as its methanol extract was examined for angiotensin-converting enzyme inhibition test where the inhibition of the angiotensin enzyme was observed at the concentration of 3.5 mg/mL indicating anti-hypertensive effect in dose-dependent manner (Park and Jhon, 2010).

8.2.3 *Phyllostachys nigra*

Phyllostachys nigra (P. nigra) is a perennial grass, also known as black bamboo. It is widely spread and grown in Southeast Asia and has been utilised medicinally for ages (Zhang et al., 2002). All parts of the species, like leaves, stems, and roots, are sources of various phytoconstituents like sterols, flavonoids, phenols, etc. which make it a potential source for medicinal exploitation (Singh et al., 2013). *P. nigra* extract obtained from the inner bark was examined on human immortalized keratinocyte (HaCaT) and human mast cell (HMC-1) cell line using ELISA and PCR techniques. Upon investigation, it resulted in downregulating the expression of inflammatory chemokine: Interleukin 6 (IL-6), Monocyte Chemoattractant Protein-1 (MCP-1), Tumor Necrosis Factor (TNF-α), Granulocyte-Macrophage Colony-Stimulating Factor (GM-CSF), and Thymus and Activation-Regulated Chemokine (TARC). Thus, the level of IgE was reduced, inhibiting the invasion of $CD4^+$, $CD8^+$ T cells, eosinophils, and mast cells which make it a potent and effective alternative medication for allergic inflammatory disease (Kim et al., 2024). The polysaccharide-PNS2A was extracted from the roots of *P. nigra* which reduced body weight by 13.2% and blood glucose by 44.5% in the diabetic rat. Also, it increased the insulin sensitivity, upregulating the synthesis pathways – hexoglycolysis, palmitate biosynthesis pathway, and degradation pathways – glutaryl-CoA and toluene degradation pathways. Additionally, it affected the gut microbiome by improving the abundance of *Lactobacillus* while reducing the amount of *Clostridia* UCG-014 and *Escherichia shigella* (Zhao et al., 2024). This species also possesses a strong anti-oxidant property as its ethanolic extract constitutes bioactive compounds like flavonoids – *p*-coumaric acid, caffeic acid, chlorogenic acid, and catechin which help in the scavenging of free radicals with an IC_{50} value of 565.63 ± 17.75 μ/mL in a 2,2-diphenyl-1-picrylhydrazyl (DPPH) test (Choi et al., 2018). In another set of studies, *P. nigra* was observed to show an anti-hypertensive effect in a dose-dependent manner when its methanolic extract was examined for inhibition of the angiotensin-converting enzyme (ACE) (Park and Jhon, 2010). *P. nigra* extracts boost the immune system against various infections, along with imparting anti-cancer properties, maintain hormonal imbalance, and cognitive enhancing properties, and contribute to improved memory. It also shows anti-allergic effects, reducing symptoms such as sneezing and itching. A brown-yellow powder extracted from the leaves of *P. nigra* is effective due to anti-oxidant, anti-aging, anti-bacterial, and anti-free radical properties (Sangeetha et al., 2015). *P. nigra* is popular in traditional treatment of respiratory disorders like asthma and bronchitis. The leaves show anti-pyretic and diuretic activity which are used for the treatment of fever, vomiting, nosebleeds, and can be taken internally for the treatment of lung infection with cough. A study showed that the leaf extract of *P. nigra* inhibits the NF-kB induced gene expression, exhibiting *in-vitro* anti-inflammatory properties (Gagliano et al., 2022). *P. nigra* species has been observed to impart anti-oxidant and anti-hypertensive properties by inhibiting ACE; also, the leaves of *P. nigra* var. *Henonis* bamboo have been utilised as a source of phenolics and flavonoids exhibiting antioxidant activity (Park and Jhon, 2010). Also, the flavone C-glucosides (orientin) have been extracted from the leaves of *P. nigra* responsible for cardioprotective activity and also block the mitochondrial apoptotic pathway thus inhibiting apoptosis (Fu et al., 2006; Zhang et al., 2008).

8.2.4 *Phyllostachys bambusoides*

Phyllostachys bambusoides, commonly known as giant timber bamboo or Japanese timber bamboo, is found in the deciduous and semi-evergreen regions of northeast India and the tropical moist deciduous forest of north and southeast India hence shows a diversified habit and habitat. They occur in different forests ranging from tropical to sub-alpine zones and are native to China but are also grown in some states of India (Nirmala et al., 2018). Shoots of *P. bambusoides* are very popular vegetables in Southeast Asian countries, while the leaves have been consumed in the form of tea for more than 1,000 years as a nutritional source and folk medicine. The leaves of this bamboo are edible with a proximate amount of vitamins, amino acids, minerals, and bioactive components but are low in fat content. The medicinal properties of *P. bambusoides* have also been explored in some research studies as its leaf and stem extracts have been observed to inhibit the process of adipogenesis in 3T3-L1 adipocytes, which are commonly used to study fat cell differentiation and play a significant role in controlling obesity (Goh et al., 2019; Kwon et al., 2017). In another set of studies, the ethyl acetate fraction (PB-EtAC) has been extracted from the leaves of *P. bambusoides* on the immune system of BALB/c mice. This indicates the administration of PB-EtAC significantly increased the titers of both IgM and IgG antibodies, generating a humoral immune response along with increasing delayed-type hypersensitivity reactions promoting enhanced cell-mediated immunity. Also, the increased levels of IFN-γ, IL-2, and IL-4 cytokines were observed to be associated with the increased phagocytic activity of macrophages and production of nitric oxide invoking both Th1 and Th2 immune responses (Kumar et al., 2014). *P. bambusoides* is also rich in various bioactive phenolic compounds, including flavonoids (luteolin, apigenin, and kaempferol), phenolic acids (caffeic acid, ferulic acid, and gallic acid), and tannins (chlorogenic acid, *p*-coumaric acid, protocatechuic acid, vanillic acid, and syringic acid). The presence of these phenolic compounds is responsible for the therapeutic potential of *P. bambusoides,* thus helping in managing diseases related to oxidative stress, inflammation, infections, and chronic conditions such as diabetes, neurodegeneration, and cancer (Kim et al., 2011; Sharma et al., 2018).

8.2.5 *Phyllostachys acuta*

Phyllostachys acuta, also known as sharp-pointed bamboo or acute bamboo, is native to China. It is characterized by its sharp-pointed shoots, culms with a diameter of 2–4 cm, height ranging from 5–10 meters, and dark green leaves which are 10–20 cm long (Li et al., 2012).

P. acuta has been traditionally used in Chinese medicine for various health benefits including anti-oxidant activity confirmed through modern research activities. The therapeutic potential of *P. acuta* has been observed in various research studies as different monosaccharides and polysaccharides have been extracted from the leaves of *P. acuta* using microwave-assisted extraction. The extracted polysaccharides have enhanced the proliferation of HepG2 cells which promote extensive anti-oxidant activities in the diphenyl picrylhydrazyl radical potential DPPH as well as the ferric-reducing antioxidant potential (FRAP) assay indicating anti-oxidant activity of *P. acuta* leaves (Zhang et al., 2016). Another investigation on *P. acuta* by Yu et al. (2024) highlights the therapeutic significance of feruloylated oligosaccharides (FOs) extracted from the hemicellulose of this bamboo. These extracted FOs effectively neutralized free radicals, including DPPH, superoxide anion, and hydroxyl radicals, while protecting cells from oxidative stress. Also, these compounds enhanced the functionality of anti-oxidant enzymes such as total antioxidant capacity (T-AOC) and superoxide dismutase (SOD), along with lowering the levels of reactive oxygen species (ROS).

8.2.6 *Phyllostachys aurea*

Phyllostachys aurea (*P. aurea*), commonly known as fish pole, golden, or walking stick bamboo, is a hardy bamboo distributed in subtropical and temperate coastal regions of Asia, America, and Africa. *P. aurea* is mainly used for amenity purposes as their flowering is rare (Miller et al., 2019).

This bamboo exhibits specific characteristics such as rapid growth, soil binding, erosion control, adaptive capability, and nutrient and water conservation properties (Nfornkah et al., 2023). The canes and leaves of this species are used in medicines and their young shoots are edible (Yormann et al., 2020). *P. aurea* leaves have been observed to synthesize silver nanoparticles (AgNPs) used for analysing anti-microbial activity against *S. aureus* and *E. coli* using the disc diffusion method. The growth of the microbes decreased with increasing concentrations of AgNPs, killing 1×10^7 CFU/mL concentration of bacteria at its lowest concentration of 20 μg/mL (Yasin et al., 2013).

8.2.7 *Phyllostachys praecox*

Phyllostachys praecox, commonly known as early bamboo, is native to East and South China and prefers moist fertile, well-drained soil for their growth. *P. praecox* has mid-green leaves and yellow/green canes of lengths up to 6 cm. The shoots are edible and tasteful, making them a popular edible bamboo (Qian et al., 2021). The major components of this delicious and healthy bamboo are cellulose, hemicellulose, and lignin which are used as natural additives in the food industry in powdered form (Yang et al., 2021). The PBSR1 polysaccharide has been extracted from *P. praceox* bamboo shoot waste which is a triple-helix galactan with 863 KD molecular weight. It showed anti-oxidant properties and immunomodulatory activity in a dose-dependent manner. It resulted in a 0.65 mg/mL EC_{50} value and highest scavenging activity of 89.2% at a concentration of 1.2 mg/mL during DPPH scavenging analysis. Additionally, evaluation through ORAC capacity resulted in 329.46 ± 12.1 μmol TE/g ORAC value which indicates that when exposed to the environment, releases the hydroxyl group which either terminates the chemical reaction or easily provides hydrogen to stabilize the free radicals, thus contributing to its anti-oxidant property (Huang et al., 2023). When high-fat diet mice were administered fresh *P. praecox* shoots powder principally consisting of cellulose, hemicellulose, and lignin (dietary fiber), it mitigated and attenuated the obesity and fatty liver disease by improving the level of triglyceride, cholesterol, and LDL-C levels that are the principal components in obesity and fatty liver disease. Also, it improved the gut microbiota by increasing the abundance of microbes like *Verrucomicrobia* and *Akkermansia* which are known to relieve obesity (Precup and Vodnar, 2019; Yang et al., 2021). Similar findings were observed when soluble dietary fiber was extracted and administered to hyperlipidemic mice which reduced the content of LDL, triglyceride, and cholesterol by 39.02%, 19.54%, and 28.33% while increasing the high-density lipoprotein cholesterol by 39.50% (Wang et al., 2019).

8.2.8 *Phyllostachys sulphurea*

Phyllostachys sulphurea, also known as sulphur bamboo, is an evergreen bamboo which grows up to 6 m, is erect, and has woody culms 30–40 mm in diameter with thin-walled internodes, and elongated rhizomes. *P. sulphurea* is used in traditional medicine to treat fever, cough, and digestive issues. The shoots of *P. sulphurea* are edible and used in Asian cuisine, while the plant is also used as an ornamental grass (Nemenyi et al., 2015). The flavonoid extract of *P. sulphurea,* isolated from its leaves, has shown extensive anti-oxidant properties due to the presence of 2,2′-azino-bis(3-ethylbenzothiazoline-6-sulfonic acid) (ABTS) free radical scavenging activity (Ouyang et al., 2019). Also, the roots of *P. sulphurea* have been consumed orally to effectively treat internal infections (Jaya et al., 2019). In a study by Dong et al. (2023), the highly soluble dietary fibres were isolated from *P. sulphurea* which imparted potential hypoglycemic activity.

8.2.9 *Phyllostachys prominens*

Phyllostachys prominens, commonly known as giant timber bamboo, is widely distributed in the south of China while its shoots are utilised as a rich, edible source. There are about 14 compounds isolated from *P. prominens,* including six flavonoids, two lignin, two phenolic glycosides, a phenolic acid, a phenylpropanoid, and amarusine, which show anti-oxidant properties preventing Hemal sclerosis, improving nutrition for tissue, and preventing aging. The anti-oxidant properties of *P. prominens* leaves have also been reported through these 14 distinct compounds, exhibiting radical scavenging activity (Xu et al., 2016).

8.2.10 *Phyllostachys glauca*

The leaves of *Phyllostachys glauca* McClure (PML), commonly known as hedge or running bamboo, have been observed to be a natural anti-oxidant source and are rich in flavonoids and phenolic acids (Fei et al., 2023; Guo et al., 2013b). The leaf extract is utilised in the form of functional food additives and extensively used in commercial products like edible oils, silicon tea, fruit and vegetable juice, bakery products, cereal, and meat products. Efforts have also been made to improve the flavour of these natural products made out of *P. glauca* leaf extract through probiotic fermentation (Mandha et al., 2021).

8.3 CONCLUSION

Several bamboo species have traditionally been consumed as food and medicine in parts of China and northeastern India. *Phyllostachys* is among the nutraceutically rich genera of bamboo, recognised for its traditional uses as medicinal grass. Species within this genus are abundant in biomolecules and phytochemicals, contributing to their potential as therapeutic sources. Despite this, multiple bioactive compounds and other medicinally important species remain unexplored. There is a vital prerequisite to explore the pharmacological properties, mechanisms, and effects of other bamboo species. However, limitations such as the presence of cyanogenic glycosides constrain their commercial usage. Furthermore, it is crucial to characterise the preparation techniques and establish sustainable utility to improve commercial utilisation of these bamboo at the global level.

REFERENCES

Abdelhameed, R.F., Habib, E.S., Ibrahim, A.K., Yamada, K., Abdel-Kader, M.S., Ahmed, S.A., Ibrahim, A.K., Badr, J.M. and Nafie, M.S., 2021. Chemical constituent profiling of *Phyllostachys heterocycla* var. *Pubescens* with selective cytotoxic polar fraction through EGFR inhibition in HepG2 cells. *Molecules*, *26*(4), 940.

Akao, Y., Seki, N., Nakagawa, Y., Yi, H., Matsumoto, K., Ito, Y., Ito, K., Funaoka, M., Maruyama, W., Naoi, M. and Nozawa, Y., 2004. A highly bioactive lignophenol derivative from bamboo lignin exhibits a potent activity to suppress apoptosis induced by oxidative stress in human neuroblastoma SH-SY5Y cells. *Bioorganic & Medicinal Chemistry*, *12*(18), 4791–4801.

Akinlabi, E.T., Anane-Fenin, K., Akwada, D.R., Akinlabi, E.T., Anane-Fenin, K. and Akwada, D.R., 2017. Bamboo taxonomy and distribution across the globe. *Bamboo: The Multipurpose Plant*, 1–37.

Chand, R., Singh, A.N. and Nirmala, C., 2017. Ethnoecological survey of underutilized plant diversity of Hamirpur district, Himachal Pradesh, India: An edibility assessment. *Environment and Ecology Research*, *5*(1), 13–29.

Choi, M.H., Jo, H.G., Yang, J.H., Ki, S.H. and Shin, H.J., 2018. Antioxidative and anti-melanogenic activities of bamboo stems (*Phyllostachys nigra* variety henosis) via PKA/CREB-mediated MITF downregulation in B16F10 melanoma cells. *International Journal of Molecular Sciences*, *19*(2), 409.

Chongtham, N., Bisht, M.S. and Haorongbam, S., 2011. Nutritional properties of bamboo shoots: Potential and prospects for utilization as a health food. *Comprehensive Reviews in Food Science and Food Safety*, *10*(3), 153–168.

Da Silva, J.G., 2014. Don't just produce more food-produce better food. *National Geographic Special Series Future Food*, *5*, 1–2.

Dong, Y., Li, Q., Guo, Y., Zhao, Y. and Cao, J., 2023. Comparison of physicochemical and in vitro hypoglycemic activity of bamboo shoot dietary fibers from different regions of Yunnan. *Frontiers in Nutrition*, *9*, 1102671.

Fei, Z., Xie, D., Wang, M., Zhang, Y., Zhang, H., Du, Q. and Jin, P., 2023. Enhanced biotransformation of bioactive components and volatile compounds of bamboo (*Phyllostachys glauca* McClure) leaf juice fermented by probiotic streptococcus thermophiles. *LWT, 173*, 114363.

Fu, X.C., Wang, M.W., Li, S.P. and Wang, H.L., 2006. Anti-apoptotic effect and the mechanism of orientin on ischaemic/reperfused myocardium. *Journal of Asian Natural Products Research, 8*(3), 265–272.

Fujimura, M., Ideguchi, M., Minami, Y., Watanabe, K. and Tadera, K., 2005. Amino acid sequence and antimicrobial activity of chitin-binding peptides, Pp-AMP 1 and Pp-AMP 2, from Japanese bamboo shoots (*Phyllostachys pubescens*). *Bioscience, Biotechnology, and Biochemistry, 69*(3), 642–645.

Gagliano, J., Anselmo-Moreira, F., Sala-Carvalho, W.R. and Furlan, C.M., 2022. What is known about the medicinal potential of bamboo? *Advances in Traditional Medicine, 22*(3), 467–495.

Goh, S., Kim, D., Choi, M.H., Shin, H.J. and Kwon, S., 2019. Effects of bamboo stem extracts on adipogenic differentiation and lipid metabolism regulating genes. *Biotechnology and Bioprocess Engineering, 24*, 454–463.

Goyal, A.K. and Brahma, B.K., 2014. Antioxidant and nutraceutical potential of bamboo: An overview. *International Journal of Fundamental and Applied Sciences (IJFAS), 3*(1), 2–10.

Guo, X.F., Yue, Y.D., Tang, F., Wang, J. and Yao, X. I., 2013a. Antioxidant properties of major flavonoids and subfractions of the extract of *Phyllostachys pubescens* leaves. *Journal of Food Biochemistry, 37*(4), 501–509.

Guo, X.-F., Yue, Y.-D., Tang, F., Wang, J., Yao, X. and Sun, J., 2013b. Simultaneous determination of seven flavonoids in Dan bamboo *Phyllostachys glauca McClure* leaf extract and in commercial products by HPLC-DAD: A new determination method of flavonoids. *Journal of Food Biochemistry, 37*, 748–757.

Huang, X., Zhang, Y., Xie, N., Cheng, J., Wang, Y., Yuan, S., Li, Q., Shi, R., He, L. and Chen, M., 2023. Molecular characterization and bioactivities of a novel polysaccharide from *Phyllostachys pracecox* bamboo shoot residues. *Foods, 12*(9), 1–20.

Ibrahim, A.K., Mehanna, E.T., Abdelhameed, R.F., Habib, E.S., Yamada, K., Ahmed, S.A., Ibrahim, A.K. and Badr, J.M., 2021. Anti-inflammatory properties of the crude extract of *Phyllostachys heterocycla* in two different models of acute inflammation in experimental rats. *Records of Pharmaceutical and Biomedical Sciences, 5*(Pharmacognosy-Microbiology), 53–59.

Isagi, Y. and Torii, A., 1997. Range expansion and its mechanisms in a naturalized bamboo species, *Phyllostachys pubescens* in Japan. *Journal of Sustainable Forestry, 6*, 127–141.

Jaya, A.M., Musa, Y., Iswoyo, H., Asmi, N. and Siregar, L.F., 2019 October. Ethnobotanical study and identification of medicinal plants based on local knowledge. In *IOP Conference Series: Earth and Environmental Science*, IOP Publishing, 343(1): 012028.

Jin, Y.C., Yuan, K. and Zhang, J., 2011. Chemical composition, and antioxidant and antimicrobial activities of essential oil of *Phyllostachys heterocycla* cv. Pubescens varieties from China. *Molecules, 16*(5), 4318–4327.

Kim, E.Y., Kim, S.B., Kim, E.J., Kim, M., Ryu, G.H., Shin, H., Lim, S.E., Sohn, Y. and Jung, H.S., 2024. *Phyllostachys nigra* Munro var alleviates inflammatory chemokine expression and DNCB-induced atopic-like dermatitis in BALB/c mice. *Journal of Ethnopharmacology, 318*, 116953.

Kim, H.J., Kang, M.A., Kim, S.H., Yim, S.H. and Lee, I.S., 2011. Bioactive phenolic constituents from the culms of *Phyllostachys bambusoides. Natural Product Sciences, 17*(4), 267–272.

Kimura, I., Kagawa, S., Tsuneki, H., Tanaka, K. and Nagashima, F., 2022. Multitasking bamboo leaf-derived compounds in prevention of infectious, inflammatory, atherosclerotic, metabolic, and neuropsychiatric diseases. *Pharmacology & Therapeutics, 235*, 108159.

Kumar, S., Sharma, G., Sidiq, T., Khajuria, A., Jain, M., Bhagwat, D. and Dhar, K.L., 2014. Immunomodulatory potential of a bioactive fraction from the leaves of *Phyllostachys bambusoides* (Bamboo) in BALB/C mice. *EXCLI Journal, 13*, 137.

Kweon, M.H., Hwang, H.J. and Sung, H.C., 2001. Identification and antioxidant activity of novel chlorogenic acid derivatives from bamboo (*Phyllostachys edulis*). *Journal of Agricultural and Food Chemistry, 49*(10), 4646–4655.

Kwon, J.H., Hwang, S.Y. and Han, J.S., 2017. Bamboo (*Phyllostachys bambusoides*) leaf extracts inhibit adipogenesis by regulating adipogenic transcription factors and enzymes in 3T3-L1 adipocytes. *Food Science and Biotechnology, 26*, 1037–1044.

Li, M.F., Sun, S.N., Xu, F. and Sun, R.C., 2012. Formic acid based organosolv pulping of bamboo (*Phyllostachys acuta*): Comparative characterization of the dissolved lignins with milled wood lignin. *Chemical Engineering Journal, 179*, 80–89.

Liese, W., & Tang, T. K. H., 2015. Properties of the bamboo culm. In *Bamboo: the plant and its uses.* Cham: Springer International Publishing. 227–256.

Liu, H., He, P., He, L., Li, Q., Cheng, J., Wang, Y., Yang, G. and Yang, B., 2018. Structure characterization and hypoglycemic activity of an arabinogalactan from *Phyllostachys heterocycla* bamboo shoot shell. *Carbohydrate Polymers*, *201*, 189–200.

Lu, B., Xia, D., Huang, W., Wu, X., Zhang, Y. and Yao, Y., 2010. Hypolipidemic effect of bamboo shoot oil (*P. Pubescens*) in Sprague–Dawley rats. *Journal of Food Science*, *75*(6), 205–211.

Mandha, J., Shumoy, H., Devaere, J., Schouteten, J.J., Gellynck, X., de Winne, A., Matemu, A.O. and Raes, K., 2021. Effect of lactic acid fermentation of watermelon juice on its sensory acceptability and volatile compounds. *Food Chemistry*, *358*, 129809.

Milani, G., Curci, F., Cavalluzzi, M.M., Crupi, P., Pisano, I., Lentini, G., Clodoveo, M.L., Franchini, C. and Corbo, F., 2020. Optimization of microwave-assisted extraction of antioxidants from bamboo shoots of *Phyllostachys pubescens*. *Molecules*, *25*(1), 215.

Miller, E., Rogers, S. and Mortimer, M., 2019. Flowering of golden bamboo (*Phyllostachys aurea*) in New Zealand. *New Zealand Garden Journal*, *22*(1), 5–9.

Mu, J., Uehara, T., Li, J. and Furuno, T., 2004. Identification and evaluation of antioxidant activities of bamboo extracts. *Forestry Studies in China*, *6*, 1–5.

Nemenyi, A., Stefanovitsne-Banyai, E., Zoltán, P.K., Hegedűs, A., Gyuricza, C., Barocsi, Z. and Helyes, L., 2015. Total antioxidant capacity and total phenolics content of *Phyllostachys* taxa shoots. *Notulae Botanicae Horti Agrobotanici Cluj-Napoca*, *43*(1), 64–69.

Nfornkah, B.N., Nath, A.J., Kaam, R., Chimi, C.D. and Mezafack, K.L., 2023. Bamboo-based forest landscape restoration: Practical lessons and initiatives to upscale in Africa. In: *Bamboo Science and Technology*. Springer Nature Singapore, Singapore, 329–356.

Nirmala, C., Bisht, M.S., Bajwa, H.K. and Santosh, O., 2018. Bamboo: A rich source of natural antioxidants and its applications in the food and pharmaceutical industry. *Trends in Food Science & Technology*, *77*, 91–99.

Ouyang, W.L., Lei, F.H., Yang, Y.J., Liu, L.L., Li, Q.Q. and Guo, A.W., 2019. Analysis of nutritional components of four bamboo leaves and antioxidant activity of flavonoid extracts in vitro. *Natural Product Research and Development*, *31*(10), 1669–1674.

Panee, J., 2015. Potential medicinal application and toxicity evaluation of extracts from bamboo plants. *Journal of Medicinal Plant Research*, *9*(23), 681.

Park, E.J. and Jhon, D.Y., 2010. The antioxidant, angiotensin converting enzyme inhibition activity, and phenolic compounds of bamboo shoot extracts. *LWT-Food Science and Technology*, *43*(4), 655–659.

Precup, G. and Vodnar, D.C., 2019. Gut *Prevotella* as a possible biomarker of diet and its eubiotic versus dysbiotic roles: A comprehensive literature review. *British Journal of Nutrition*, *122*(2), 131–140.

Qian, Z., Sun, X., Gao, J. and Zhuang, S., 2021. Effects of bamboo (*Phyllostachys praecox*) cultivation on soil nitrogen fractions and mineralization. *Forests*, *12*(8), 1109.

Rana, K.R., Chongtham, N. and Bisht, M.S., 2022. Evaluation of proximate composition, vitamins, amino acids, antioxidant activities and bioactive compounds of young edible bamboo (*Phyllostachys mannii* Gamble). *Current Research in Nutrition and Food Science Journal*, *10*(1), 321–333.

Sangeetha, R., Diea, Y.K.T., Chaitra, C., Malvi, P.G. and Shinomol, G.K., 2015. The amazing bamboo: A review on its medicinal and pharmacological potential. *Indian Journal of Nutrition*, *2*(1), 1–7.

Santosh, O., Bajwa, H.K., Bisht, M.S. and Chongtham, N., 2021. Application of bamboo in the food and pharmaceutical industry. In: *Biotechnological Advances in Bamboo*. Springer, Singapore, 401–429.

Seo, C.S. and Song, K.H., 2021. Phytochemical characterization for quality control of *Phyllostachys pubescens* Leaves using high-performance liquid chromatography coupled with diode array detector and tandem mass detector. *Plants*, *11*(1), 50.

Shang, Y.F., Kim, S.M. and Um, B.H., 2014. Optimisation of pressurised liquid extraction of antioxidants from black bamboo leaves. *Food Chemistry*, *154*, 164–170.

Sharma, V., Devi, T.P., Nirmala, C. and Bisht, M.S., 2018. Bamboo shoots: An untapped source of essential nutrients and bioactive compounds. *Third International Conference on Research Development Application Science Engineering and Management*, *7(8)*, 247–255.

Shen, D.Y., Song, H.L., Zou, T.T., Wan, S.Y. and Li, M.K., 2022. Characterization of odor-active compounds in Moso bamboo (*Phyllostachys pubescens* Mazel) leaf via gas chromatography-ion mobility spectrometry, one-and two-dimensional gas chromatography-olfactory-mass spectrometry, and electronic nose. *Food Research International*, *152*,110916.

Singh, R., Monga, A.K. and Souravh, B., 2013. *Phyllostachys nigra* (Lodd.) Munro: A systematic review. *Inventi Rapid: Planta Activa*, *2*, 1–9.

Son, E., Kim, S.H., Yang, W.K., Kim, D.S. and Cha, J., 2017. Antiplatelet mechanism of an herbal mixture prepared from the extracts of *Phyllostachys Pubescens* leaves and *Prunus Mume* fruits. *BMC Complementary and Alternative Medicine*, *17*(1), 541–552.

Song, K.H., Seo, C.-S., Yang, W.-K., Gu, H.-O., Kim, K.-J. and Kim, S.-H., 2021. Extracts of *Phyllostachys pubescens* leaves represses human steroid 5-alpha reductase type 2 promoter activity in BHP-1 cells and ameliorates testosterone-induced benign prostatic hyperplasia in rat model. *Nutrients*, *13*, 1–13.

Suga, A., Takaishi, Y., Goto, S., Munakata, T., Yamauchi, I. and Kogure, K., 2003. Two lignan dimers from bamboo stems (*Phyllostachys edulis*). *Phytochemistry*, *64*(5), 991–996.

Tanaka, A., Kim, H.J., Oda, S., Shimizu, K. and Kondo, R., 2011. Antibacterial activity of Moso bamboo shoot skin (*Phyllostachys pubescens*) against Staphylococcus aureus. *Journal of Wood Science*, *57*, 542–544.

Tanaka, A., Shimizu, K. and Kondo, R., 2013. Antibacterial compounds from shoot skins of Moso bamboo (*Phyllostachys pubescens*). *Journal of Wood Science*, *59*, 155–159.

Tang, Y.M., Li, A.P., Xiao, J.P., Li, D.Y. and Wang, L., 2021. Effects of bamboo shoots (*Phyllostachys edulis*) dietary fibers prepared by different processes on the adsorption characteristics of polyphenols. *Journal of Food Biochemistry*, *45*(5), 13721.

Tao, C., Wang, Y., Zhang, X., Li, L., Wu, Y., Han, X., Jiang, X. and Lv, Z., 2019. Mechanism of action of essential oils extracted from bamboo (*Phyllostachys heterocycla* cv. Pubescens) leaves: Chemical composition and antimicrobial activity against four food-related microorganisms. *BioResources*, *14*(1), 1419–1434.

Tao, C., Wu, J., Liu, Y., Liu, M., Yang, R. and Lv, Z., 2018. Antimicrobial activities of bamboo (*Phyllostachys heterocycla* cv. Pubescens) leaf essential oil and its major components. *European Food Research and Technology*, *244*, 881–891.

Tilman, D., 1999. Global environmental impacts of agricultural expansion: The need for sustainable and efficient practices. *Proceedings of the National Academy of Sciences*, *96*(11), 5995–6000.

Tundis, R., Augimeri, G., Vivacqua, A., Romeo, R., Sicari, V., Bonofiglio, D. and Loizzo, M.R., 2023. Anti-inflammatory and antioxidant effects of leaves and sheath from bamboo (*Phyllostachys edulis* J. Houz). *Antioxidants*, *12*(6), 1239.

Verma, S., Mili, M., Hada, V., Kalita, P., Hashmi, S.A.R., Rathore, S.K.S. and Srivastava, A.K., 2022. Bamboo leaves: Sustainable feedstock in pharmaceuticals, food, and agricultural sector. *Environmental Technology Reviews*, *11*(1), 266–281.

Wang, C.H., Ni, Z.J., Ma, Y.L., Thakur, K., Zhang, J.G., Prasad, C. and Wei, Z.J., 2019. Antioxidant and hypolipidemic potential of soluble dietary fiber extracts derived from Bamboo shoots (*Phyllostachys praecox*). *Current Topics in Nutraceutical Research*, *17*(2), 195–205.

Wang, S., Tang, F., Yue, Y., Yao, X., Wei, Q. and Yu, J., 2013. Simultaneous determination of 12 coumarins in bamboo leaves by HPLC. *Journal of AOAC International*, *96*(5), 942–946.

Wedler, J., Daubitz, T., Schlotterbeck, G. and Butterweck, V., 2014. In vitro anti-inflammatory and wound-healing potential of a *Phyllostachys edulis* leaf extract–identification of isoorientin as an active compound. *Planta Medica*, *80*(18), 1678–1684.

Wróblewska, K.B., de Oliveira, D.C., Grombone-Guaratini, M.T. and Moreno, P.R.H., 2018. Medicinal properties of bamboos. *Pharmacognosy-Medicinal Plants*, 1–18.

Xiao, Z., Zhang, Q., Dai, J., Wang, X., Yang, Q., Cai, C., Mao, J. and Ge, Q., 2020. Structural characterization, antioxidant and antimicrobial activity of water-soluble polysaccharides from bamboo (*Phyllostachys pubescens* Mazel) leaves. *International Journal of Biological Macromolecules*, *142*, 432–442.

Xu, X.B., Jiang, H., Sun, J., Tang, F., Guo, X.F. and Wang, J., 2016. Chemical constituents and antioxidant properties of *Phyllostachys prominens* Gramineae (WY Xiong) leaf extracts. *Tropical Journal of Pharmaceutical Research*, *15*(3), 569–575.

Yang, J., Wu, L., Yang, H. and Pan, Y., 2021. Using the major components (cellulose, hemicellulose, and lignin) of *Phyllostachys praecox* bamboo shoot as dietary fiber. *Frontiers in Bioengineering and Biotechnology*, *9*, 1–8.

Yasin, S., Liu, L. and Yao, J., 2013. Biosynthesis of silver nanoparticles by bamboo leaves extract and their antimicrobial activity. *Journal of Fiber Bioengineering and Informatics*, *6*(6), 77–84.

Yong-Chun, J., Lin, Y. and Yuan, K., 2012. A novel high-performance liquid chromatography fingerprint approach to discriminate *Phyllostachys pubescens* from China. *Pharmacognosy Magazine*, *8*(29), 42.

Yormann, G.E., Rúgolo, Z.E. and Apóstolo, N.M., 2020. Culms of *Bambusa vulgaris* cv. vittata, *Arundinaria simonii* and *Phyllostachys aurea* (Poaceae, Bambusoideae): Characterization of vascular bundles and fibers. *Flora*, *263*, 151523.

Yu, X., Cao, H., Liu, X., Liu, J., Lyu, Y., Wang, D. and Wei, M., 2024. Isolation and characterization of feruloylated oligosaccharides from *Phyllostachys acuta* and in vitro antioxidant activity. *Carbohydrate Research*, *542*, 109193.

Zhang, X., Li, M., Zhong, L., Peng, X. and Sun, R., 2016. Microwave-assisted extraction of polysaccharides from bamboo (*Phyllostachys acuta*) leaves and their antioxidant activity. *BioResources*, *11*(2), 5100–5112.

Zhang, Y., Jiao, J., Liu, C., Wu, X. and Zhang, Y., 2008. Isolation and purification of four flavone C-glycosides from antioxidant of bamboo leaves by macroporous resin column chromatography and preparative high-performance liquid chromatography. *Food Chemistry*, *107*(3), 1326–1336.

Zhang, Y., Wu, X.Q. and Yu, Z.Y., 2002. Comparison study on total flavonoid content and anti-free redical activity of the leaves of bamboo, *Phyllostachys* nigra, and *Ginkgo bilabo*. Zhongguo Zhong yao za zhi= Zhongguo Zhongyao Zazhi= *China Journal of Chinese Materia Medica*, *27*(4), 254–257.

Zhao, K., Wu, X., Han, G., Sun, L., Zheng, C., Hou, H., Xu, B.B., El-Bahy, Z.M., Qian, C., Kallel, M. and Algadi, H., 2024. *Phyllostachys nigra* (Lodd. ex Lindl.) derived polysaccharide with enhanced glycolipid metabolism regulation and mice gut microbiome. *International Journal of Biological Macromolecules*, *257*, 128588.

9 The Health Benefits of Biosilica from Bamboo

Ankita Acharya and Moulima Das

9.1 INTRODUCTION

Since ancient times, bamboo has been used as a building material due to its strength and flexibility; it is a natural building material found in tropical to mild temperate regions. Bamboo has been rediscovered since the 1980s as an inexpensive, sustainable resource and several bamboo building materials have been developed for modern needs. On top of that, bamboo presents ecological advantages as well. Bamboo contains several metabolites. Phenolic compounds were the most common secondary metabolites reported for bamboo species, followed by coumarins and alkaloids (Gagliano et al., 2022). The most common bioassay was for antioxidant capability. Although bamboos are known to have a diversity of flavonoids, additional groups of components should be further researched (Joshi et al., 2022). A few bamboo species have had their *in-vivo* biological potential investigated. As a result, bamboo contains commercially important secondary metabolite classes with potential medical applications. However, additional studies on phytochemistry and *in-vivo* testing are still needed to further evaluate their advantages to human health (Panee, 2015). With an increased interest in sustainable energy generation, agricultural leftovers have lately been discovered as viable raw materials for producing high-value goods. As one of the most fundamental raw materials associated with the Industrial Revolution, silica has continually contributed to technological advancement. Agricultural waste has a high silica concentration, is inexpensive, and is of comparable quality, making it an ideal silica source with the potential for mass manufacturing. Agricultural waste contains silica in concentrations ranging from 50% to 90%, making it an economical supply of silica (Yusof Nur Izzati Amira et al., 2023). Bamboo leaves are an agricultural waste resource, and bamboo leaf ash contains around 75.90% to 82.86% silica (Saldi et al., 2021). Numerous research has been reported on silica generation from different agricultural wastes, including corn cobs, maize husks, rice husks, baggages, and bamboo stalks. Furthermore, some research used bamboo leaves to produce silica. The open burning of agricultural waste has resulted in significant pollution, highlighting the necessity for this technique. The vast volume of garbage might become a new source of silica production for the country. Silica extraction from bamboo leaves is a result of the need to manage enormous volumes of agricultural waste and extract more value-added goods from them (Hethnawi et al., 2021). Silica is a key component of granite, marble, sand, soil, and a variety of other minerals. Silicate minerals are categorized into two types: amorphous and crystalline. Silica has a great variety of industrial uses due to its amorphous nature rather than its crystalline form (Mahltig et al., 2008). The utilization and extraction of silica from bamboo leaves find uses in a range of sectors, including the ceramics, cement, and glass industries, as well as the biomedical and rubber industries (Bui et al., 2012). Several advanced industrial applications can benefit from silica extracted in nanoscale sizes, such as rubber fillers, ceramics, polymer composites, corrosion inhibition, photocatalysis, nanomaterials, etc. (Teacă et al., 2019).

9.2 BAMBOO AS A SOURCE OF BIOSILICA

Bamboo, a common plant species in tropical and subtropical countries, is a valuable local economic resource. First and foremost, being a member of the Poaceae family and Bambusoideae subfamily, bamboo is fast-growing and plays a very important role in the livelihoods of people at the local level (Olawale, 2020). Bamboos occupy more than 0.8 percent of the world's forest area of 31 million

DOI: 10.1201/9781003485933-9

hectares in China, Brazil, and India. This is a popular source of cellulose-based nanofibers due to its rapid growth and high cellulose content. Malaysia has over 50 bamboo species, with 14 commercially used (Silviana and Bayu, 2018). Bamboo is an optimal supplier of cellulose-based nanofibers. Bamboo species in Malaysia are popular due to their size, culm walls, and ease of cultivation. They work well in factories. With fewer trees available, there is a need to find new materials for building (Gautam et al., 2023). Bamboo is a good choice because it is strong like wood, grows quickly, is environmentally friendly, and does not contribute to climate change. Bamboo can also store carbon, making it a great replacement for wood. Factors influencing silica nanoparticle production include affordability, high silica content, equivalence, energy content, and feasibility for large-industrial purposes (Muthu et al., 2017). As the demand grows, people need a way to make things that do not cost too much, are good for the environment, and can last a long time. Picking the right starting material for making silica is very important for making good-quality products that can last. Daily, people generate large amounts of waste from various origins, such as farms, industrials, and households. This garbage is derived from major plants such as rice, wheat, bamboo leaves, sugarcane, and corn (Silva et al., 2023). Due to its absorption of soil, this waste is silica-rich and leads to environmental problems. For developing silica, the study examines various waste materials, such as rice husk, wheat straw, bamboo leaves ash, and sugarcane bagasse (Vaibhav et al., 2015). Bamboo leaves are used to make bamboo charcoal and wrappings for food are a biodegradable waste. Early work explored their use in cement as pozzolanic materials. Depending on the maturation of the leaves, their growth conditions, and provenance, silica content varies in bamboo leaves but can be chemically processed to yield amorphous silica after being burnt (Olawale, 2020; Olawale et al., 2021)

9.3 HEALTH BENEFITS OF BAMBOO-DERIVED BIOSILICA

9.3.1 Bone and Joint Health

Bamboo silica is rich in several essential vitamins and minerals. It contains vitamin B_6, which is vital to energizing people and maintaining a healthy brain. It has phosphorus to keep bones and teeth strong, and magnesium that helps muscles work properly and nerves to function well (Tundis et al., 2023). Bamboo silica has abundant minerals that are necessary in bones. These minerals play a role in forming and strengthening connective tissues, resulting in sturdier and healthier bones (Canavan et al., 2017). Eating bamboo can be good, especially as people get older and bones become less dense. This might lower the chance of getting a bone disease called osteoporosis (Chongtham et al., 2014). Bamboo silica has natural substances that help reduce swelling, which can cause issues like joint pain and inflammation. Eating things with anti-inflammatory qualities, such as bamboo silica, can be especially helpful for people with conditions that cause inflammation. So, including bamboo in their diet might help keep people healthy by controlling inflammation levels (Jin and Yuan, 2012).

9.3.2 Skin and Hair Health

Bamboo-derived biosilica, a natural substance with the ability to protect against sunlight, has been discovered to block the creation of melanin, much like skin-brightening substances such as arbutin and kojic acid. This might aid in diminishing dark patches and uneven skin tone, resulting in a glowing, consistent skin appearance (Choi et al., 2013). It also helps control the production of oil on the skin and reduces breakouts. Brazilian bamboo plants are getting noticed for their ethanolic extracts made from biosilica, showing promise for use in skincare products (Kalyan et al., 2023). Bamboo silica essential oils are famous for their ability to fight aging and protect against damage, which makes them perfect for skin care. These oils are rich in antioxidants and have the property to kill germs and bacteria, resulting in fresh, clear skin for individuals (Chongtham and Bisht, 2020).

Biosilica is one of the minerals that assists in hair treatment, which is why it can be found in most shampoos and conditioners as it improves health. It promotes hair growth, strengthens hair roots, and nourishes the scalp, leading to thicker and voluminous-looking hair. It even eliminates itchiness and irritation of the scalp, as well as dandruff, while its well-calming effects help soothe scalp problems (Afrin et al., 2012; Carey et al., 2009).

9.3.3 Detoxification and Immune Support

Biosilica from bamboo plants, especially Bamboo Extract (BEX), can help lessen the harmful effects of fat in cells. Akt is an important enzyme found in complex cells, which plays a key role in both normal human body functions and diseases. When Akt is activated, it affects various body processes such as staying alive, growing, multiplying, taking in sugar, forming new blood vessels, and processing food (Kumar et al., 2014). Akt signaling controls lipid metabolism by phosphorylating and inhibiting glycogen synthase kinase 3, which promotes the degradation of sterol regulatory element binding proteins (SREBPs). BEX decreases total Akt and p-Akt levels in lipid production in normal conditions. In lipotoxic conditions, BEX increases protein levels, promoting cell survival and inhibiting apoptosis. This highlights BEX's significant impact on cell signaling pathways, influenced by Free Fatty Acid (FFA) content in the environment (Gagliano et al., 2022). The study indicates that a natural substance has an inhibitory impact on lipoapoptosis, pointing to BEX as a possible phytotherapeutic method for T2DM-associated lipotoxicity. Additional *in-vivo* testing in animal models will show BEX's bioavailability and metabolism (Ballhorn et al., 2016). Biosilica's biological reactions involve the stimulation of the immune system, involving both innate and adaptive immunity. Silica is detected as a pathogen-associated molecular pattern by innate immune receptors, leading to the death of key cells and macrophages (Chang et al., 2013). During stimulation, macrophages produce cytokines that stimulate fibroblasts, leading to the development of dendritic cells, which serve as the link between the innate and adaptive immune systems. This results in fibro-hyaline tissue development (Chen et al., 2018).

9.4 MECHANISMS OF ACTION OF BIOSILICA FROM BAMBOO SPECIES

Biosilica from *Phyllostachys edulis* – Maozhu or moso, is the fastest-growing bamboo species. It has a high content of polyphenols and flavonoids in biosilica derived from the leaves of Phyllostachys edulis. This extract, BEX, has cell-protective properties against lipotoxicity caused by high palmitic acid (Huang and Young, 2019). BEX downregulates the overproduction of inflammatory mediators like IL-6 and MCP-1 induced by palmitic acid. Flavonoids such as tricin and 7-O-methyl tricin are primarily responsible for this effect by inhibiting pathways regulating inflammatory responses, including NFkappaB and AP-1 (Wu et al., 2015). BEX appears to influence the brain and cancer development. A major component of this biosilica is a compound called 3-O-caffeoyl-one-methylquinic acid. This substance increases the production of a protein called heme oxygenase-1 and other Nrf2-dependent phase II detoxifying genes, protects against peroxide-induced cytotoxicity, and prevents photocarcinogenesis by apoptotic elimination of tumor protein 53 mutant and DNA-repair defective cells (Fang et al., 2018). Biosilica contains chlorogenic acid–related substances to shield against cell damage. The oils extracted from the biosilica contain a chemical known as cis-3-hexanol, which is both cell-protective and antibacterial (Kalderis et al., 2023). Biosilica derived from the oil of bamboo shoots, especially those from a variety named *Phyllostachys edulis*, reduces cholesterol, LDL (bad cholesterol level), triglycerides, phytosterol, as well as lipoprotein lipase enzyme. It also assists the fatty liver and raises the quantity of cholesterol expelled in feces from rats fed a high-cholesterol diet plan. This oil contains a chemical known as acetylcholine which is critical for communication between animals and insects, but it remains to be seen if it might be useful in humans as nutrition or as medicine (Li et al., 2022).

Biosilica from *Phyllostachys nigra* – Black bamboo, one representative of the *Phyllostachys* family, contains flavonoids, potent antioxidants. Biosilica of this kind contains flavone c-glucosides and *p*-coumaric acid, which remain in the colon for lengthy periods and may help one eliminate damaging free radicals (Dlamini et al., 2022). A substance called luteolin 6-C-(6″-O-trans-caffeoylglucoside), found in biosilica, helps protect eye nerve cells from damage caused by harmful molecules and boosts the activity of a key enzyme that might help prevent problems linked to diabetes. This substance could also help in preventing complications related to diabetes (Iroegbu and Ray, 2021). Orientin, a substance from black bamboo that helps protect against cell damage, has been discovered to help relax blood vessels by stopping the release of calcium inside cells and the entry of calcium from outside (Dauletbek et al., 2021). It also prevents cell death caused by low oxygen and then reoxygenation in the heart and heart muscle cells. The part of *P. nigra* biosilica that is like a solvent, called acetone, has been shown to help leukemia cells become more normal and stop the production of a substance called IL-12 in immune cells called macrophages (Liu et al., 2022).

Biosilica from *Sasa senanensis* – Kumaizasa bamboo, a type of bamboo used in traditional Japanese medicine, has been discovered to have several health benefits. An alkaline extract, known as "Sasa Health," from the biosilica was found to slow down the growth of breast tumors in mice with spontaneous or Her2/NeuN-induced tumorigenesis, influence spontaneous motor activity in both intact and gonadectomized mice, and help reduce some negative effects of not having enough food (Feng et al., 2021). The extract was found to be effective in treating various health conditions. This extract was found to have anti-tumor effects, inhibiting the growth of DMBA-induced mammary tumors and inoculating sarcoma and melanoma cells (Huang et al., 2015). The extract's effectiveness was attributed to immunopotentiation, with 1,3-β-glucan activity (Figure 9.1).

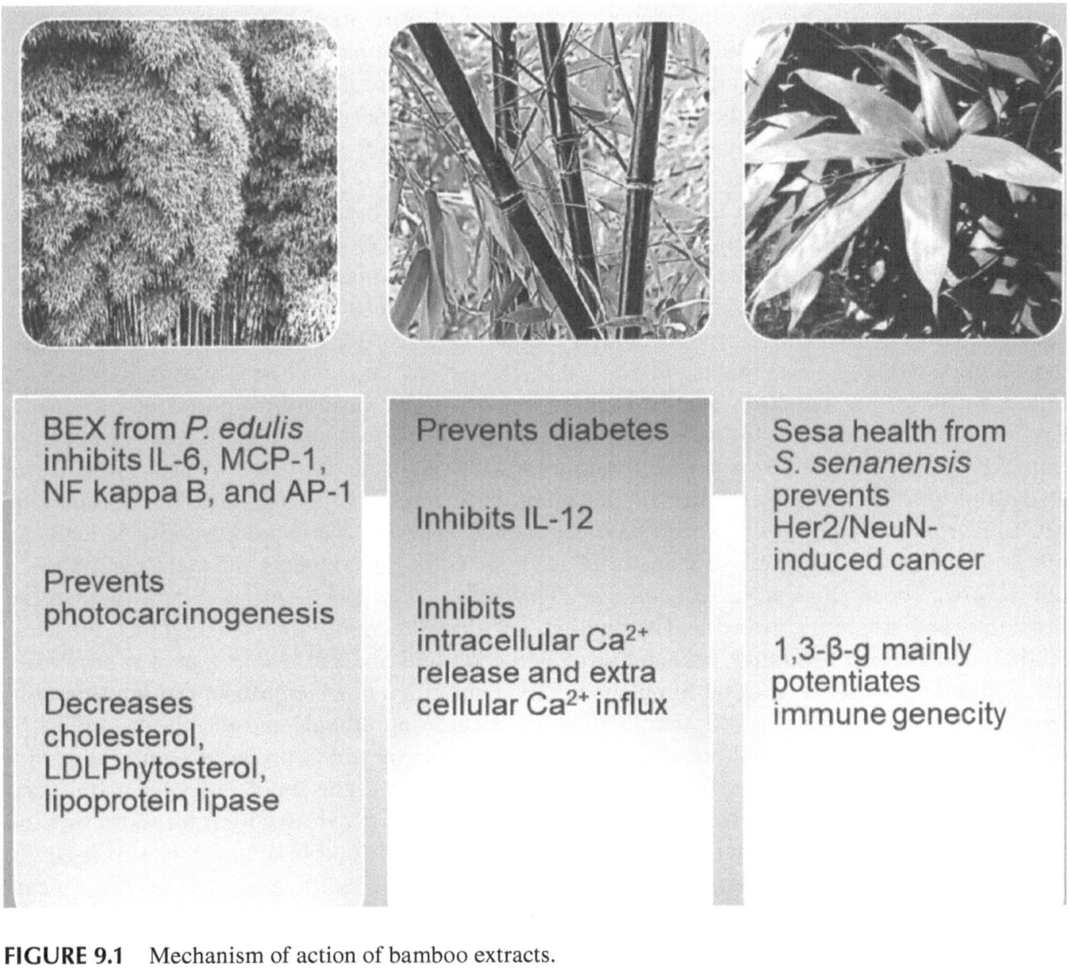

FIGURE 9.1 Mechanism of action of bamboo extracts.

9.5 INNOVATIVE APPLICATIONS

9.5.1 Supplementation

Biosilica, a vital mineral for bone, skin, and hair health, is naturally present in the body but diminishes with age. Supplements can help replenish biosilica levels. The type of biosilica used in a bone supplement is crucial, with horsetail, bamboo, and diatomaceous earth being the most common sources (Xue et al., 2023). Horsetail and bamboo are excellent sources, while diatomaceous earth contains other minerals. As people age, bones become brittle and fracture-prone, making biosilica supplements a superb addition to any health regimen to improve bone density and strength (Du et al., 2023; Mimendi et al., 2022). Some of the biosilica growth supplements are listed below.

a. Orgono Living Silica
b. Dr. Mercola Silica Collagen Builder Dietary Supplement
c. Living Silica Plant Based Collagen Booster

Patients using Living Silica for weeks show improvement in bone health. The easy-to-take liquid formula is vegan, without artificial flavors or preservatives, and has no unpleasant aftertaste. Living Silica's proprietary blend offers high absorption rates and has been laboratory tested for quality and safety. Despite its higher price point than other bone health supplements, its benefits make it worth the investment, ensuring full utilization (Chongtham et al., 2014). Dr. Mercola Silica Collagen Builder Dietary Supplement's patented formula maximizes organic silica absorption, crucial for collagen production and nutrient absorption. It offers maximum benefits from each capsule, although results may take time. The supplement is slightly expensive compared to other silica supplements and may be difficult for some to swallow (Chongtham et al., 2011). Living Silica Plant Based Collagen Booster is a liquid supplement with a proprietary blend of silicon, known for its high absorption rate, promoting collagen production and supporting bone and joint health. It is easy to take and absorb, ensuring optimal body utilization (Chongtham et al., 2007). Living biosilica is a vegan-friendly plant-based supplement free from common allergens like gluten, dairy, and soy, making it an ideal choice for those with dietary restrictions. Living Silica, a product with a slightly bitter taste, can be enhanced by mixing it with juice or water (Basumatary et al., 2017). Despite its large size, it contains 16.9 fl oz of liquid, sufficient for a month's supply, making it suitable for those with limited storage space. LABO Nutrition Bioactive Silica is a vegan-friendly, gluten-free supplement that significantly improves bone health. Its synergistic formula includes vitamins D_3 and K_2, ensuring calcium absorption in bones and blood vessels (Chauhan et al., 2016). The supplement also contains MCT oil from coconut oil, known for its health benefits. LABO Nutrition Bioactive Silica is an ultra-collagen booster/stimulator that stimulates bone collagen synthesis and stabilization without the use of beef, chicken, or fish collagen. This helps support bone mineral density, increase flexibility and strength, restore and strengthen cartilage, tendons, ligaments, joints, muscles, teeth, blood vessels, and intestines, and promotes good oral and gut health (Satya et al., 2012). The Eidon Liquid Silica Mineral Supplement has reportedly shown significant improvement in hair and nail strength and health, thanks to its all-natural ingredients and the absence of artificial preservatives or colors. Although the taste may not be everyone's preference, mixing it with water or juice can make it more palatable (Granato et al., 2010). The supplement is a natural and effective way to support bones, hair, and skin, making it a worthwhile option for those seeking a natural and effective solution (Park and Jhon, 2010). The Cellfood Essential Silica Anti-Aging Formula has been found to significantly improve bone and joint health, as well as overall appearance. The liquid formula is easy to take and has a pleasant taste, making it a great addition to daily routines (Bhatt et al., 2005). Although it may take time to see noticeable results, the wait is worth it. The product is tested and trusted, manufactured in an FDA-compliant facility, and is gluten-free, thiaminase-free, and non-GMO. Overall, it is highly recommended for those looking to improve their health and appearance (Pandey et al., 2012).

9.5.2 Functional Foods

Bamboo shoots contain a high amount of silica. Bamboo shoots, collected during the middle of the growing season, are suitable for consumption due to their high height and development. They can be processed into fresh or fermented forms into traditional cooking dishes including the countries like China, Korea, Japan, Thailand, and Taiwan (Satya et al., 2010). Some of the popular dishes of bamboo silica are depicted in Table 9.1.

Bamboo shoots are cleaned, shredded, and treated with water or boiled to remove acidity (Zhao et al., 2014). In Australia and New Zealand, bamboo shoot pieces are slated for 8–10 minutes after boiling. The shoots are also used in Thai curry, soup, and Vietnamese broths. Súp bún mang gá, a Vietnamese broth, is popular for breakfast. Canned bamboo shoots are exported to countries like the USA, the United Kingdom, Australia, Singapore, Canada, and New Zealand (Chen et al., 2023). Locals consume bamboo shoots after boiling them at various temperatures and durations to eliminate the acrid flavor and poisonous components. They are utilized in pickles, snacks, papads, and other fried foods in rural parts of the Western Ghats (Yang et al., 2021). Northeast India, consisting of eight states and covering 18.4 million hectares, is a treasure house of bamboo, contributing over 66% of India's total resources. The region is characterized by diverse ethnic communities, cultures, and religions, and is home to 16 edible bamboo species, including *Arundinaria callosa, Bambusa nutans, Bambusa pallida, Dendrocalamus hamiltonii, Dendrocalamus giganteus, Dendrocalamus brandisii, Melocanna baccifera, Melocanna bambusoides, Dendrocalamus hookeri, Dendrocalamus sikkimensis, Dendrocalamus strictus,* and *Phyllostachys manni* (Mustafa et al., 2021).

TABLE 9.1
Traditional Bamboo Recipes Used in Different Countries

Sl. No.	Name of the Recipe	Ingredients of the Dish	Country Food	References
1.	Gulai rebung	Bamboo shoots with coconut milk	Indonesia	Ben-zhi et al. (2005)
2.	Labong	Bamboo shoots in coconut milk with chillies	Philippines	Ben-zhi et al. (2005)
3.	Alu tama	Fermented bamboo shoots with potato	Nepal	Zhao et al. (2014)
4.	Bamboo salad	Chopped bamboo	Thailand and Vietnam	Yeasmin et al. (2015)
5.	Ma khua proh	Bamboo shoots with white pumpkin	Thailand	Yeasmin et al. (2015)
6.	Dom jud nomai	Bamboo shoot with pork ribs	Vietnam	Chen et al. (2023)
7.	Ekung and Soibum	Fermented bamboo shoots	Arunachal Pradesh, India	Haldipur and Srividya (2021)
8.	Apatani	Bamboo curry with meat, fish, or vegetables	Tribal area in India	Haldipur and Srividya (2021)
9.	Godhak	fermented bamboo shoots with bananas, fish, salt, chili, and garlic	Tripura, India	Chongtham et al. (2018)
10.	Mesu	Bamboo pickle	Sikkim and Darjeeling, India	Chongtham et al. (2018)
11.	Kardi	Soaked bamboo shoots	Western India	Chen et al. (2023)

9.5.3 Cosmetics

The popularity of bamboo has grown due to its environmentally friendly properties and is now used in clothing and furniture, making it a versatile and sustainable alternative to traditional materials (Rathour et al., 2022). Bamboo-based biosilica is environmentally friendly and provides many benefits for skin health and beauty. It has antioxidants, which protect the skin from free radicals and premature aging, making it ideal for acne-prone skin. It also has antibacterial properties, making it a suitable ingredient for acne-prone products (Binfield et al., 2022). It soothes and calms red or tender skin which makes it great for sensitive skin. It is rich in antioxidants, germ-fighting properties, and a natural source of silica: bamboo silica. Silica also plays a significant role in maintaining the health of skin, hair, and nails. It also helps make collagen, a protein that helps keep skin stretchy and resilient (Liese and Weiner, 1996). Bamboo silica, with its antioxidant properties, helps to promote the production of collagen, which is an essential protein that keeps skin firm and youthful. With age, the body produces less collagen contributing to fine lines, wrinkles, and sagging skin. Bamboo extract-based skincare products boost collagen production and hydrate the skin, which helps make wrinkles and lines appear less defined and encourages plump skin and firmness (Dwivedi et al., 2019). This is what makes bamboo silica a perfect ingredient for skin care. It has a calming and soothing effect, which is why it works great on sensitive skin. and can help maintain skin balance and calmness, especially for those prone to irritation from skincare products (Canavan et al., 2017).

9.6 CONCLUSION

Bamboo silica is a gourmet superfood that is an integral part of the traditional diet consumed by many tribal communities across the globe. It is sustainable as a food source, both processed and fresh. To make sure it is safe to eat and keeps its nutrients, it needs to be processed correctly. The world's need for renewable resources and the increasing number of people on Earth have made it difficult to keep natural resources in good condition. Bamboo silica offers a solution to meet current and future needs; however, there is comparatively little interest in this underutilized crop, which has enormous export potential and is a valuable nutraceutical. Therefore, researchers should focus on utilizing bamboo silica for sustainable food production. The amalgamation of traditional and contemporary technologies can help produce affordable food for poor farmers and consumers. This involves minimal cultivation and conservation of soil-conserving crops. Large-scale bamboo production can significantly contribute to rural economies and boost export wages without disrupting the biodiversity.

REFERENCES

Afrin, T., Tsuzuki, T., Kanwar, R. K., & Wang, X. 2012. The origin of the antibacterial property of bamboo. *Journal of the Textile Institute 103*(8): 844–849.

Ballhorn, D. J., Rakotoarivelo, F. P., & Kautz, S. 2016. Coevolution of cyanogenic bamboos and bamboo lemurs on Madagascar. *PLoS One 11*(8): e0158935.

Basumatary, A., Kumar Middha, S., Usha, T., Kumar Basumatary, A., Kumar Brahma, B., & Goyal, A. K. 2017. Bamboo shoots as a nutritive boon for Northeast India: An overview. *3 Biotech 7*: 1–6.

Ben-Zhi, Z., Mao-Yi, F., Jin-Zhong, X., Xiao-Sheng, Y., & Zheng-Cai, L. 2005. Ecological functions of bamboo forest: Research and application. *Journal of Forestry Research 16*: 143–147.

Bhatt, B. P., Singh, K., & Singh, A. 2005. Nutritional values of some commercial edible bamboo species of the North Eastern Himalayan Region, India. *Journal of Bamboo and Rattan 4*(2): 111–124.

Binfield, L., Britton, T. L., Dai, C., & Innes, J. 2022. Evidence on the social, economic, and environmental impact of interventions that facilitate bamboo industry development for sustainable livelihoods: A systematic map protocol. *Environmental Evidence 11*(1): 33.

Bui, L. A. T., Chun, T. C., Chao, L. H., & Wei, S. W. 2012. Effect of silica forms in rice husk ash on the properties of concrete. *International Journal of Minerals, Metallurgy and Materials 19*(3): 252–258.

Canavan, S., Richardson, D. M., Visser, V., Le Roux, J. J., Vorontsova, M. S., & Wilson, J. R. U. 2017. The global distribution of bamboos: Assessing correlates of introduction and invasion. *AoB Plants 9*(1): 1–18.

Carey, W., Dasi, J., Rao, N., & Gottumukkala, K. 2009. Anti-inflammatory activity of methanolic extract of *Bambusa vulgaris* leaves. *International Journal of Green Pharmacy 3*(3): 234–238.

Chang, W. J., Chang, M. J., Chang, S. T., & Yeh, T. F. 2013. Chemical composition and immunohistological variations of a growing bamboo shoot. *Journal of Wood Chemistry and Technology 33*(2): 144–155.

Chauhan, O. P., Unni, L. E., Kallepalli, C., Pakalapati, S. R., & Batra, H. V. 2016. Bamboo shoots: Composition, nutritional value, therapeutic role and product development for value addition. *International Journal of Food and Fermentation Technology 6*(1): 1–12.

Chen, G., Bu, F., Chen, X., Li, C., Wang, S., & Kan, J. 2018. Ultrasonic extraction, structural characterization, physicochemical properties and antioxidant activities of polysaccharides from bamboo shoots (*Chimonobambusa quadrangularis*) processing by-products. *International Journal of Biological Macromolecules 112*: 656–666.

Chen, Z., Pan, X., Hu, L., Ji, H., Yu, X., & Shao, J. F. 2023. A comparative evaluation of chemical composition and nutritional value of bamboo rice and major cereals reveals the potential utility of bamboo rice as functional food. *Food Chemistry: X 18*: 100723.

Choi, S., Park, M. S., Lee, Y. R., Lee, Y. C., Kim, T. W., Do, S. G., Kim, D. S., & Hwa Jeon, B. 2013. A standardized Bamboo leaf extract inhibits monocyte adhesion to endothelial cells by modulating vascular cell adhesion protein-1. *Nutrition Research and Practice 7*(1): 9–14.

Chongtham, N., Bisht, M. S., Bajwa, H. K., & Santosh, O. 2018. Bamboo: A rich source of natural antioxidants and its applications in the food and pharmaceutical industry. *Trends in Food Science & Technology 77*: 91–99.

Chongtham, N., Bisht, M. S., & Laishram, M. 2014. Bioactive compounds in bamboo shoots: Health benefits and prospects for developing functional foods. *International Journal of Food Science & Technology 49*(6): 1425–1431.

Chongtham, N., & Bisht, M. S. 2020. Bamboo shoot : Superfood for Nutrition, health and medicine. In *Bamboo Shoot*. CRC Press.

Chongtham, N., Bisht, M. S., & Haorongbam, S. 2011. Nutritional properties of bamboo shoots: Potential and prospects for utilization as a health food. *Comprehensive Reviews in Food Science and Food Safety 10*(3): 153–168.

Dauletbek, A., Li, H., Xiong, Z., & Lorenzo, R. 2021. A review of mechanical behavior of structural laminated bamboo lumber. *Sustainable Structures 1*(1): 1–18.

Dlamini, L. C., Fakudze, S., Makombe, G. G., Muse, S., & Zhu, J. 2022. Bamboo as a valuable resource and its utilization in historical and modern-day China. *BioResources 17*(1): 1926–1938.

Du, J., Yang, K. L., Yuan, Z. Q., Liu, Z. M., Li, X. Y., Liu, S. J., … Wu, R. M. 2023. Effect of physical treatment methods on the properties of natural bamboo materials. *Construction and Building Materials 394*: 132170.

Dwivedi, A. K., Kumar, A., Baredar, P., & Prakash, O. 2019. Bamboo as a complementary crop to address climate change and livelihoods–Insights from India. *Forest Policy and Economics 102*: 66–74.

Fang, C. H., Jiang, Z. H., Sun, Z. J., Liu, H. R., Zhang, X. B., Zhang, R., & Fei, B. H. 2018. An overview on bamboo culm flattening. *Construction and Building Materials 171*: 65–74.

Feng, Q., Huang, Y., Ye, C., Fei, B., & Yang, S. 2021. Impact of hygrothermal treatment on the physical properties and chemical composition of Moso bamboo (*Phyllostachys edulis*). *Holzforschung 75*(7): 614–625.

Gagliano, J., Anselmo-Moreira, F., Sala-Carvalho, W. R., & Furlan, C. M. 2022. What is known about the medicinal potential of bamboo? *Advances in Traditional Medicine 22*(3): 467–495.

Gautam, N., Rajesh, Y., Kale, N., Jagtap, M., Chaudhari, H., & Pansare, S. 2023 December. Silica extraction from bamboo leaves using alkaline extraction method. *Materials Today: Proceedings 111*: 155–160.

Granato, D., Branco, G. F., Nazzaro, F., Cruz, A. G., & Faria, J. A. F. 2010. Functional foods and nondairy probiotic food development: Trends, concepts, and products. *Comprehensive Reviews in Food Science and Food Safety 9*(3): 292–302.

Haldipur, A. C., & Srividya, N. 2021. A comparative evaluation of in vitro antihyperglycemic potential of bamboo seed rice (*Bambusa arundinacea*) and Garudan samba (*Oryza sativa*): An integrated metabolomics, enzymatic and molecular docking approach. *Journal of Cereal Science 99*: 103200.

Hethnawi, A., Hashlamoun, K., Sessarego, S., Chehelamirani, M., & Nassar, N. N. 2021. Simultaneous removal of silica and TOC from steam assisted gravity drainage (SAGD) produced water using iron-hydroxide-coated walnut shell filter media. *Journal of Water Process Engineering 43*: 102016.

Huang, J. K., & Young, W. B. 2019. The mechanical, hygral, and interfacial strength of continuous bamboo fiber reinforced epoxy composites. *Composites Part B: Engineering 166*: 272–283.

Huang, X. D., Hse, C. Y., & Shupe, T. F. 2015. Study of Moso bamboo's permeability and mechanical properties. *Emerging Materials Research 4*(1): 130–138.

Iroegbu, A. O. C., & Ray, S. S. 2021. Bamboos: From bioresource to sustainable materials and chemicals. *Sustainability 13*(21): 12200.

Jin, Y., & Yuan, K. 2012. Studies on the functional components and bioactivity and the relativity of bamboo shoots and shells. *Applied Mechanics and Materials 108*: 314–319.

Joshi, U. R., Roy, R., & Satsangi, P. G. 2022. Investigation of bamboo leaves as an alternative source of silica: Extraction, characterization and its application as an adsorbent for methylene blue sequestration. *ChemistrySelect 7*(21): e202200011.

Kalderis, D., Seifi, A., Trang, T. K., Tsubota, T., Anastopoulos, I., Manariotis, I., … Khataee, A. 2023. Bamboo-derived adsorbents for environmental remediation: A review of recent progress. *Environmental Research 224*: 115533.

Kalyan, N., Santosh, O., Indira, A., Kuhad, A., & Chongtham, N. 2023. Therapeutic aspects of bamboo for wound healing. In: *Bamboo Science and Technology*. Springer Nature Singapore, Singapore, 237–264.

Kumar, S., Sharma, G., Sidiq, T., Khajuria, A., Jain, M., Bhagwat, D., & Dhar, K. L. 2014. Immunomodulatory potential of a bioactive fraction from the leaves of *Phyllostachys bambusoides* (Bamboo) in BALB/C mice. *EXCLI Journal 13*: 137.

Li, Z. Z., Luan, Y., Hu, J. B., Fang, C. H., Liu, L. T., Ma, Y. F., … Fei, B. H. 2022. Bamboo heat treatments and their effects on bamboo properties. *Construction and Building Materials 331*: 127320.

Liese, W., & Weiner, G. 1996. Ageing of bamboo culms. A review. *Wood Science and Technology 30*(2): 77–89.

Liu, L., Yuan, Z., Fan, X., Pan, C., & Li, X. 2022. A review of interfacial bonding mechanism of bamboo fiber reinforced polymer composites. *Cellulose 29*(1): 83–100.

Mahltig, B., Swaboda, C., Roessler, A., & Böttcher, H. 2008. Functionalising wood by nanosol application. *Journal of Materials Chemistry 18*(27): 3180–92.

Mimendi, L., Lorenzo, R., & Li, H. 2022. An innovative digital workflow to design, build and manage bamboo structures. *Sustainable Structures 2*(1): 1–16.

Mustafa, A. A., Derise, M. R., Yong, W. T. L., & Rodrigues, K. F. 2021. A concise review of *Dendrocalamus* asper and related bamboos: Germplasm conservation, propagation and molecular biology. *Plants 10*(9): 1897.

Muthu, M., Ramachandran, D., Hasan, N., Jeevanandam, M., Gopal, J., & Chun, S. 2017. Unprecedented nitrate adsorption efficiency of carbon-silicon nano composites prepared from bamboo leaves. *Materials Chemistry and Physics 189*: 12–21.

Olawale, O. 2020. Bamboo leaves as an alternative source for silica in ceramics using box Benhken design. *Scientific African 8*: e00418.

Olawale, O., Akinyemi, B., Ogundipe, S. J., Abayomi, S. T., & Adekunle, A. A. 2021. Response surface methodology for silicon production from bamboo leaves. *IOP Conference Series: Earth and Environmental Science 665*(1): 012075.

Pandey, A. K., Ojha, V., & Choubey, S. K. 2012. Development and shelf-life evaluation of value added edible products from bamboo shoots. *American Journal of Food Technology 7*(6): 363–371.

Panee, J. 2015. Potential medicinal application and toxicity evaluation of extracts from bamboo plants. *Journal of Medicinal Plant Research 9*(23): 681.

Park, E. J., & Jhon, D. Y. 2010. The antioxidant, angiotensin converting enzyme inhibition activity, and phenolic compounds of bamboo shoot extracts. *LWT-Food Science and Technology 43*(4): 655–659.

Rathour, R., Kumar, H., Prasad, K., Anerao, P., Kumar, M., Kapley, A., … Singh, L. 2022. Multifunctional applications of bamboo crop beyond environmental management: An Indian prospective. *Bioengineered 13*(4): 8893–8914.

Saldi, G. D., Louvat, P., Schott, J., & Gaillardet, J. 2021. The pH dependence of the isotopic composition of boron adsorbed on amorphous silica. *Geochimica et Cosmochimica Acta 308*: 1–20.

Satya, S., Bal, L. M., Singhal, P., & Naik, S. N. 2010. Bamboo shoot processing: Food quality and safety aspect (a review). *Trends in Food Science and Technology 21*(4): 181–189.

Satya, S., Singhal, P., Bal, L. M., & Sudhakar, P. 2012. Bamboo shoot: A potential source of food security. *Mediterranean Journal of Nutrition and Metabolism 5*(1): 1–10.

Silva, L. H. P., de Paiva, F. F. G., Tamashiro, J. R., & Kinoshita, A. 2023. Potential of bamboo leaf ash as supplementary binder materials-a systematic literature review. *Journal of Building Engineering 71*: 106547.

Silviana, S., & Bayu, W. J. 2018. Silicon conversion from bamboo leaf silica by magnesiothermic reduction for development of Li-ion baterry anode. *Matec Web of Conferences 156*: 05021.

Teacă, C. A., Roşu, D., Mustaţă, F., Rusu, T., Roşu, L., Roşca, I., & Varganici, C. D. 2019. Natural bio-based products for wood coating and protection against degradation: A review. *BioResources 14*(2): 4873–4901.

Tundis, R., Augimeri, G., Vivacqua, A., Romeo, R., Sicari, V., Bonofiglio, D., & Loizzo, M. R. 2023. Anti-inflammatory and antioxidant effects of leaves and sheath from bamboo (*Phyllostacys edulis* J. Houz). *Antioxidants 12*(6): 1239.

Vaibhav, V., Vijayalakshmi, U., & Roopan, S. M. 2015. Agricultural waste as a source for the production of silica nanoparticles. *Spectrochimica Acta Part A: Molecular and Biomolecular Spectroscopy 139*: 515–520.

Wu, C., Zhou, S., Li, R., Wang, D., & Zhao, C. 2015. Reactivity improvement of bamboo dissolving pulp by xylanase modification. *BioResources 10*(3): 4970–4977.

Xue, X., Zhou, W., Sayed, U., Feng, Z., Li, H., Li, Y., ... Lorenzo, R. 2023. Design and construction of "Bamboo Cubic " facade with laminated bamboo lumber. *Sustainable Structures 3*(2): 1–15.

Yang, W. T., Gong, X. X., Ji, H., & Shao, J. F. 2021. Qualitative and quantitative characterization of nutrient content and morphology in seeds of bamboo, rice, and wheat. *Journal of Cereal Science 101*: 103273.

Yeasmin, L., Ali, M. N., Gantait, S., & Chakraborty, S. 2015. Bamboo: An overview on its genetic diversity and characterization. *3 Biotech 5*: 1–11.

Yusof Nur Izzati Amira, M., Izzati, N., Nadiah, A., Amini Mohd Hazim, M., Ramle Sitti Fatimah, M., Mohammad Haaziq, A., & Hidayani, J. 2023. Utilisation of silica from bamboo leaves ash and modification with M-Toluoyl chloride in shellac; It's microbial effect on Pulai wood. *Materials Today: Proceedings*.

Zhao, H., Wang, L., Dong, L., Sun, H., & Gao, Z. 2014. Discovery and comparative profiling of microRNAs in representative monopodial bamboo (*Phyllostachys edulis*) and sympodial bamboo (*Dendrocalamus latiflorus*). *PLoS One 9*(7): e102375.

FURTHER READING

Chongtham, N., Elangbam, D., & Sharma, M. L. 2007. Changes in nutrient components during ageing of emerging juvenile bamboo shoots. *International Journal of Food Sciences and Nutrition 58*(8): 612–618.

10 *Lophatherum gracile*
A Medicinally Active Bamboo

Viral Patel, Mithun Sing Rajput, Manan Raval,
Sneha Makwana, and Ravish Patel

ABBREVIATIONS

ABTS	2,2′-diazide-bis (3-ethylbenzothiazoline-6-sulfonic acid) di-ammonium salt
ACE2	Acetyl cholinesterase
ATP	Adenosine triphosphate
DPPH	2,2-diphenyl-1-pyridinohydrazinyl
FRAP	Ferric reducing antioxidant power
GPX	Glutathione peroxidase
HPLC-MS	High-performance liquid chromatography coupled with mass spectroscopy
HUVECs	Human umbilical vein endothelial cells
JNK	c-Jun N-terminal kinases
LC-MS	Liquid chromatography–mass spectrometry
L. gracile	*Lophatherum gracile*
MAPKs	Mitogen-activated protein kinase
MDA	Malondialdehyde
mRNA	Messenger ribonucleic acid
NF-κB	Nuclear factor kappa-B
PMA	Phorbol myristate acetate
RANKL	Receptor activator of nuclear factor kappa-B
ROS	Reactive oxygen species
RSV	Respiratory syncytial virus
SARS-CoV-2	Severe acute respiratory syndrome-related coronavirus
SOD	Super-oxide dismutase
UHPLC	Ultra-high-performance liquid chromatography

10.1 INTRODUCTION

Lophatherum gracile Brongn. (*L. gracile*) is a bamboo primarily found in humid, tropical regions including India, south China, and Japan. It belongs to the Graminae family, with Bambuseae as the sub-family and is a perennial plant. In China, the leaves of *L. gracile* are also known as "Dan Zhu Ye" and the leaf extract is a household remedy for urinary tract inflammation, ardour urinae, hydrodipsia, and pyreticosis. For more than 2,000 years, the dried leaves and stems of the plant have been used in Chinese homes as an important food ingredient as well as traditional medicine. The ancient documents have supported the effectiveness of *L. gracile* for diuresis, irritation, fever, thirst, pain, and urinary tract inflammation (Chen et al., 2023).

According to the plant anatomy described in *Flora of China*, the average height of *L. gracile* varies from 40 to 100 cm. The shape of the root is a small spindle type, while the leaves are lanceolate in structure and approximately 20 cm long and 2.5 cm wide. The plant holds a wide range of active constituents – phenolic acids, polysaccharides flavonoids, and triterpenes. The herb also contains phytochemicals like alkaloids, amino acids, and minerals. The presence of phytochemicals imparts various

 DOI: 10.1201/9781003485933-10

biological properties as reported both *in-vivo* and *in-vitro* as anti-inflammatory, antipathogenic, antimicrobial, antioxidant, anticancer, and liver protection. In addition, the leaf extract of *L. gracile* has been reported to show hyperglycemic activities and astringent properties. The versatility of the herb holds applications as a food supplement, too. The extract of the herb has been traditionally used as a beverage in combination with tea and alcohol. Nowadays, the extract of *L. gracile* has been used as a natural food additive, as a super-food, and also holds importance in the cosmetic industry (Li et al., 2008).

With the information available on different constituents, the functions and applications of *L. gracile* are patchy and hence, there is a need to give a comprehensive understanding of such a versatile herb. Many potential benefits of *L. gracile* are yet to be discovered. Herein, the chapter details recent studies that considered different phytoconstituents, along with supportive *in-vitro* and *in-vivo* data revealing the importance of the herb, and are summarized, which will provide a constructive reference for future studies on *L. gracile*.

10.2 ACTIVE PHYTOCONSTITUENTS OF *L. GRACILE*

L. gracile holds plentiful beneficial phytoconstituents. There is considerable literature suggesting the presence of flavonoids, polysaccharides, phenolic acids, etc. (as shown in Figure 10.1). Among the long list of phytoconstituents, only flavonoids, phenolic acids, and triterpenoids have been characterized by different researchers across the globe.

Chemically, flavonoids consist of polyphenolic entities extensively found in fresh fruits and vegetables. Flavonoids have been reported to show many health-promoting functions. The leaf and stem of *L. gracile* possess nearly 50% of the flavonoids; however, their content may slightly vary with the geographical cultivation, climatic zone, the soil texture, and other habitats surrounding its cultivation. The extract of *L. gracile* was analyzed using high-performance liquid chromatography coupled with mass spectroscopy (HPLC-MS) to estimate the composition of the constituents. The results identified the presence of 44 different flavonoid monomers; surprisingly, most were glycoside derivatives of quercetin, luteolin, and kaempferol. Further, it was observed by Zhang et al. (2021), that some of the flavonoids – orientin, saponaretin, rutin, and isoorientin – were frequently detected from the extracts of *L gracile* collected from diverse geographical locations, serving as indicators of *L. gracile*. Another group of scientists probed the phenolic fingerprint of *L. gracile* collected from diverse locations. They reported around 20 different compounds in different studies; however, the key components have been identified as orientin, chlorogenic acid, and saponaretin, which could be used as markers for quality control of *L. gracile* (Liu et al., 2022; Tang et al., 2015). Other flavonoids reportedly present in *L. gracile* include vitexin, tricin 7-O-β-D-glucopyranoside, tricin 7-Oneohesperidoside, isovitexin, swertiajaponin, and isoorientin. Besides flavone glycoside, some C-glucosides present in *L. gracile* include lutonarin and swertisin (Liu et al., 2022).

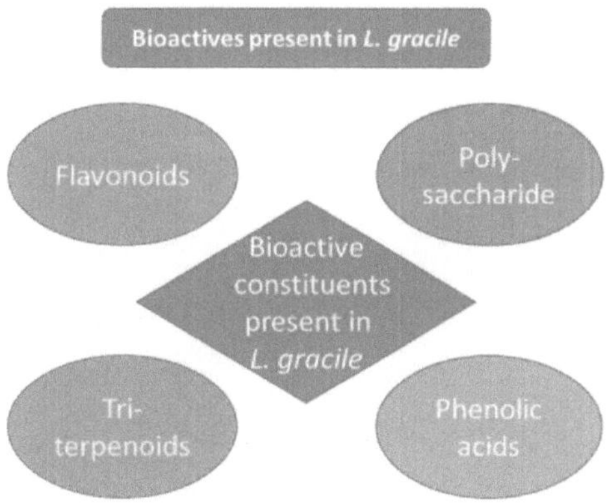

FIGURE 10.1 Bioactive constituents of *L. gracile*.

The majority of all edible plants contain phenolic acids as one of the bioactive constituents. It has been reported that *L. gracile*, irrespective of its origin, consists of 14 phenolic acids, among which chlorogenic acid is present in the highest content. Many of the phenolic acids are derivatives of quinic acid or caffeic acid. The presence of *p*-coumaric acid in *L. gracile* has also been reported in some literature. Tang et al. (2015) identified 10 bioactive components, including coumaroylquinic acid. These constituents possess strong antioxidant properties. Many research studies concluded with a suggestion that *L. gracile* is an indispensable source of natural chlorogenic acid (Quan et al., 2002).

A natural macromolecule, polysaccharide was testified in *L. gracile,* comprised of both cellulose and non-cellulose. The non-cellulose polysaccharide derived from *L. gracile* was referred to as BLF80, and entailed glucose, galactose, fructose, mannose, and xylose. The potential antioxidant activity of these polysaccharides has been confirmed (Yan et al., 2015). Similarly, the presence of soluble dietary fibers (around 30.03%) was confirmed in *L. gracile* stems and leaves, corroborating its water-holding capacity (Fan et al., 2022).

Besides the main constituents of other phytoconstituents, *L. gracile* also holds bountiful other components like triterpenoids together with arundoin, cylindrin, friedelin, and taraxerol. Xu Jie-Kun et al. reported the presence of salcolin A, salcolin B, afzelin, tricin, luteolin, tricin 7-O-β-D-glucopyranoside, tricin 7-O-neohesperidoside, β-(*p*-methoxyphenyl) acrylic acid, daucosterol, and β-sitosterol in *L. gracile* (Guleria et al., 2013; Li et al., 2008).

10.3 BIOLOGICAL IMPORTANCE OF ACTIVE CONSTITUENTS IN *L. GRACILE*

L. gracile has been identified as a medicinal plant with a wide spectrum of biological activity owing to different phytoconstituents present in different parts of the plant (presented in Figure 10.2). Table 10.1 reports a few studies, along with *in-vitro* and *in-vivo* data.

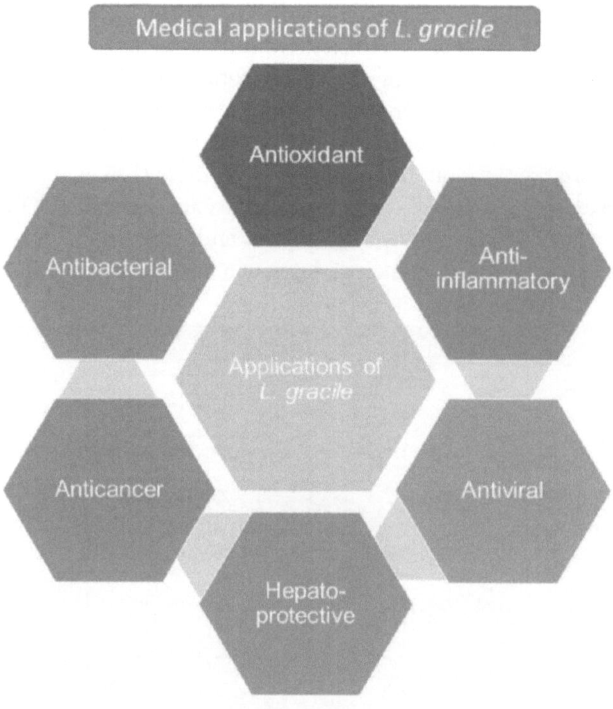

FIGURE 10.2 Medical applications of *L. gracile* extracts.

TABLE 10.1
***In-vitro* and *In-vivo* Effects of *L. gracile* Extracts**

Active Constituent	Biological Activity	IC50/Dose	Proposed MOA	Reference
Flavonoid	Antioxidant	50 mg/mL, 100 mg/mL	Reduces ROS production and Increase in SOD	(Jia et al., 2021)
Poly saccharide	Antioxidant	400 mg/kg	Increase in SOD and GPX activity. Reduce ROS production	(Ge et al., 2014)
Ethanolic leaf extract	Antioxidant	Proportional to dose	Significant antioxidant effect observed in-vitro	(Zhang et al., 2021)
Ethanol extract	Anti-inflammatory	30 μg/mL	JNK inhibition and reduction in Ca^{+2} levels	(Lai et al., 2021)
Flavanoid-isoorientin	Anti-inflammatory	30 μg/mL	Reduces ROS and superoxide anions	(Chen et al., 2023)
Extract/Flavanoid	Antibacterial		*In-vitro* showed antibacterial effect against S. aureus, E. coli, B. subtilis antibacterial effects in vitro (S. aureus, E. coli, B. subtilis	(Xue et al., 2009)
Ethanol extract	Antiviral	20 μg/mL	Inhibits synthesis of mRNA and protein expression of IL-1β, TNF-α and IFN-ɣ	(Chen et al., 2019)
Flavanoid/Isoorientin	Antiviral	30 μg/mL	Inhibits ACE2 overexpression	(Chen et al., 2023)
Ethanol extract	Hepato-protective	0.45 mg/mL	Inhibits hydroxyl radicals and MDA levels with increase in SOD activity	(Ma et al., 2024)
Ethanol extract	Hepato-protective	0.3 kg/mL	Increases antioxidant enzyme activity	(Qingfeng et al., 2016)
Chloroform extract	Anticancer	177.85 μg/mL	Inhibits cell differentiation and proliferation	(Istiqomah et al., 2015)
Ethanol extract	Anticancer	50, 100 mg/kg	Inhibits metastasis and angiogenesis	(Kim et al., 2016)

10.3.1 Antioxidant Effects

Ultra-high-performance liquid chromatography (UHPLC), combined with quadrupole Orbitrap mass spectrometry (Q-Orbitrap/MS) over the phenolic extract of *L. gracile,* led them to identify 14 flavonoids and 7 phenolic acids. The antioxidant activity of these isolated components was estimated by 1,1-diphenyl-2-trinitrophenylhydrazine (DPPH), 2,2'-diazide-bis (3-ethylbenzothiazoline-6-sulfonic acid), di-ammonium salt (ABTS), and ferric reducing antioxidant power (FRAP) assay. The outcomes suggest IC_{50} values for DPPH and ABTS assays stretching from 5.2 to 16.1 mg/mL and 1.2 to 2.8 mg/mL correspondingly, while with the FRAP assay, the $FeSO_4$ concentrations were observed in the range of 1.84–4.20 mmol/L. They further suggested pharmacodynamics activity to be conducted to support the data (Liu et al., 2022). Another study on the flavonoid extract of *L. gracile* tested on myocardial ischemia-reperfusion injury in rats showed that the extract ameliorated the injury by augmenting the activity of super-oxide dismutase (SOD) and glutathione peroxidase (GPX) in both cardiac myocytes and serum, later fast-tracking the scavenging of reactive oxygen species (ROS) and plummeting the oxidative impairment in cardiac myocytes (Jia et al., 2021). In line with these findings, Sharma and Mehdi (2023) reported that *L. gracile* can be used to treat antiaging caused by oxidative changes. Ding et al. (2024), through their study, provided the evidence of antioxidant properties of *L. gracile* leaves by scavenging the free radicals of 2,2-diphenyl-1-pyridinohydrazinyl (DPPH). *In-vitro* antioxidant assay conducted by Qingfeng et al. (2016) on the polysaccharide extract of *L. gracile* revealed a significant inhibitory effect on superoxide radicals. The same extract showed strong DPPH radical foraging activities, hydroxyl radical foraging activities, and the ferrous ion chelating effect was also prominent.

The researcher proposed the use of *L. gracile* as a natural antioxidant and can be used as supplementary food for humans. Similarly, Huang et al. (2015) inspected the antiaging effect of *L gracile* polysaccharide in the experimental mice model. They reported a significant increase of 22.18%, 46.28%, and 25.60%, respectively, in the spleen, thymus index, and neuron number as compared with the model group. The study also reported a remarkable increase in SOD activity and GPX activity in liver serum and brain tissue; also, there was a decrease in malondialdehyde (MDA) content, highlighting the good antioxidant effects both *in vitro* and *in vivo*. Li et al. (2008), in their study, collected the *L. gracile* from 21 different regions and studied the extracts for antioxidant potential via DPPH activity and ABTS scavenging rates. They reported DPPH activity in the range of 28% to 95.97% and ABTS scavenging activity rates between 32.20% and 89.58%. Since the antioxidant property is directly linked with the total flavonoid amount, the higher flavonoid contents would enhance DPPH and ABTS activity to as much as 40%. The results were supported by findings of Ma et al. (2020), where cynaroside isolated from *L. gracile* indicated DPPH scavenging activity at IC_{50} of 0.43 μM. The studies support the idea that *L. gracile* can be a potential antioxidant agent for both the food and pharmaceutical industries.

10.3.2 Anti-inflammatory Activity

Inflammation is the body's immune reaction to any pathogenic attack; however, prolonged inflammation might reduce the metabolic functioning of the body. *L. gracile* has been shown to have strong anti-inflammatory potential. A study confirmed that *L. gracile* has repressed inflammation in LGE-exhibited anti-inflammatory activities in N-formyl-methionyl-leucyl-phenylalanine (fMLF)–induced human neutrophils by modulating JNK and calcium mobilization pathways, independently (Lai et al., 2021).

10.3.3 Antibacterial and Antiviral Activities

Research has supported the antimicrobial potential of *L. gracile*. The strong antiviral activity of four flavone glycosides was examined *in vitro* against respiratory syncytial virus (RSV) by Wang et al. (2012). In one study, several flavone C-glycosides isolated from the leaves of *L. gracile* were studied to treat RSV using HEp-2 cell lines. The study showed a dose-dependent anti-RSV activity with an IC_{50} value of 20 μg/mL beside RSV long strain and an IC_{50} value of 25 μg/mL against RSV-A2 strain. Also, oral administration of the extract considerably reduced the viral load with a reduction in lesions on lung tissues, *in vivo* (Chen et al., 2019). In another study steered by Chen et al. (2023), the anti-inflammatory responses were attenuated and replication of the virus was inhibited by the bioactive present in *L. gracile* in severe acute respiratory syndrome-related coronavirus (SARS-CoV-2) models. They reported the prominent inhibitory effect of flavoLG and isoorientin on SARS-CoV-2 pseudovirus infection by specifically interfering with its attachment to ACE2 (angiotensin converting enzyme-2). In addition, four glycosyl-flavones isolated from *L. gracile* showed antibacterial activities against *Staphylococcus aureus* (Zeng et al., 2013). More recently, the antibacterial activity of luteolin in multidrug-resistant *Escherichia coli* has been shown where downregulation of multiple metabolic pathways was seen causing a decline in ATP (adenosine triphosphate) synthesis, structure degeneration, cell wall damage, downregulation of mRNA, and content leakage by luteolin (Ding et al., 2024). Such results deduce that *L. gracile* has the potential to serve as an antiviral and antibacterial agent.

10.3.4 Hepatoprotective Effects

The liver is a vital organ of the human body. Certain substances, like alcohol, chemicals, drugs, and other exogenous substances, have hazardous effects on the liver; hence, liver-protecting agents aid in preventing liver damage. Several plants have shown evidence of potential hepatoprotective properties. The hepatoprotective property of *L. gracile* against alcohol-induced damage was studied using HepG2 cells. The results showed a higher survival rate with low ROS and MDA content in *L. gracile*–treated groups, where the extract in the range of 0.23–0.45 mg/mL remarkably improved the alcohol-induced oxidative impairment in HepG2 cells and their viability rates spiked from 59.23% to 67.76%, suggesting a hepatoprotective behavior of *L. gracile* by improving the antioxidant capacity of liver cells (Ma et al., 2024). Comparable annotations were testified by Zhao et al. in a study conducted by means of *in-vitro* digestion and alcohol-induced oxidative damage models. The results suggested an advanced network pharmacological analysis of *L. gracile* extract along with its liquid chromatography–mass spectrometry (LC-MS) analysis studied by Zheng et al. (2023) revealed that the flavanoids ameliorated metabolism in the liver and inflammation via JNK and nuclear factor κB in addition to the PI3K/AKT pathway. This study was further elaborated in diabetic mice and it was found to improve liver damage at a dose of 35 mg/kg. These outcomes advocate that *L. gracile* could be explored for hepatoprotective functions.

10.3.5 Anticancer Activity

Secondary plant metabolites were found to inhibit cell division and metastasis in cancer cells. An extensive study with water, chloroform, hexane, and 80% ethanol extract of *L. gracile* suggested the preeminent negative effect with an IC_{50} value of 177.8 μ/mL with chloroform extract when studied with breast cancer T47D cells (Istiqomah et al., 2015). In addition, Kim et al. (2016) stated that daily oral doses of 50 and 100 mg/kg repressed metastatic colonization in B16F10 cells in the lungs of genetically modified mice with no obvious side effects. They also reported significant inhibition of p38 and NF-κB caused as a result of phorbol myristate acetate (PMA-) induced stimulation by *L. gracile* extract. Kim et al. reported, thus, the studies suggest that *L. gracile* extract can inhibit tumor growth and its metastasis.

10.4 OTHER REPORTED BIOACTIVITIES

L. gracile has been traditionally used as a herbal medicine, with its roots originating from China. Various biological properties like antiviral, anti-inflammatory, anticancer, antiviral, antibacterial, and neutrophilic inflammatory effects were reported by researchers working with herbal flora and fauna. According to traditional Chinese medicine, the herbal formula consisting of decoction of 10 different herbs, including *L gracile*, exerted anti-inflammatory, antiallergic, antiangiogenic, and antioxidant properties, suggesting their potential for the management of atopic dermatitis, eczema, and other skin diseases, with no signs of toxicity (hepatic, renal, and haematological) (Yan et al., 2020). Similar therapeutic effects of *L. gracile* have been confirmed in different research-based studies.

The ethanolic and hexane extract of *L. gracile* exhibit inhibition of *Plasmodium falciparum*, showing the possibility of its usage as an inhibitor of malarial transmission (Wulandari et al., 2018). Lee et al. (2022) examined the effect of water extract of *L. gracile* leaves on osteoclast differentiation and bone loss. They identified inhibition of osteoclastogenesis, a process of differentiation of osteoclasts, by quashing the RANKL-induced stimulation of MAPKs and NF-κB. Also, the *L. gracile* extract prevented bone loss in OVX-induced postmenopausal osteoporosis in mouse models, thus suggesting bone-protective effects of *L. gracile* extract. The plant extract of *L. gracile* also shows lactate-dehydrogenase activity, causing lactate metabolism while reducing the influence on homeostasis of the body upon exercise, hence suggesting antifatigue effects in experimental mice models (Deng et al., 2021).

10.5 CONCLUSION AND FUTURE PROSPECTS

L. gracile has garnered attention owing to its diverse bioactivities. The presence of a wide range of phytoconstituents and their biological implications have implied that *L. gracile* is a favorable herb for the development of various formulations, incorporating flavonoids, phenolic acids, triterpenoids, and polysaccharides. However, a challenge comes to developing technologies to control the quality of the herbs obtained from different geographical sources. There is also a need to pay attention to the constituents present in minor quantities, as they have shown potential effects even in small quantities. While there are ample studies on understanding the bioactivity of phytoconstituents present in *L. gracile*, there has been the least focus has been on the structural-activity-relation of the phytoconstituents, which calls for directed efforts. These studies can be followed by clinical evaluation under protocol-specific clinical studies. Overall, there is a huge scope of investigation with *L. gracile* to unveil its pathway as a potential medical agent.

REFERENCES

Chen, L.F., Zhong, Y.L., Luo, D., Liu, Z., Tang, W., Cheng, W., Xiong, S., Li, Y.L. and Li, M.M. 2019. Antiviral activity of ethanol extract of *Lophatherum gracile* against respiratory syncytial virus infection. *Journal of Ethnopharmacology 242*: 111575.

Chen, Y.L., Chen, C.Y., Lai, K.H., Chang, Y.C. and Hwang, T.L. 2023. Anti-inflammatory and antiviral activities of flavone C-glycosides of *Lophatherum gracile* for COVID-19. *Journal of Functional Foods 101*: 105407.

Deng, Y., Huang, D. and Yue, T. 2021. Purification technology of polysaccharides from *Lophatherum gracile* Brongn. by Macroporous resin adsorption and its effect on athletic endurance of mice. *Journal of Science Technology of Food Industry 42*(14): 169–174.

Ding, Y., Wen, G., Wei, X., Zhou, H., Li, C., Luo, Z., Ou, D., Yang, J. and Song, X. 2024. Antibacterial activity and mechanism of luteolin isolated from *Lophatherum gracile* Brongn. against multidrug-resistant *Escherichia coli. Frontiers in Pharmacology 15*: 1430564.

Fan, H., Liu, F.G., Wang, Y.T., Li, Y.X. and Liu, X.B. 2022. Ball milling combined with alkaline-assisted enzymatic extraction improved the physicochemical and functional properties of soluble dietary fiber from Herba Lophatheri. *Food Science 43*(24): 74–82.

Ge, Q., Huang, J., Mao, J.W., Gong, J.Y., Zhou, Y.F. and Huang, J.X. 2014. Optimization of total polysaccharide extraction from Herba Lophatheri using RSM and antioxidant activities. *International Journal of Biological Macromolecules 67*: 37–42.

Guleria, S., Tiku, A.K., Singh, G., Koul, A., Gupta, S. and Rana, S. 2013. In vitro antioxidant activity and phenolic contents in methanol extracts from medicinal plants. *Journal of Plant Biochemistry and Biotechnology 22*: 9–15.

Huang, S.J., Yin, A.W., Gong, D. and Li, T.F. 2015. Anti-aging effects of polysaccharides from *Lophatherum gracile* brongn. *Modern Food Science and Technology 31*(11): 51–55.

Istiqomah, A., Muti'ah, R. and Hayati, E.K. 2015. Anticancer activity against breast cancer cells T47D and identification of its compound from extracts and fractions of leaves bamboo grass (*Lophaterum gracile B.*). *ALCHEMY: Journal of Chemistry 4*(1): 6–16.

Jia, J.Y., Zang, E.H., Lv, L.J., Li, Q.Y., Zhang, C.H., Xia, Y., Zhang, L., Dang, L.S. and Li, M.H. 2021. Flavonoids in myocardial ischemia-reperfusion injury: Therapeutic effects and mechanisms. *Chinese Herbal Medicines 13*(1): 49–63.

Kim, A., Im, M., Gu, M.J. and Ma, J.Y. 2016. Ethanol extract of Lophatheri Herba exhibits anti-cancer activity in human cancer cells by suppression of metastatic and angiogenic potential. *Scientific Reports 6*(1): 36277.

Lai, K.H., Chen, P.J., Chen, C.C., Yang, S.H., El-Shazly, M., Chang, Y.C., Wu, Y.H., Wu, Y.H., Wang, Y.H., Hsieh, H.L. and Hwang, T.L. 2021. *Lophatherum gracile* Brongn. attenuates neutrophilic inflammation through inhibition of JNK and calcium. *Journal of Ethnopharmacology 264*: 113224.

Lee, S.J., Jang, S.A., Kim, S.C., Ryuk, J.A. and Ha, H. 2022. *Lophatherum gracile* Bronghiart suppresses receptor activator of nuclear factor kappa-b ligand-stimulated osteoclastogenesis and prevents ovariectomy-induced osteoporosis. *International Journal of Molecular Sciences 23*(22): 13942.

Li, H.B., Wong, C.C., Cheng, K.W. and Chen, F. 2008. Antioxidant properties in vitro and total phenolic contents in methanol extracts from medicinal plants. *LWT-Food Science and Technology 41*(3): 385–390.

Liu, X., Wang, Y., Ge, W., Cai, G., Guo, Y. and Gong, J. 2022. Spectrum–effect relationship between ultra-high-performance liquid chromatography fingerprints and antioxidant activities of *Lophatherum gracile* Brongn. *Food Science & Nutrition 10*(5): 1592–1601.

Ma, N.H., Guo, J., Xu Chen, S.H., Yuan, X.R., Zhang, T. and Ding, Y. 2020. Antioxidant and compositional HPLC analysis of three common bamboo leaves. *Molecules 25*(2): 409.

Ma, Y.L., Wu, Z.F., Li, Z., Wang, Y., Shang, Y.F., Thakur, K. and Wei, Z.J. 2024. In vitro digestibility and hepato-protective potential of *Lophatherum gracile* Brongn. leave extract. *Food Chemistry 433*: 137336.

Qingfeng, H., Li, Y., Liu, J., Zhang, P., Yan, S., He, X. and Zhang, A. 2016. Hepatoprotective activity of *Lophatherum gracile* leaves of ethanol extracts against carbon tetrachloride-induced liver damage in mice. *International Journal of Pharmacology 12*(4): 387–393.

Quan, C., Lijun, W. and Lijun, R. 2002. Chemical studies on the constituents of *Lophatherum gracile* Brongn. *Shenyang yao ke da xue xue bao Journal of Shenyang Pharmaceutical University 19*(4): 257–259.

Sharma, V. and Mehdi, M.M. 2023. Oxidative stress, inflammation and hormesis: The role of dietary and lifestyle modifications on aging. *Neurochemistry International 164*: 105490.

Tang, Q., Shao, M., Wang, Y., Zhao, H., Fan, C., Huang, X., Li, Y. and Ye, W. 2015. Simultaneous determination of 10 bioactive components of *Lophatherum gracile* Brongn by HPLC-DAD. *Journal of Chromatographic Science 53*(6): 963–967.

Wang, Y., Chen, M., Zhang, J., Zhang, X.L., Huang, X.J., Wu, X., Zhang, Q.W., Li, Y.L. and Ye, W.C. 2012. Flavone C-glycosides from the leaves of *Lophatherum gracile* and their in vitro antiviral activity. *Planta Medica 78*(01): 46–51.

Wulandari, E., Yuliani, D., Hayati, E.K. and Muti'ah, R. 2018. Aktivitas antimalaria Ekstrak Kasar Etanol dan Fraksi n-Heksana Rumput Bambu (*Lophatherum gracile* B.) secara in vitro. *ALCHEMY: Journal of Chemistry 6*(1): 18–23.

Xue, Y.Q., Song, J., Ye, S.P. and Yuan, K. 2009. Separation, identification and its antibacterial activity of glycosylflavones in *Lophatherum gracile* Brongn. *West China Journal of Pharmaceutical Sciences 24*(3): 218–220.

Yan, F., Li, F., Liu, J., Ye, S., Zhang, Y., Jia, J., Li, H., Chen, D. and Mo, X. 2020. The formulae and biologically active ingredients of Chinese herbal medicines for the treatment of atopic dermatitis. *Biomedicine & Pharmacotherapy 127*: 110142.

Yan, X., Ye, R. and Chen, Y. 2015. Blasting extrusion processing: The increase of soluble dietary fiber content and extraction of soluble-fiber polysaccharides from wheat bran. *Food Chemistry 180*: 106–115.

Zeng, P., Zhang, Y., Pan, C., Jia, Q., Guo, F., Li, Y., Zhu, W. and Chen, K. 2013. Advances in studying of the pharmacological activities and structure–activity relationships of natural C-glycosylflavonoids. *Acta Pharmaceutica Sinica B 3*(3): 154–162.

Zhang, P., Chun, Z., Shao, Q., Fu, L., Luo, Y., Gu, D. and Chen, R. 2021. Evaluation of the phytochemicals and antioxidant activity of *Lophatherum gracile* Brongn based on chemical fingerprinting by HPLC with electrochemical detection. *Journal of Separation Science 44*(20): 3777–3788.

Zheng, J.H., Lin, S.X., Li, X.Y., Shen, C.Y., Zheng, S.W. and Chen, W.B. 2023. Flavonoids from *Lophatherum gracile* Brongn. Ameliorate liver damages in high-fat diet and streptozotocin-induced diabetic mice by regulating PI3K/AKT and NF-Kappa B pathways. *Journal of Food Biochemistry 2023*(1): 5273349.

11 Ethnobotanical and Pharmacological Insights into *Bambusa* Genus

Babita Joshi, Aribam Indira, Nikita Kalyan, Nirmala Chongtham, and Ashwani Koul

11.1 INTRODUCTION

Bamboo is a significant health-promoting plant used as traditional food and medicine in Asian countries. Several Chinese, Indian, Indo-Persian, and Tibetan traditional pharmacopeias have documented the health benefits of various bamboo parts, such as shoots, leaves, culms, rhizomes, seeds, etc. in preventing and treating several ailments (Cheng et al., 2023; Indira et al., 2023a). In recent years, these bamboo parts have been reported to possess various health benefiting properties, including anticarcinogenic, antidiabetic, antimicrobial, antiulcer, anti-inflammatory, and immunostimulating properties (Chongtham and Bisht, 2020; Indira, 2023b; Kalyan et al., 2023). These medicinal properties are attributed to pharmaceutical compounds present in bamboo, such as phenols, phenolic acids, flavonoids, polysaccharides, sterols, terpenoids, alkaloids, and many more (Bajwa et al., 2021). Various value-added products, functional foods, nutraceuticals, and medicinal formulations prepared with the fresh and processed parts of bamboo are in great demand among consumers (Joshi et al., 2023; Santosh et al., 2023; Santosh et al., 2024).

Bambusa is one of the largest genera of bamboo belonging to the family Poaceae and subfamily Bambusoideae, with over 100 species (Benjamin et al., 2023). The morphological features of bamboo species belonging to the *Bambusa* genus include treelike physiological traits, hollow and woody stems with thick walls, dominant primary branch with several secondary branches, petiolate foliage, and extensive rhizome network (Banik, 2016). *Bambusa* species are spread across the tropical and subtropical regions of south Asia, Africa, and South America (Bahru and Ding, 2021). They provide a livelihood to the local population and help the nations to achieve sustainable goals, including eradication of hunger, good health, and sustainable economic livelihood (Chongtham and Bisht, 2020). *Bambusa* species, including *Bambusa arundinacea, B. balcooa, B. bambos, B. nutans, B. textilis, B. tulda,* and *B. vulgaris* are rich in bioactive compounds such as polyphenols, flavonoids, phytosterols, steroids, polysaccharides, terpenoids, alkaloids, tannins, glycosides, and fatty acids (Nirmala et al., 2018; Wang et al., 2024; Zhang et al., 2024). The utilization of *Bambusa* species in traditional medicine of several nations dates back years ago and therefore, documentation of traditional knowledge along with scientific evaluation to validate the health claims is needed. The present chapter compiles the ethnobotanical and pharmacological activities of *Bambusa* species that would encourage further research into the widespread use and therapeutic applications of various indigenous bamboo species of the *Bambusa* genera.

11.2 ETHNOBOTANY AND TRADITIONAL MEDICINAL APPLICATION OF *BAMBUSA* SPP.

The application of various *Bambusa* species in folk and traditional medicinal systems of Asian countries are well known to treat a variety of ailments (Table 11.1). In India, *B. arundinacea* and *B. vulgaris* are the most commonly used bamboo species in traditional medicine (Benjamin et al., 2023). The

 DOI: 10.1201/9781003485933-11

TABLE 11.1
Ethnobotanical and Traditional Medicinal Uses of *Bambusa* Spp.

Bambusa spp.	Country	Region	Local Names	Part Used	Medicinal Preparation and Application	Diseases Treated	References
Bambusa arundinacea	Bangladesh	Kushtia district	Bans	Leaves, stems, and roots	–	Cough, cold, leukoderma, joint pain	Sharkar et al., 2013
	Bangladesh	–	Gora, Kewal, Songough, Otosi	Juvenile shoots	–	Stomach upset, antimalarial	Saalu, 2016
	India	Tirunelveli Hills of Western Ghats; Pachamalai Hills of Tamil Nadu	Moongil	Seeds and seed oil	Paste; oral	Rheumatism	Ayyanar and Ignacimuthu, 2011; Prabhu et al., 2021
	India	Kani tribe of Kanyakumari district of Tamil Nadu	–	Seeds	–	Enhance fertility and reproductive ability	Panda and Padhy, 2007
	India	Pudukottai District, Tamil Nadu	Moongil	Young leaves; terminal buds	Along with turmeric leaves; grounded with *Aloe vera* to form paste	Fractured bones	Anbarashan et al., 2011
	India	District of Jammu and Kashmir	Bans	Roots	Root decoction taken orally	Kidney stone	Bhatia et al., 2014
	India	Atoor Village, Kanyakumari District, Tamil Nadu	Mungil	Leaves	Oral administration of leaf juice mixed with honey	Cough in children	Jeeva and Femila, 2012
	India	Kanyakumari District, Tamil Nadu	Mulmunkil	Roots and leaves	Decoction of roots; infusion of leaves	Skin diseases	Johnsy et al., 2012
	India	Andhra Pradesh	Veduru	Leaves	Decoction, 30–50 mL taken twice a day orally	Asthma	Savithramma et al., 2007
	India	Kanyakumari District, Tamil Nadu	–	Whole plant	Paste of whole plant with turmeric and *Areca catechu* is used	Contusion, swellings	Sukumaran et al., 2014
	India	Kanyakumari District, Tamil Nadu	–	Young shoots	Poulticed, extracted, and topical application	Dislodgment of worms and ulcers; bandage for varmam	Sukumaran et al., 2014
	India	Kanyakumari District, Tamil Nadu	–	Leaf buds	Decoction	Discharge of menses	Sukumaran et al., 2014
Bambusa balcooa	India	Tripura	–	–	–	–	Majumder et al., 2014
	India	Moran tribe of Tinsukia district of Assam	–	Leaves	Decoction	Diabetes	Kalita and Phukan, 2010

(Continued)

TABLE 11.1 (Continued)
Ethnobotanical and Traditional Medicinal Uses of *Bambusa* Spp.

Bambusa spp.	Country	Region	Local Names	Part Used	Medicinal Preparation and Application	Diseases Treated	References
Bambusa bambos	Bangladesh	Rangamati Hill District, Chittagong Hill Tracts	Kanta bans	Leaves, stems, roots, and sprouts	–	Asthma, bleeding disorder, cold, cough, emmenagogue, inflammation, laxative, and leukoderma	Rudra et al., 2021
	India	Palamalai Region of Eastern Ghats	Moongil	Seeds	Topical application of seed paste	Rheumatis	Silambarasan and Ayyanar, 2015
	India	Savandurga Forest, Bangalore Rural District, Karnataka	Bidiru	Grains	Oral consumption as rice prepared from grains	Various diseases	Udayan et al., 2005
	Thailand	Chawang District, Nakhon Si Thammarat Province	–	Leaves	Powdered	Constipation	Neamsuvan et al., 2016
	Thailand	Chawang District, Nakhon Si Thammarat Province	–	Leaves	Bolus	Bilharzia	Neamsuvan et al., 2016
	Thailand	Khao Luang Region, Nakhon Si Thammarat	Phai Pa	Young stems	Decoction; potion; oral	Leprosy and lipoma	Srisawat et al., 2016
Bambusa multiplex	Bangladesh	Chittagong Hill Tracts	Thirwa	Leaves and roots	Oral administration of leaf juice; topical application of paste prepared with roots	Fevers, abscesses, and itches	Hossan et al., 2009
Bambusa oldhamii	Bangladesh	Chittagong Hill Tracts	O-anna	Leaves and roots	Oral administration of leaf juice; topical application of paste prepared with roots	Fevers, abscesses, and itches	Hossan et al., 2009
Bambusa pervariabilis	China	Sansui County	Zhu Ru	–	–	Clearing heat and stop bleeding	Luo et al., 2020
Bambusa tulda	Bangladesh	Muktipara, Chuadanga Sadar, Chuandanga District	Bans	Soft inner core of stems	Oral administration of decoction with parts of other medicinal plants	Cardiovascular disorders and weakness of heart	Rahmatullah and Biswas, 2012
	Bangladesh	Rangamati Hill District, Chittagong Hill Tracts	Bans	Leaves	–	Arthritis and rheumatism	Rudra et al., 2021
	India	Assam	Bans	Roots	Oral administration of decoction	Piles and constipation	Sharma et al., 2012
	India	Assam	Ejo	–	–	Tetanus infection	Sharma et al., 2012
	India	Adi tribe of Arunachal Pradesh	–	–	–	abortifacient, tetanus, stomach pain, and steroidal drug production	Sharma and Borthakur, 2008
Bambusa vulgaris	China	Xishuangbanna, Yunnan Province	–	Culm	–	Edema	Wang et al., 2002

	China	Xishuangbanna, Yunnan Province	–	–	Boiled with *Nicolsonia renifolia*, *Dillenia indica*, and *Mahonia* spp.	Hepatitis	Wang et al., 2002
	India	Tiruvallur District, Tamil Nadu	–	–	–	Type 2 diabetes	Esakkimuthu et al., 2016
	India	Kathiyavadi Village, Vellore District, Tamil Nadu	Mughil	Leaves	Oral administration of extract	Reduced sperm count	Sathiyaraj et al., 2012
	India	Subhartipuram, Meerut District, Uttar Pradesh	–	Young shoots	Cooked shoots	Rheumatis and malaria	Singh et al., 2020
	India	Subhartipuram, Meerut District, Uttar Pradesh	–	Barks	–	Astringent, emmenagogue, malaria, fevers, and heart problems	Singh et al., 2020
	India	Subhartipuram, Meerut District, Uttar Pradesh	–	Leaves	Boiled extract	Clean-out dilation and expulsion after parturition	Singh et al., 2020
	Indonesia	Pulang Pisau Regency, Central Kalimantan-Indonesia	Bambu Kuning	Stems	Oral administration of decoction	Malaria	Novaryatiin and Indah, 2019
	Indonesia	Karangwangi Village, Cianjur, West Java	Haur Koneng	–	Oral administration	Liver disease and gonorrhea	Partasasmita et al., 2017
	Indonesia	Karangwangi Village, Cidaun Sub-District of Cianjur, West Java	Haur Koneng	Buds	–	Cancer	Setiawati et al., 2017
	Laos	Paksan District, Bolikhamsai Province	Mai Sang Kham	Roots	Oral administration of boiled roots with the root of Mak Feuang Phu	Kidney stones	Libman et al., 2006
	Philippines	Batan Island of Batanes	Kawayan	Stems	Oral administration of decoction; topical application after burnt with coconut oil	Allergy	Abe and Ohtani, 2013
	Philippines	Central Philippines	Kawayan	Leaves	Topical application of boiled leaves	Postpartum wash	Ong and Kim, 2015
	Sri Lanka	–	–	Leaves	Oral administration of leaf extract	Dengue fever	Kumarasinghe et al., 2014
Bambusa spp.	Cambodia	Phnom Penh; Kandal; Takeo; Kampong Speu	Russey Srok	Stem	Oral administration of decoction	Liver disorders	Chassagne et al., 2017
	India	Chhattisgarh	Bans	Leaves	Oral administration of leaf juice	Cancer	Jain and Jain, 2010
	India	Chhattisgarh	–	Seeds	Mixed with honey	Cancer	Jain and Jain, 2010
	India	Chhattisgarh	–	Bark	Oral administration of decoction	Cancer	Jain and Jain, 2010
	Malaysia	Kampung Menunggui, Kota Belud District	Buluh	Juvenile shoots	Oral administration of boiled shoots	Postpartum diet	Awang-Kanak et al., 2018

leaf extract of *B. arundinacea* has been traditionally used for the treatment of diabetes, edema ulcer, pyrexia, tetter, hepatotoxicity, and hypertension (Cheng et al., 2023). The decoction of tender *B. balcooa* leaves are used for the treatment of diabetes by the Moran tribe of the Tinsukia district of Assam (Kalita and Phukan, 2010). The Kani tribe of the Kanyakumari district of Tamil Nadu, India, used the seeds of *B. arundinacea* to enhance fertility and reproductive ability (Panda and Padhy, 2007), and were applied as a paste to treat rheumatism (Ayyanar and Ignacimuthu, 2011). The tribes also used the seeds of *B. bambos* for the treatment of rheumatism (Silambarasan and Ayyanar, 2015). Sharma and Borthakur (2008) reported that *B. tulda* is used by the Adi tribe of Arunachal Pradesh for treatments including abortifacient, tetanus, stomach pain, and steroidal drug production. Devi (2018) reported that bamboo shoot extract has been used traditionally to treat wounds and flake-free hair in Manipur. *B. nutans* and *B. tulda* shoots in the form of paste or decoction are used for healing wounds, skin diseases, and dog bites. The leaf paste of *B. tulda* is applied to heal dog bites, while the boiled decoction of fermented shoots is used to cure tumors, meningitis, and ringworm. The decoction of *B. vulgaris* roots is considered diuretic, tonic, depurative, laxative, and cooling and is used for the treatment of burning sensations, arthralgia, debility, dysuria, and arka poisoning (Sharma, 2018). Different components of *B. arundinacea* have been utilized for the treatment of ailments such as ringworm, bleeding gums, joint pain, skin disorders, liver cirrhosis, and abdominal tumors (Zubair et al., 2013).

In Sri Lanka, the leaf extract of *B. vulgaris* has been a traditional remedy to treat dengue fever (Kumarasinghe et al., 2014). The leaves of *B. vulgaris* have been reported to be used for Ayurvedic treatments of diabetes, inflammations, paralysis, ulcers, and skin disorders (Lodhi et al., 2016; Macharla et al., 2012). In Bangladesh, the Rakhain tribe uses the leaves and roots of *B. multiplex* to treat fevers and skin itches (Hossan et al., 2009), while the root of *B. arundinacea* is used for joint pain treatment (Sharkar et al., 2013). The leaf, stem, root, and sprouts of *B. bambos* are used to treat various ailments, including asthma, colds, coughs, inflammation, laxatives, and leukoderma in Bangladesh (Rudra et al., 2021). In Malaysia, the juvenile shoots of *Bambusa* spp. are used as a postpartum diet (Awang-Kanak et al., 2018). The leaves of *B. bambos* in Thailand are used to treat constipation and bilharzia, while the decoction of young shoots are used for leprosy and lipoma (Neamsuvan et al., 2016; Srisawat et al., 2016). In Sansui County of China, *Bambusa pervariabilis* is used for clearing heat and stopping bleeding, while in the Yunnan province, the culms of *B. vulgaris* are used for curing edema (Luo et al., 2020; Wang et al., 2002). The stems and buds of *B. vulgaris* in Indonesia have been used to cure several ailments, including malaria, liver disease, gonorrhea, and cancer (Novaryatiin and Indah, 2019; Partasasmita et al., 2017; Setiawati et al., 2017). There are several bamboo species belonging to the genus *Bambusa* that are integral parts of traditional and folklore medicine of Asian countries and, therefore, it is very important to validate the health benefits of bamboo and its potential in the nutraceutical industry.

11.3 PHARMACOLOGICAL IMPORTANCE OF *BAMBUSA* SPP.

The bamboo species of the *Bambusa* genus show a variety of pharmacological activities (Figure 11.1) and their potential in pharmaceutical and nutraceutical industries.

11.3.1 Antioxidant Activity

The antioxidant potential of *Bambusa* spp. has been reported by several researchers. The methanol extract of *B. arundinacea* exhibited DPPH (1,1-diphenyl-2-picrylhydrazyl) radical scavenging activity of 68.24% was observed at 100 μg/mL (Sharma, 2017). Badwaik et al. (2014) reported the antioxidant potential of fresh shoots of *B. balcooa* to be 26.62% of DPPH radical scavenging activity. Further, Badwaik et al. (2015) carried out an *in-vitro* antioxidant assay using DPPH free radical scavenging assay of the shoots of *B. balcooa*, *B. pallida,* and *B. tulda.* The antioxidant activity of these species ranged from 19.17% to 27.12%. Both acetone and aqueous extracts of *B. vulgaris* demonstrated dose-dependent antioxidant properties by scavenging hydrogen peroxide and DPPH radicals (Goyal et al., 2013). The aqueous extract *B. balcooa* exhibited a higher yield in DPPH and

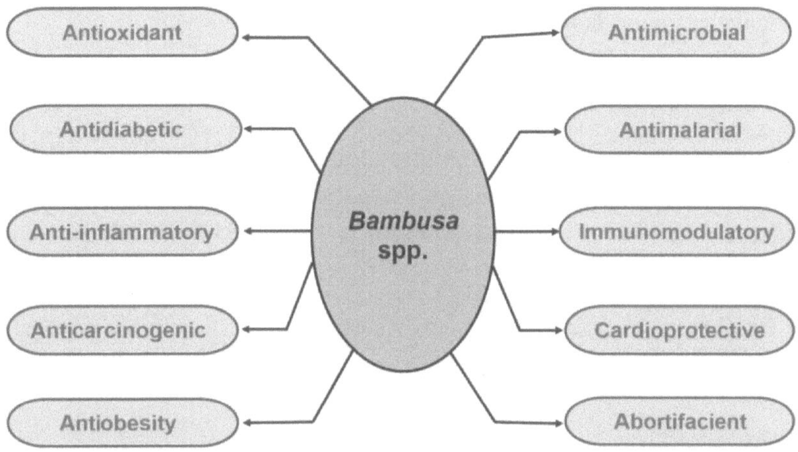

FIGURE 11.1 Pharmacological activities of *Bambusa* spp.

FRAP assays compared to the methanol and acetone extracts (Goyal et al., 2017). Joshi et al. (2024) reported a higher antioxidant activity of leaf and fermented shoots of *B. balcooa,* as indicated by results of FRAP and DPPH assays. The IC_{50} value of the *B. balcooa* leaf and fermented shoot for DPPH assay was 73.92 µg/mL and 147.48 µg/mL, respectively, while for FRAP it was 89.61 mg/mL and 80.90 mg/mL, respectively. The antioxidant activity assessment of fresh, boiled, and fermented shoots of *B. nutans* revealed that fermented shoots have the highest DPPH free radical scavenging activity and FRAP analysis with 51.8% and 54.35% higher activity compared to the fresh shoot (Indira et al., 2024b). The results of *in-vivo* experiments showed that administering fermented-shoots extract resulted in a 31.85% increase in GSH levels and an 11.12% decrease in LPO levels. Indira et al. (2022) evaluated the DPPH and FRAP activity of dried bamboo leaves of various bamboo species including *B. balcooa, B. nutans, B. polymorpha, B. tulda,* and *B. vulgaris,* with some commercially available herbal teas and reported a significant antioxidant potential of leaves of the *Bambusa* species.

11.3.2 Anti-inflammatory Effect

The methanolic extract of *B. arundinacea* leaves demonstrates superior efficacy against immunologically induced paw edema and carrageenan-induced edema in albino rats compared to standard ulcerogenic medications (Muniappan and Sundararaj, 2003). The anti-inflammatory effects of ethyl acetate and aqueous fraction of *B. vulgaris* leaf extracts in carrageenan-induced inflammation of paw edema in wistar albino rats was reported by Lodhi et al. (2016). The results showed that both fractions significantly reduced the percent inhibition of paw edema, and the ethyl acetate fraction–treated groups showed higher inhibition than the aqueous fraction in a dose-related manner. Another study by Carey et al. (2009) showed that 400 mg/kg *B. vulgaris* extract reduced significantly the acetic acid–induced permeability of peritoneal capillaries. Furthermore, the hydroalcoholic extract of the *B. arundinacea* shoot exhibited strong anti-inflammatory properties by stabilizing the membrane of human red blood cells (HRBCs) and utilizing the protein denaturation technique (Vanitha et al., 2016). Singla et al. (2017) showed that nanobiocomposites in film and ointments synthesized using cellulose nanocrystals isolated from *B. bambos* leaves have significantly enhanced the *in-vivo* skin tissue repair by decreasing the production of inflammatory cytokines and increasing fibroblast proliferation, angiogenesis, tissue neo-epithelization, and regeneration.

11.3.3 Anticarcinogenic Effect

The anticarcinogenic properties of the *B. bambos* leaf extracts in chloroform and hydro-alcohol were evaluated using the trypan blue dye exclusion method on Daltons' Lymphoma Ascites and Ehrlich's Ascites carcinoma cell lines (Muneerudeen et al., 2013). The chloroform extract showed superior performance and exhibited no significant signs of hemolysis, indicating that the extracts are not harmful to normal erythrocytes. Silver nanoparticles synthesized with the leaf samples of *B. arundinaceae* and *B. nutans* were reported to have effective anticarcinogenic activities against PC3 and Vero cancer cell lines (Kalaiarasi et al., 2015). Jayarambabu et al. (2021) demonstrated that zinc oxide nanoparticles derived from *B. arundinacea* exhibited anticancer activity against MCF-7 cell lines, as a decrease in cell viability was observed in a concentration-dependent manner.

11.3.4 Antidiabetic Effect

The aqueous extract of leaves of *B. arundinacea* revealed hypoglycemic effects in both euglycemic and hyperglycemic rats (Joshi et al., 2009). The oral dose of 1000 mg/kg in hyperglycemic rats after 3 hours and 500 mg/kg in euglycemic rats after 30 minutes effectively showed the blood sugar lowering effect. The antidiabetic effects of *B. vulgaris* petroleum ether extract at doses of 100, 200, and 400 mg/kg significantly lowered the fasting blood glucose levels in hyperglycemic mice by 32.8%, 45.8%, and 55.3%, respectively, when compared to control animals (Senthilkumar et al., 2011). Senthilkumar (2012) further reported a 13.50% increase in glucose uptake characteristics in the hemidiaphragm of isolated mice due to *B. vulgaris* methanolic extract as compared to the control group. The *in-vitro* hypoglycemic effect of bamboo leaf extract under various pH levels was investigated by Middha and Usha (2012) and was reported that *B. vulgaris* has a hypoglycemic effect across all pH levels, whereas *B. balcooa* and *B. pallida* exhibited the highest activity in acidic conditions. Ahmed et al. (2013) demonstrated that the 28-day treatment of *B. arundinacea* in combination with Qurs Tabasheer significantly reduces serum glucose levels. The significant antihyperglycemic potential of the aerial part of *B. vulgaris* was attributed to the presence of saponins and alkaloids in the extract (Haque et al., 2015). Goyal et al. (2017) reported that the administration of the aqueous extract *B. balcooa* at doses of 100 and 200 mg/kg significantly reduced the fasting blood sugar and glycated hemoglobin levels, while increasing the plasma insulin levels in alloxan-induced diabetic rats compared to the control group. Similarly, the treatment of leaf extracts of *B. tulda* in alloxan-induced diabetic rats showed a significant decrease in blood glucose, serum triglycerides, and total cholesterol levels (Dey et al., 2015). The LC-MS analysis of extracts revealed the presence of *p*-hydroxybenzoic acid and salicylic acid that could have been attributed to the hyperglycemic activity. The antidiabetic potential demonstrated by zinc oxide (ZnO) nanoparticles produced from *B. arundinacea* was higher than the crude extract, as indicated by a lowered blood glucose and increased insulin production from the islet of Langerhans (Jayarambabu et al., 2021).

11.3.5 Antiobesity Effect

The leaf extract of *B. multiplex* demonstrated the antiobese activity in both *in-vitro* and *in-vivo* models (Li et al., 2021). The study revealed that the extract treatment lowered the pancreatic lipase level, inhibited the α-glucosidase, and reduced the absorption of lipids and fats, along with the increased excretion of fat via excreta. The methanolic leaf extract of *B. textilis* also was reported to exhibit antiobesity activities *in vivo* in a high-fat diet rat model (Liu et al., 2016). The study demonstrated that treatment with 200 and 400 mg/kg b.w. of methanolic extract significantly decreased the levels of TC, TG, and LDL-C while elevating the HDL-C level in the treatment groups compared to the group fed a high-fat diet.

11.3.6 Cardioprotective Effect

The crude leaf extract of *B. blumeana* demonstrated the cardioprotective activity against isoproterenol-induced myocardial infarction in Sprague-Dawley rats (Rosales et al., 2013). The results showed activities of serum marker enzymes, including lactate dehydrogenase, serum aspartate aminotransferase, and creatine kinase–MB, were significantly improved in rats pretreated with leaf extract.

11.3.7 Immunomodulatory Effect

The juice of young shoots of *B. vulgaris,* when administered in rats at a dose of 400 mg/kg and 800 mg/kg b.w., modulated both functional as well as immunological measures in a dose-dependent manner. The splenocyte count was significantly increased, whereas the lymphocyte to neutrophil ratio, platelet counts, and bone marrow counts were significantly reduced in the animals treated with bamboo juice. A lower functional response for phagocytic activity was also reported, proving the immunomodulatory activity (Thathsarani et al., 2021).

11.3.8 Antimicrobial Effect

The ethanolic extract of *B. arundinacea* at concentrations of 25, 50, 100, and 150 μl effectively showed the antibacterial activity against *S. aureus* and *Bacillus subtilis* in a dose-dependent manner (Jayarambabu et al., 2021). Sandhiya et al. (2013) also reported the antimicrobial activity of *B. arundinacea* leaves against *E. coli, Pseudomonas aeruginosa, B. subtilis,* and *S. aureus.* The study revealed that hydroalcohol and ethanol extracts were active against all screened bacterial strains with a maximum zone of inhibition against *E. coli* (21 mm and 19 mm, respectively). The bactericidal action of both aqueous and ethanolic extracts of *B. vulgaris* at concentrations range from 2.5 mg/mL to 80 mg/mL was reported by Bolou et al. (2011). Wasnik and Tumane (2014) studied the antibacterial activity of solvent extracts (methanol, acetone, ethanol, petroleum ether, and n-Hexane) of *B. bambos* leaves against multiple drug-resistant bacteria by the agar well diffusion method. The results showed that ethanolic extract has the highest inhibitory activity against *Citrobacter* spp., *Proteus mirabilis,* and *Streptococcus* spp., with an 18-mm zone of inhibition. The ethanolic leaf extracts of *B. tuldoides* at the concentration of 20 g/40 mL extracts exhibited greater antibacterial activity compared to the sorghum liquors and fermented maize (Gabriel-Ajobiewe et al., 2014). Boiled *B. vulgaris* shoot extracts (using solvents water, ethanol, ethyl acetate, and hexane) exhibited a stronger antifungal activity with a minimum inhibitory concentration (MIC) of 0.01 to 2.50 mg/mL compared to their antibacterial activity (MIC: 0.31–2.50 mg/mL) (Kong et al., 2020). Saducos (2021) reported that *B. blumeana* extracts have an antifungal effect comparable to a 1 mg/mL concentration of fluconazole (a pharmaceutical antifungal agent).

The silver nanoparticles prepared by using *B. arundinacea* leaf extract demonstrated a higher antibacterial activity against *E. coli, S. aureus,* and *B. subtilis,* with a maximum zone of inhibition of 21.0 mm, 20.0 mm, and 19.0 mm, respectively, when compared with both silver nitrate and streptomycin alone (Patel et al., 2020). Similarly, Naradala et al. (2021) reported antibacterial activity of copper nanoparticles prepared using an aqueous leaf extract of *B. arundinacea* against *B. subtilis, E. coli, S. aureus,* and *Proteus vulgaris.* The nanoparticles demonstrated a high degree of antibacterial activity for *E. coli* and *B. subtilis,* whereas moderate activity was observed in the case of *P. vulgaris* and *S. aureus.* The active constituents identified as stigmasterol and dihydrobrassicasterol, with side-chain structures responsible for inhibiting the microbial growth.

11.3.9 Antimalarial Activity

The leaves of *B. vulgaris* in dried and fresh form in combination with *Citrus aurantifolia* and *Polyanthia longifolia* leaves were used to treat malaria in Bosomtwe and Sekyere East Districts of

Ghana (Komlaga et al., 2015). In another study, Komlaga et al. (2016) demonstrated that the ethyl acetate and petroleum ether fraction of *B. vulgaris* exhibited the significant antiplasmodial activity against the *Plasmodium falciparum* 3D7 strain with IC_{50} values below 1 μg/mL. Additionally, the ethanolic extract of *B. vulgaris* demonstrated highly potent and specific antimalarial activity against *P. falciparum*, with an IC_{50} of 4.7 μg/mL and a selectivity index of 28.9. The methanolic extract of *B. arundinacea* showed a significant efficacy against *Plasmodium berghei*, with 26.0% suppression of plasmodial infection in infected mice model (Esmaeili et al., 2015).

11.3.10 Abortifacient Activity

The aqueous extracts of *B. vulgaris* leaves possess abortifacient properties, as oral administration of the extract at a dose of 50 mg/kg and 500 mg/kg increased abortion rates by 60% and 100%, respectively, and reduced fetus survival rates to 29% and 0%, respectively (Yakubu and Bukoye, 2009). Both dosages also resulted in higher postimplantation loss and resorption index, along with reduced serum levels of progesterone, luteinizing hormone, and follicle-stimulating hormones. Yakubu et al. (2009) also reported detrimental impact of aqueous extracts of *B. vulgaris* leaves in pregnant Dutch rabbits at doses of 250 mg/kg and 500 mg/kg body weight. The administration of the extract showed toxic effects on liver and kidney functions, as indicated by elevated blood levels of bilirubin, creatinine, uric acid, bicarbonate ions, sodium, and potassium.

11.3.11 Other Health Benefits

The leaf extract of *B. vulgaris* protects the nephron against pro-oxidants treated kidney homogenate (Adeyanju et al., 2019). The oral administration of fresh and fermented shoot extracts of *B. nutans* demonstrated the hepatoprotective activity in the mice model (Indira et al., 2024a). Leaf extracts of *B. vulgaris* have been shown to suppress anxiety (Del Rosario et al., 2012; Fitri et al., 2020). Triterpenoids extracted from the culm shavings of *B. tuldoides* also exhibited antifatigue activity in mice as administration of the extract improved the length of swimming and climbing with no effect on the body weight of mice. Moreover, the serum urea nitrogen and blood lactic acid level decreased (Zhang et al., 2006).

11.4 CONCLUSION

The *Bambusa* genus consists of several species with the potential to serve as an affordable and widely available resource in pharmaceutical, nutraceutical, and biomedical industries. The large bioresource of *Bambusa* spp. would provide the raw material for medicinal preparations for prevention, management, and treatment of prevailing diseases, such as heart disease, obesity, and cancer. Previous research has shown that *Bambusa* spp. offers various health benefits and has great potential as a source for pharmaceutical components due to its high content of nutrients and bioactive compounds. However, further extensive research investigations are needed to validate the effectiveness of traditional health claims; effective extraction of bioactive components; and the development of safe and low-cost nutraceuticals, healthy food, and pharmaceutical formulations.

ACKNOWLEDGMENTS

The authors would like to extend heartfelt gratitude to University Grant Commission, New Delhi, India, and Council of Scientific and Industrial Research, New Delhi, India, for their contribution in the execution and completion of this work.

REFERENCES

Abe, R., & Ohtani, K. 2013. An ethnobotanical study of medicinal plants and traditional therapies on Batan Island, the Philippines. *Journal of Ethnopharmacology 145*(2): 554–565.

Adeyanju, O., Akomolafe, S. F., Atoki, V., & Oboh, G. 2019. The leaves *of Bambusa vulgaris (L.)* ameliorate. Pro-oxidants induced nephrotoxicity in rats-in vitro. *FUTA Journal of Research in Sciences 15*(1): 102–12.

Ahmed, F. A., & Ali, R. F. 2013. Bioactive compounds and antioxidant activity of fresh and processed white cauliflower. *BioMed Research International 2013*(1): 367819.

Anbarashan, M., Parthasarthy, N., & Padmavathy, A. 2011. Ethno-floristic survey in sacred groves, Pudukkottai district, Tamil Nadu-India. *Journal of Medicinal Plants Research 5*(3): 439–443.

Awang-Kanak, F., Bakar, M. F. A., Mohamed, M., & Norazlimi, N. A. 2018. Utilization of natural resources: Preliminary study on ethnopharmacological application of 'ulam'or traditional vegetables among Sama-Bajau of Kampung Menunggui, Kota Belud, Sabah. *AIP Conference Proceedings 2016*(1): 020029-1–020029-7.

Ayyanar, M., & Ignacimuthu, S. 2011. Ethnobotanical survey of medicinal plants commonly used by Kani tribals in Tirunelveli hills of Western Ghats, India. *Journal of Ethnopharmacology 134*(3): 851–864.

Badwaik, L. S., Borah, P. K., Borah, K., Das, A. J., Deka, S. C., & Sharma, H. K. 2014. Influence of fermentation on nutritional compositions, antioxidant activity, total phenolic and microbial load of bamboo shoot. *Food Science and Technology Research 20*(2): 255–262.

Badwaik, L. S., Gautam, G., & Deka, S. C. 2015. Influence of blanching on antioxidant, nutritional and physical properties of bamboo shoot. *Journal of Agricultural Sciences–Sri Lanka 10*(3): 140–150.

Bahru, T., & Ding, Y. 2021. A review on bamboo resource in the African region: A call for special focus and action. *International Journal of Forestry Research 2021*(1): 8835673.

Bajwa, H. K., Santosh, O., & Nirmala, C. 2021. Bamboo shoot for food and nutritional security. *Journal of Pharmacognosy and Phytochemistry 10*(4S): 24–30.

Banik, R. L. 2016. *Silviculture of South Asian Priority Bamboos.* Springer Singapore, 21–115.

Benjamin, M. A. Z., Saikim, F. H., Ng, S. Y., & Rusdi, N. A. 2023. A comprehensive review of the ethnobotanical, phytochemical, and pharmacological properties of the genus *Bambusa. Journal of Applied Pharmaceutical Science 13*(5): 001–022.

Bhatia, H., Sharma, Y. P., Manhas, R. K., & Kumar, K. 2014. Ethnomedicinal plants used by the villagers of district Udhampur, J&K, India. *Journal of Ethnopharmacology 151*(2): 1005–1018.

Bolou, G. E. K., Bagré, I., Ouattara, K., & Djaman, A. J. 2011. Evaluation of the antibacterial activity of 14 medicinal plants in Côte d'Ivoire. *Tropical Journal of Pharmaceutical Research 10*(3): 335–340.

Carey, W. M., Dasi, J. M. B., Rao, N. V., & Gottumukkala, K. M. 2009. Anti-inflammatory activity of methanolic extract of *Bambusa vulgaris* leaves. *International Journal of Green Pharmacy 3*(3): 234–238.

Chassagne, F., Deharo, E., Punley, H., & Bourdy, G. 2017. Treatment and management of liver diseases by Khmer traditional healers practicing in Phnom Penh area, Cambodia. *Journal of Ethnopharmacology 202*: 38–53.

Cheng, Y., Wan, S., Yao, L., Lin, D., Wu, T., Chen, Y., Zhang, A., & Lu, C. 2023. Bamboo leaf: A review of traditional medicinal property, phytochemistry, pharmacology, and purification technology. *Journal of Ethnopharmacology 306*: 116166.

Chongtham, N., & Bisht, M. S. 2020. *Bamboo Shoot: Superfood for Nutrition, Health and Medicine.* CRC Press.

Del Rosario, A., McDermott, M. M., & Panee, J. 2012. Effects of a high-fat diet and bamboo extract supplement on anxiety-and depression-like neurobehaviours in mice. *British Journal of Nutrition 108*(7): 1143–1149.

Devi, P. T. 2018. Edible bamboos of Manipur: analysis of nutrients, anti-nutrients and bioactive compounds in young shoots. A thesis submitted to Panjab University, Chandigarh, India.

Dey, S., Middha, S. K., Usha, T., Brahma, B. K., & Goyal, A. K. 2015. In-vitro antioxidant profiling of *Bambusa tulda* Roxb. aqueous methanolic leaf extract growing in the forests of Kokrajhar district. BTAD, Assam, India. *International Journal of Fundamental and Applied Science 4*: 81–86.

Esakkimuthu, S., Mutheeswaran, S., Arvinth, S., Paulraj, M. G., Pandikumar, P., & Ignacimuthu, S. 2016. Quantitative ethnomedicinal survey of medicinal plants given for cardiometabolic diseases by the non-institutionally trained siddha practitioners of Tiruvallur district, Tamil Nadu, India. *Journal of Ethnopharmacology 186*: 329–342.

Esmaeili, S., Ghiaee, A., Naghibi, F., & Mosaddegh, M. 2015. Antiplasmodial activity and cytotoxicity of plants used in traditional medicine of Iran for the treatment of fever. *Iranian Journal of Pharmaceutical Research 14*: 103.

Fitri, A., Asra, R., & Rivai, H. 2020. Overview of the traditional, phytochemical, and pharmacological uses of gold bamboo (*Bambusa vulgaris*). *World Journal of Pharmacy and Pharmaceutical Sciences 9*(8): 21.

Gabriel-Ajobiewe, R. A. O., Cyril-Olutayo, M. C., Olumekun, V. O., & Akinbileje, F. J. 2014. Effect of *Bambusa tuldoides* cv. ventricosa leaf extracted with fermented steep liquors of maize and sorghum on some pathogenic organisms. *Malaysian Journal of Microbiology 10*(4): 280–289.

Goyal, A. K., Middha, S. K., & Sen, A. 2013. *Bambusa vulgaris* Schrad. ex JC Wendl. var. vittata Riviere & C. Riviere leaves attenuate oxidative stress-An in vitro biochemical assay. *Indian Journal of Natural Products and Resources 4*(4): 436–440.

Goyal, A. K., Middha, S. K., Usha, T., & Sen, A. 2017. Analysis of toxic, antidiabetic and antioxidant potential of *Bambusa balcooa* Roxb. leaf extracts in alloxan-induced diabetic rats. *3 Biotech 7*: 1–11.

Haque, A. M., Das, A. K., Bashar, S. S., Al-Mahamud, R., & Rahmatullah, M. 2015. Analgesic and antihyperglycemic activity evaluation of *Bambusa vulgaris* aerial parts. *Journal of Applied Pharmaceutical Science 5*(9): 127–130.

Hossan, M. S., Hanif, A., Khan, M., Bari, S., Jahan, R., & Rahamatullah, M. 2009. Ethnobotanical survey of the Rakhian tribe inhabiting the Chittagong hill tracts region of Bangladesh. *American-Eurasian Journal of Sustainable Agriculture 3*: 172–180.

Indira, A., Joshi, B., Koul, A., & Chongtham, N. 2024a. Comparative hepato-ameliorative effects of *Bambusa nutans* fresh and fermented shoot extracts on STZ induced diabetic LACA mice. *Talanta 274*: 126035.

Indira, A., Joshi, B., Oinam, S., Koul, A., & Chongtham, N. 2023a. Assessment of preliminary subacute oral toxicity of *Bambusa nutans* aqueous extract based on histoarchitectural and biochemical alterations in LACA mice. *Comparative Clinical Pathology 32*(2): 253–265.

Indira, A., Joshi, B., Oinam, S., Koul, A., & Chongtham, N. 2023b. Potential of bamboo in the prevention of diabetes-related disorders: Possible mechanisms for prevention. In: *Bamboo Science and Technology*. Springer Nature Singapore, Singapore, 89–124.

Indira, A., Santosh, O., Koul, A., & Nirmala, C. 2022. Comparative assessment of the antioxidant potential of bamboo leaves, along with some locally and commercially consumed beverages in India. *Advances in Bamboo Science 1*: 100007.

Indira, A., Shahar, B., Joshi, B., & Chongtham, N. 2024b. Assessment of bioactive compound variations and in-vitro and in-vivo antioxidant activity in edible fresh and processed *Bambusa nutans* shoot through FTIR, GC/MS and HPLC analyses. *Food Chemistry 452*: 139552.

Jain, R., & Jain, S. K. 2010. Traditional medicinal plants as anticancer agents from Chhattisgarh, India: An overview. *International Journal of Phytomedicine 2*(3): 186–196.

Jayarambabu, N., Rao, L. S., Akshaykranth, A., Rao, T. V., & Kumar, R. R. 2021. Study of optical and structural properties of natural bamboo fiber powder prepared by ball milling method. *The European Physical Journal Plus 136*(10): 989.

Jeeva, S., & Femila, V. 2012. Ethnobotanical investigation of Nadars in Atoor village, Kanyakumari District, Tamil Nadu, India. *Asian Pacific Journal of Tropical Biomedicine 2*(2): S593–S600.

Johnsy, G., Sargunam, S. D., & Kaviyarasan, V. 2012. Indigenous knowledge of medicinal plants used for the treatment of skin diseases by the Kaani tribe of Kanyakumari District. *International Journal of Pharmacy and Pharmaceutical Sciences 4*(1): 309–313.

Joshi, B., Indira, A., Kalyan, N., Koul, A. & Chongtham, N. 2024. Assessment of *in vitro* antioxidant potential of *Bambusa balcooa* crude extracts and their safety evaluation based on biochemical alteration in C57bL/6 mice. *Food and Humanity 3*: 100335.

Joshi, B., Indira, A., Oinam, S., Koul, A., & Chongtham, N. 2023. Fermented bamboo shoots: A potential source of nutritional and health supplements. In: *Bamboo Science and Technology*. Springer Nature Singapore, 210–236.

Joshi, R. K., Patil, P. A., Mujawar, M. H. K., Kumar, D., & Kholkute, S. D. 2009. Hypoglycemic activity of *Bambusa arundinacea* leaf aqueous extract in euglycemic and hyperglycemic Wistar rats. *Pharmacologyonline 3*: 789–795.

Kalaiarasi, K., Prasannaraj, G., Sahi, S. V., & Venkatachalam, P. 2015. Phytofabrication of biomolecule-coated metallic silver nanoparticles using leaf extracts of in vitro-raised bamboo species and its anticancer activity against human PC3 cell lines. *Turkish Journal of Biology 39*(2): 223–232.

Kalita, D., & Phukan, B. 2010. Some ethnomedicines used by the Tai Ahom of Dibrugarh district, Assam, India. *Indian Journal of Natural Products and Resources 1*(4): 507–511.

Kalyan, N., Santosh, O., Indira, A., Kuhad, A., & Chongtham, N. 2023. Therapeutic aspects of bamboo for wound healing. In: *Bamboo Science and Technology*. Springer Nature Singapore, 237–264.

Komlaga, G., Agyare, C., Dickson, R. A., Mensah, M. L. K., Annan, K., Loiseau, P. M., & Champy, P. 2015. Medicinal plants and finished marketed herbal products used in the treatment of malaria in the Ashanti region, Ghana. *Journal of Ethnopharmacology 172*: 333–346.

Komlaga, G., Cojean, S., Dickson, R. A., Beniddir, M. A., Suyyagh-Albouz, S., Mensah, M. L., Agyare, C., Champy, P., & Loiseau, P. M. 2016. Antiplasmodial activity of selected medicinal plants used to treat malaria in Ghana. *Parasitology Research 115*: 3185–3195.

Kong, C. K., Tan, Y. N., Chye, F. Y., & Sit, N. W. 2020. Nutritional compositions, biological activities, and phytochemical contents of the edible bamboo shoot, *Dendrocalamus asper*, from Malaysia. *International Food Research Journal 27*(3): 54–54.

Kumarasinghe, N., Perera, P. K., Lankeshwara, L. K., Subasinghe, H. G., & Abertahane, W. 2014. A Sri Lankan traditional medicine (*Bamboo swarasa*) to control Dengue symptoms and signs – A preliminary observation. Published in 2nd International Conference on Ayurveda, Unani, Siddha and Traditional Medicine and Ayu Expo -2014 at Institute of Indigenous Medicine, University of Colombo, Rajagiriya, Sri Lanka.

Li, Y., Chen, B., Cao, H. Y., Li, J. E., Chen, L. L., & Zhang, Q. F. 2021. Pancreatic lipase inhibitory activity of *Bambusa multiplex* cv. Fernleaf leaf extract in vitro and in vivo. *Food & Function 12*(16): 7440–7447.

Libman, A., Bouamanivong, S., Southavong, B., Sydara, K., & Soejarto, D. D. 2006. Medicinal plants: An important asset to health care in a region of Central Laos. *Journal of Ethnopharmacology 106*(3): 303–311.

Liu, M. H., Ko, C. H., Ma, N., Tan, P. W., Fu, W. M., & He, J. Y. 2016. Chemical profiles, antioxidant and anti-obesity effects of extract of *Bambusa textilis* McClure leaves. *Journal of Functional Foods 22*: 533–546.

Lodhi, S., Jain, A. P., Rai, G., & Yadav, A. K. 2016. Preliminary investigation for wound healing and anti-inflammatory effects of *Bambusa vulgaris* leaves in rats. *Journal of Ayurveda and Integrative Medicine 7*(1): 14–22.

Luo, B., Ahmed, S., & Long, C. 2020. Bamboos for weaving and relevant traditional knowledge in Sansui, Southwest China. *Journal of Ethnobiology and Ethnomedicine 16*(1): 9.

Macharla, S. P., Goli, V., & Nath, A. R. 2012. Antidiabetic activity of Bambusa arundinaceae Root Extracts on Alloxan Induced Diabetic Rats. *Asian Journal of Research in Pharmaceutical Science 2*(2): 73–75.

Majumder, J., Bhattacharjee, P. P., Datta, B. K., & Agarwala, B. K. 2014. Ethno-medicinal plants used by Bengali communities in Tripura, northeast India. *Journal of Forestry Research 25*: 713–716.

Middha, S. K., & Usha, T. 2012. An in-vitro new vista to identify hypoglycemic activity. *International Journal of Fundamental and Appllied Science 1*(2): 27–29.

Muneerudeen, J., Himanshu, J., Gururaja, M. P., Swapna, D., Lekshmi, P. V., & Jipnomon, P. 2013. Anticancer potential of *Bambusa bambos* leaf extracts. *International Research Journal of Pharmacy 4*(4): 205–208.

Muniappan, M., & Sundararaj, T. 2003. Antiinflammatory and antiulcer activities of *Bambusa arundinacea. Journal of Ethnopharmacology 88*(2–3): 161–167.

Naradala, J., Allam, A., Tumu, V. R., & Rajaboina, R. K. 2021. Antibacterial activity of copper nanoparticles synthesized by *Bambusa arundinacea* leaves extract. *Biointerface Research in Applied Chemistry 12*: 1230–1236.

Neamsuvan, O., Phumchareon, T., Bunphan, W., & Kaosaeng, W. 2016. Plant materials for gastrointestinal diseases used in Chawang district, Nakhon Si Thammarat province, Thailand. *Journal of Ethnopharmacology 194*: 179–187.

Nirmala, C., Bisht, M. S., Bajwa, H. K., & Santosh, O. 2018. Bamboo: A rich source of natural antioxidants and its applications in the food and pharmaceutical industry. *Trends in Food Science & Technology 77*: 91–99.

Novaryatiin, S., & Indah, I. 2019. The medicinal plants used in Anjir Pulang Pisau, Central Kalimantan-Indonesia. *Pharmacognosy Journal 11*(6): 1572–1579.

Ong, H. G., & Kim, Y. D. 2015. Herbal therapies and social-health policies: Indigenous Ati Negrito Women's dilemma and reproductive healthcare transitions in the Philippines. *Evidence-Based Complementary and Alternative Medicine 2015*(1): 491209.

Panda, T., & Padhy, R. N. 2007. Sustainable food habits of the hill-dwelling Kandha tribe in Kalahandi district of Orissa. *Indian Journal of Traditional Knowledge 6*(1): 103–105.

Partasasmita, R., Anâ, A. Z. I. F. A. H., Iskandar, J., mutaqin, A. Z., Annisa, A., & Ratningsih, N. 2017. Karangwangi peopleâ€™ s local knowledge of bamboo and its role: Implications for management of cultural keystone species. *Biodiversitas Journal of Biological Diversity 18*(1): 275–282.

Patel, H. V., Mangrola, A. V., Bariya, H. S., & Patel, J. D. 2020. Antibacterial and larvicidal activity of biologically synthesized silver nanoparticles from *Bambusa arundinaceae* leaves extract. *Asian Journal of Biological and Life Sciences 9*(1): 43.

Prabhu, S., Vijayakumar, S., Yabesh, J. M., Prakashbabu, R., & Murugan, R. 2021. An ethnobotanical study of medicinal plants used in Pachamalai hills of Tamil Nadu, India. *Journal of Herbal Medicine 25*: 100400.

Rahmatullah, M., & Biswas, K. R. 2012. Traditional medicinal practices of a Sardar healer of the Sardar (Dhangor) community of Bangladesh. *The Journal of Alternative and Complementary Medicine 18*(1): 10–19.

Rosales, S. L. O., Anog, B. F. A., Blanza, R. M., Cuartero, M. A. D., Landicho, M. C., Talledo, E. G. D., Cabanela, R. A., Dumaoal, O. S. R., & Magbojos, C. R. 2013. Cardioprotective activity of *Bambusa blumeana* Schultes leaf crude extract against isoproterenol induced myocardial infarction in Sprague-Dawley rats. *The Steth 7*: 41–57.

Rudra, S., Islam, K. N., Rahman, M. M., & Uddin, S. B. 2021. Medicinal plant diversity and their therapeutic uses in selected village common forests in Chittagong hill tracts, Bangladesh. *Journal of Herbs, Spices & Medicinal Plants 27*(1): 83–107.

Saalu, L. C. 2016. Nigerian Folklore medicinal plants with potential antifertility activity in males: A scientific appraisal. *Research Journal of Medicinal Plants 10*(3): 201–227.

Saducos, A. G. 2021. Antimycotic potential of Kawayang tinik against pathogenic fungal species. *Plant Science Today 8*(2): 403–409.

Sandhiya, S., Subhasree, N., Shivapriya, S., Agrawal, A., & Dubey, G. P. 2013. In vitro antioxidant and antimicrobial potential of *Bambusa arundinacea* (Retz.) Willd. *International Journal of Pharmacy and Pharmaceutical Sciences 5*(2): 359–362.

Santosh, O., Bajwa, H. K., Indira, A., Bano, S., Dolma, N., Bisht, M. S., & Chongtham, N. 2023. Novel food product development through food-to-food fortification with nutrient and bioactive compound-rich Bamboo shoot. In: *Bamboo Science and Technology*. Springer Nature, Singapore, 67–87.

Santosh, O., Nirmala, C., Bajwa, H. K., Bisht, M. S., Indira, A., & Bano, S. 2024. Fortification of noodles with processed bamboo shoots: Evaluating the impact on nutritional composition, antioxidant activity and sensory characteristics. *Advances in Bamboo Science 8*: 100089.

Sathiyaraj, K., Sivaraj, A., Thirumalai, T., & Senthilkumar, B. 2012. Ethnobotanical study of antifertility medicinal plants used by the local people in Kathiyavadi village, Vellore District, Tamil Nadu, India. *Asian Pacific Journal of Tropical Biomedicine 3*(2): S1285–S1288.

Savithramma, N., Sulochana, C., & Rao, K. N. 2007. Ethnobotanical survey of plants used to treat asthma in Andhra Pradesh, India. *Journal of Ethnopharmacology 113*(1): 54–61.

Senthilkumar, M. K., Sivakumar, P., Changanakkattil, F., Rajesh, V., & Perumal, P. 2011. Evaluation of antidiabetic activity of *Bambusa vulgaris* leaves in streptozotocin induced diabetic rats. *International Journal of Pharmaceutical Sciences and Drug Research 3*(3): 208–210.

SenthilKumar, M. K. 2012. Pharmacognostical, phytochemical and pharmacological screening for *Bambusa vulgaris* (Gramineae) and *Pandanus odoratissimus* (Pandanaceae). Ph D. Thesis.

Setiawati, T., Mutaqin, A. Z., Irawan, B., Anamillah, A., & Iskandar, J. 2017. Species diversity and utilization of bamboo to support lifes the community of Karangwangi Village, Cidaun Sub-District of Cianjur, Indonesia. *Biodiversitas Journal of Biological Diversity 18*(1): 58–64.

Sharkar, P., Rahman, M. M., Haque Masum, G. Z., Nayeem, M. A., Hossen, M. M., & Azad, A. K. 2013. Ethnomedicinal importance of the plants in villages in Kushtia Sador and Mirpur Upozila, Bangladesh. *Journal of Herbs, Spices & Medicinal Plants 19*(4): 401–417.

Sharma, A. 2017. Evaluation of antioxidant potential of Drakshavaleha: A polyherbal ayurvedic formulation. *Asian Journal of Pharmaceutics 11*: 04.

Sharma, T.P., & Borthakur, S.K. 2008. Ethnobotanical observations on Bamboos among Adi tribes in Arunachal Pradesh. *Indian Journal of Traditional Knowledge 7*(4):594–597.

Sharma, U. K., Pegu, S., Hazarika, D., & Das, A. 2012. Medico-religious plants used by the Hajong community of Assam, India. *Journal of Ethnopharmacology 143*(3): 787–800.

Sharma, V. (2018). Changes in bioactive components and anti-nutrients during processing of bamboo shoots. Ph. D. Thesis.

Silambarasan, R., & Ayyanar, M. 2015. An ethnobotanical study of medicinal plants in Palamalai region of Eastern Ghats, India. *Journal of Ethnopharmacology 172*: 162–178.

Singh, R., Upadhyay, S. K., Rani, A., Kumar, P., & Kumar, A. 2020. Ethanobotanical study of Subhartipuram, Meerut, Uttar Pradesh, India. II. Diversity and pharmacological significance of shrubs and climbers. *International Journal of Pharmaceutical Research 12*(2): 383–393.

Singla, R., Soni, S., Kulurkar, P. M., Kumari, A., Mahesh, S., Patial, V., Padwad, Y. S., & Yadav, S. K. 2017. In situ functionalized nanobiocomposites dressings of bamboo cellulose nanocrystals and silver nanoparticles for accelerated wound healing. *Carbohydrate Polymers 155*: 152–162.

Srisawat, T., Suvarnasingh, A., & Maneenoon, K. 2016. Traditional medicinal plants notably used to treat skin disorders nearby Khao Luang mountain hills region, Nakhon Si Thammarat, Southern Thailand. *Journal of Herbs, Spices & Medicinal Plants 22*(1): 35–56.

Sukumaran, S., Brintha, T. S., Subitha, P., Sheebha, Y. A., & Jeeva, S. 2014. Usage of medicinal plants by two cultural communities of Kanyakumari district, Tamil Nadu, South India. *Journal of Chemical and Pharmaceutical Research 6*(8): 67–79.

Thathsarani, N., Jayasinghe, C. D., Jayawardena, U., & Nilakarawasam, N. 2021. Effect of oral administration of the fresh juice of Bamboo (*Bambusa vulgaris*) young shoots on enumeration of bone marrow cells, platelets, splenocyte and phagocytic activity of peritoneal macrophages of rats. *Phytomedicine Plus 1*(3): 100059.

Udayan, P. S., George, S., Tushar, K. V., & Balachandran, I. 2005. Some common medicinal plants used by the Nayaka community, Savandurga forest of Magadi taluk, Bangalore rural district, Karnataka, India. *Journal of Natural Remedies 5*(1): 35–40.

Vanitha, B., Kumar, R. K., Duraiswamy, B., Gnanavel, G., & Neelamma, G. 2016. Estimation of total phenol and total flavanoid content with antioxidant and anti-inflammatory activity of *Bambusa arundinacae* (bamboo shoot) grown in Nilgiris. *Der Pharma Chemica 8*: 307.

Wang, K., Hong, L. T., & Rao, V. R. 2002. Diversity and indigenous utilization of bamboo in Xishuangbanna, Yunnan Province, Southwest China. *CABI Databases 1*(3): 263–273.

Wang, M., Yu, A., Hu, W., Zhang, Z., Wang, Z., Meng, Y., Yang, B., & Kuang, H. 2024. Extraction, purification, structural characteristic, health benefit, and product application of the polysaccharides from bamboo shoot: A review. *International Journal of Biological Macromolecules 271*(2): 132581.

Wasnik, D. D., & Tumane, P. M. 2014. Antibacterial activity of *Bambusa bambose* L. against multiple drug resistant (MDR) bacteria isolated from clinical specimen. *International Journal of Pharmaceutical Sciences 25*(1): 215–18.

Yakubu, M. T., & Bukoye, B. B. 2009. Abortifacient potentials of the aqueous extract of *Bambusa vulgaris* leaves in pregnant Dutch rabbits. *Contraception 80*(3): 308–313.

Yakubu, M. T., Bukoye, B. B., Oladiji, A. T., & Akanji, M. A. 2009. Toxicological implications of aqueous extract of *Bambusa vulgaris* leaves in pregnant Dutch rabbits. *Human & Experimental Toxicology 28*(9): 591–598.

Zhang, Y., Wu, L., Li, Y., Yang, J., Yang, H., Zhao, Y., & Chen, G. 2024. Bamboo shoot and its food applications in last decade: An undervalued edible resource from forest to feed future people. *Trends in Food Science & Technology 146*: 104399.

Zhang, Y., Yao, X., Bao, B., & Zhang, Y. 2006. Anti-fatigue activity of a triterpenoid-rich extract from Chinese bamboo shavings (Caulis bamfusae in taeniam). *Phytotherapy Research 20*(10): 872–876.

Zubair, M., Bibi, Z., Rizwan, K., Rasool, N., Zahoor, A. F., & Riaz, M. 2013. In vitro antimicrobial and haemolytic studies of *Bambusa arundinaceae* leaves. *Journal of Applied Pharmaceutical Science 3*(4): 111–115.

12 The Medicinal Potential of *Sasa senanensis*

Prior Art and Future Explorations

Ravjot Kaur, Jitender Madan, Joohee Pradhan, Nagendra Singh Chauhan, and Ashish Baldi

12.1 INTRODUCTION

Humanity has always relied on nature to provide necessities like clothing, food, housing, medicine, spices, fertilizers, and means of transportation (Gratani et al., 2008). The plants have been utilized for medicinal purposes since ancient times. The traditional medical system of Ayurveda remains practiced in many societies. Because of their nutritional value and therapeutic qualities, medicinal plants have been used extensively in traditional medicine (Baldi and Bansal, 2020; Singh et al., 2021a). Bamboo is a large, woody, perennial grass that is a member of the order Monocotyledon and the group Angiosperms. One minor subfamily, *Centrothecoideae*, and five big subfamilies, *Arundinoideae, Pooideae, Chloridodeae, Panicoideae,* and *Bambusoideae,* make up the grass family *Poaceae* (also known as *Gramineae*) (Pandey, 1994) There are around 60–70 genera and 1,200–1,500 species of bamboo around the globe. Approximately 50% of these species are found in Asia, mostly in the Indo-Burmese region (Wong, 1995). Bamboo shoots have evolved into a lucrative global industry, generating billions of dollars annually. China stands as the largest supplier of bamboo shoots to the global market, earning approximately US$130 million per year from exports alone. Around 200 bamboo species are suitable for producing edible and tasty bamboo shoots, with Moso bamboo (*Phyllostachys edulis*) and Kumaisasa (*Sasa senanensis*) emerging as the top choices in the international market. Bamboo is becoming famous around the world as a nutritious and healthy food option because of its lack of residual toxicity; low-fat content; and abundance of edible fiber and essential mineral elements such as potassium, calcium, manganese, zinc, chromium, copper, and iron, with lower levels of phosphorus and selenium (Bystriakova et al., 2003). Bamboo has gained remarkable traction as an herbal medicine within the global healthcare systems. Many species of bamboo have demonstrated therapeutic potential and have been utilized since ancient times. The enduring utilization and scientifically validated benefits of these bamboo varieties underscore their significant medicinal value. Bamboo has played an important role in human civilization throughout history and continues to contribute to the well-being of contemporary society. For instance, *Bambusa* leaves are used as a traditional treatment for ailments like coughs, fevers, and leprosy. Research has revealed that extracts from *B. arundinacea* possess anti-inflammatory, antiulcer, antifertility, antimicrobial, and hypoglycemic properties (Nazreen et al., 2011).

Sasa senanensis (*SS*), also known as Kumaisasa, a Japanese native dwarf bamboo plant, has a long-standing tradition of being utilized for various purposes. Its leaves are traditionally employed as tea, healthy food ingredients, or in folk medicine practices. For instance, they are often used for wrapping sushi and Chiaki (rice cake) to prevent decay. *SS* is an evergreen bamboo that reaches a height of 1.8 meters (6 feet), maintaining foliage throughout the year. This species is hermaphroditic, possessing both male and female reproductive organs, and is pollinated by wind. *SS* thrives in various soil types, including light (sandy), medium (loamy), and heavy (clay) soils, and is adaptable to a range of pH levels, from mildly acidic to neutral and mildly alkaline soils. *SS* can grow in semi-shaded environments, such as light woodlands, and prefers moist soil conditions (Lu et al., 2006). *Sasa* plant

DOI: 10.1201/9781003485933-12

dried leaves have been used medicinally, for herbal tea, and in functional foods (Chuyen et al., 1983). SASA-Health®, an 'alkaline extract' made from *SS* leaves, is classified as an over-the-counter drug in group III and is anticipated to be less dangerous than group II Kampo medicines (Zhou et al., 2009). *SS* is said to be effective in treating stomatitis, weariness, reduced appetite, halitosis, and body odor; however, there is not enough scientific data to back up these claims because there are not any appropriate biomarkers (Ono et al., 2011). In studies, *SS* has demonstrated anti-inflammatory, antibacterial, antiviral (Sakagami et al., 2008), anti-UV (Sakagami et al., 2010), anticancer, antioxidant, and radical-scavenging properties, along with its synergistic effects with vitamin C, CYP3A4-inhibitory activity, lichenoid dysplasia, bone health, immune regulator, and antihalitosis activity. *SS* shares several biological characteristics with LCC, including significant anti-HIV, anti-UV, and synergistic activity with vitamin C, as anticipated from their availability under alkaline conditions.

The objective of this compilation is to investigate the medicinal qualities and therapeutic potential of the bamboo variety, *SS*, by analyzing several scientific studies and research findings. The chapter examines *SS* extracts' antitumor, antiviral, antifatigue, and other health-improving qualities, offering insights into their mechanisms of action and possible applications in conventional and modern medicine. Furthermore, it aims to enhance the importance of further investigation into the bioactive components found in bamboo, the related formulations, and the integration of treatments derived from bamboo into all-encompassing healthcare strategies for the benefit of present and future generations. Challenges associated with current state of art for further therapeutic exploration are also discussed to present as base report for researchers.

12.2 BOTANICAL DESCRIPTION (*Sasa senanensis* PFAF Plant Database, n.d.)

Taxonomy and classification (ACIR Community, n.d.)

Kingdom:	Plantae
Phylum:	Tracheophyta
Class:	Magnoliopsida
Order:	Poales
Family:	*Poaceae*
Genus:	*Sasa*
Species:	*senanensis*

Habitat (*Sasa senanensis*: Useful Temperate Plants, n.d.)

Woodlands and damp hollows in mountains, ranging from East Asia, Far East Russia, and Japan.

Weed Potential:	Yes
Habit:	Evergreen bamboo
Height:	1.80 m
Pollinators:	Wind
Cultivation Status:	Wild

12.3 CULTIVATION DETAILS

SS (Figure 12.1) thrives in the cool temperate zone of its habitat range and demonstrates resilience to temperatures as low as –20°C. It prefers partial shade and flourishes in well-draining, humus-rich loam soil with adequate moisture during the growing season. Thriving in thin woodland environments, this species exhibits notable resistance to honey fungus.

FIGURE 12.1 *Sasa senanensis* plant.

SS has an interesting way of growing. Every year, every plant produces several new stems, and in the first year of growth, these stems grow to their greatest height. Only the growth of additional side branches and leaves may be considered subsequent expansion in the stem. In certain mature tropical species, during peak growth periods, new stems can grow rapidly each day to a height of up to 30 meters. *SS* and other bamboo are plant species that grow the fastest on the globe because of their quick growth rate. *SS* and other plants of the family are usually monocarp, meaning they live for a few years before going through a flowering phase that lasts one to three years and is frequently followed by senescence. During this time, the plants release a large number of seeds. Nonetheless, flowering patterns differ; some species flower sporadically, meaning that plants bloom every year without dying in between, while others experience gregarious flowering events, in which every plant in a species flower at the same time (*Sasa senanensis:* Useful Temperate Plants, n.d.).

12.4 MORPHOLOGY

SS, an evergreen bamboo, can reach heights of 200 cm or more. Its upright, woody culms typically measure around 3 to 5 mm in diameter and feature thin-walled internodes. The elongated rhizomes give the plant a running growth habit, allowing it to produce new canes. Rapidly, it establishes an open thicket characterized by slender, upright culms adorned with a loose canopy of large, spreading leaves (*Sasa senanensis:* Useful Temperate Plants, n.d.).

12.5 CHEMICAL CONSTITUENTS

SS mainly composed of polysaccharides, although it also contains low molecular weight compounds such as p-coumaric acid, ferulic acid, vanillin, apigenin, luteolin, tricin, syringaresinol, phenylpropanoid, p-hydroxybenzoic acid, *p*-hydroxybenzaldehyde, and 3,4-dihydroxybenzaldehyde (Matsuta et al., 2011a,b; Yagi et al., 2013). The chemical structures of representative chemical constituents are given in Figure 12.2.

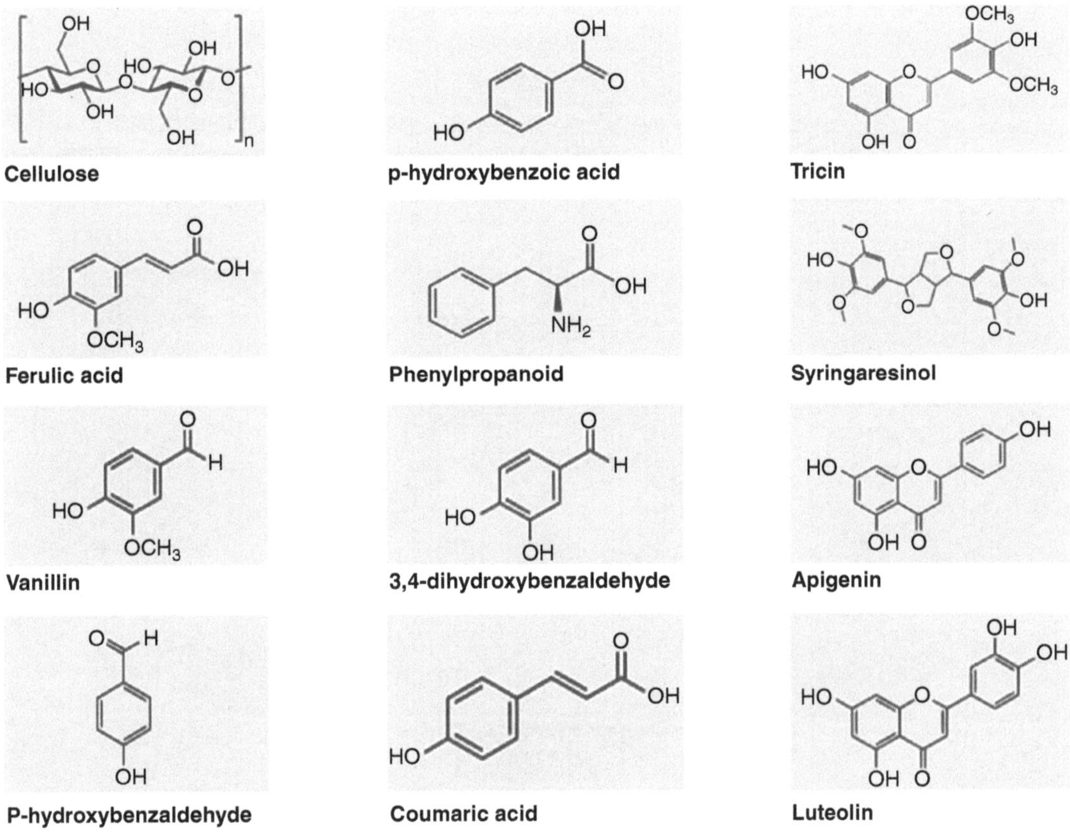

FIGURE 12.2 Major chemical constituents of *Sasa senanensis*.

12.6 PHARMACOLOGICAL ACTIVITIES

As per published literature, *SS* has been explored for various pharmacological activities, as detailed below, and has shown significant potential for therapeutic benefits (Figure 12.3).

12.6.1 Antioxidant Properties

Significant antioxidant activity in *SS* extracts was shown in reducing oxidative damage in mice caused by restraint stress. It is important to remember, though, that the efficacy of some functional molecules is determined by their activity, which includes tissue distribution and absorption. As a result, whereas *SS* extract may show strong antioxidant activity, this does not necessarily correspond to its action *in vivo*. However, *SS* extract is a promising, readily available redox modulator that may also be a useful natural antioxidant source. This characteristic might help stop different oxidative stressors from progressing (Hollman, 2004; Khatun et al., 2013).

12.6.2 Antimicrobial Activity

SS exhibited a bacteriostatic effect, rather than bactericidal, on *F. nucleatum* and *P. intermedia*, bacteria associated with periodontal diseases. While the inhibition of bacterial growth was not complete, *SS* nearly eradicated hydrogen sulfide and methyl mercaptan from the bacterial cultures. This implies that *SS* might metabolize or capture these volatile compounds, in addition to its antibacterial properties. Consequently, *SS* appears to have potential applications in dentistry as a mouthwash.

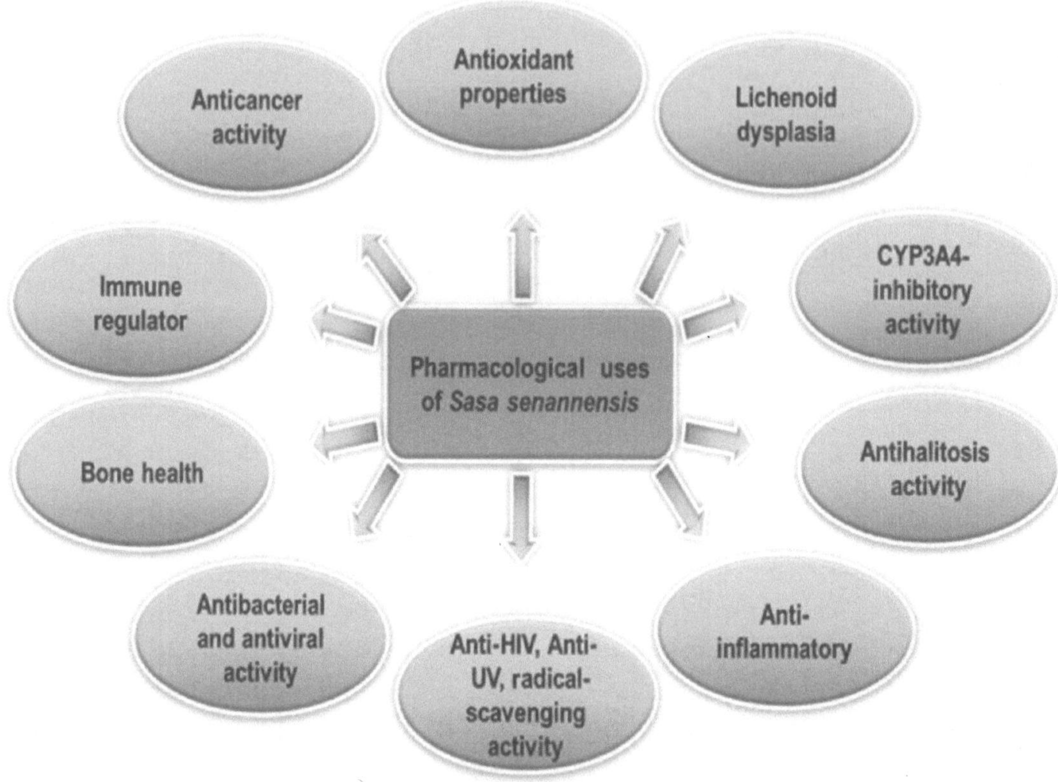

FIGURE 12.3 Pharmacological uses of *Sasa senanensis.*

SS demonstrated protective effects on human T-cell leukemia molt-4 cells against the cytopathy effect induced by HIV infection, exhibiting greater anti-HIV activity compared to tannins and flavonoids, and similar efficacy to natural and synthetic lignins. Additionally, *SS* shielded madin-darby canine kidney cells from the cytopathic effects of influenza virus infection. Furthermore, *SS* synergistically augmented the superoxide anion and hydroxyl radical-scavenging activity of vitamin C (Sakagami et al., 2008).

12.6.3 Anticancer Activity

Meshima Max (MM), an extract blend containing *Phellinus linteus*, leaves from *SS*, and *Inonotus obliquus*, a type of mushroom, may prevent cancer from starting and tumor progression. MM may therefore have the ability to prevent cancer as supported by a hypothesis that has been corroborated by research using both transplanted and carcinogen-induced tumor models. The most significant element of innate immunity, macrophage activation was suggested as the cause of this impact. MM showed very little cytotoxicity, particularly against healthy cells. MM could therefore develop into a different type of supplemental nutrient for cancer patients as well as a post-chemotherapy maintenance strategy (Fang et al., 2020).

Another study revealed that the 'antitumor' efficacy of the leaf extract of *SS* was assessed using various mouse tumor models, including S-180, C38, and Meth-A. Oral administration of the extracts commenced after the size of tumor reached approximately 6 mm, with concentrations of 0.05% or higher utilized. Results indicated significant suppression of tumor growth in both the S-180 and C38 tumor models. Furthermore, the treatment group exhibited significantly prolonged overall survival compared to the control group. Activation of macrophages and NK cells by the extracts proposed that their antitumor effect was mediated through immunopotentiation (Seki et al., 2010).

SS was utilized in traditional medicine for many years. Several investigations highlighted various biological properties of an alkaline extract obtained from *SS* leaves called 'Sasa Health'. When mice with spontaneous or Her2/NeuN-induced mammary carcinogenesis were fed this extract, their breast tumor development and incidence were suppressed (Ren et al., 2004; Tsunoda et al., 1998). Additionally, it mitigated certain negative effects of restricted eating and affected the motor activity of both intact and gonadectomized mice (Nagasawa et al., 2001a,b).

In 2010, hot water extract from *Sasa* leaves demonstrated antitumor effects. According to Seki and Maeda (2010), this extract successfully prevented the development of melanoma and sarcoma cells that had been injected, as well as the formation of breast tumors that were generated by 7,12-dimethylbenz(α)anthracene. Immunopotentiation was attributed to the extract's antitumor efficaciousness, with 1,3-β-glucan found as the likely key immune-potentiating ingredient (Seki et al., 2010).

12.6.4 Anti-inflammatory Activity

The 'alkaline' extract of *SS* demonstrates a plethora of biological activities. The extract affects the activity of macrophages. *SS* blocks the formation of nitric oxide by lipopolysaccharide-activated mouse macrophage-like RAW264.7 cells (Auluck et al., 2008). Western blot and reverse transcription polymerase chain reaction (RT-PCR) analyses revealed about inhibition stemmed from suppressing the expression of inducible nitric oxide synthase at both the protein and mRNA levels. Electron spin resonance spectroscopy depicts *SS* scavenged the nitric oxide radical produced by Calbiochem A nitric oxide in a dose-dependent manner (Brown et al., 2008).

To further ascertain its anti-inflammatory activity, the activity of prostaglandin E_2 production and the enzymes involved in the arachidonic acid pathway were affected by *SS* and was revealed in the study. *SS* effectively inhibited prostaglandin E_2 production by lipopolysaccharide-stimulated RAW264.7 cells, albeit slightly less than its inhibition of nitric oxide production. Moreover, *SS* inhibited cyclooxygenase expressions at both the protein and mRNA levels, albeit slightly less compared to its inhibition of nitric oxide synthase expression. The concentration of arginine in *SS* is less compared to the culture medium. These findings proposed that *SS* exhibits a modest anti-inflammatory activity (Sakagami et al., 2010).

12.6.5 Antihalitosis Activity

In a study, a small test was performed to figure out the activity of *SS*. Twelve people participated in the test. They brushed their teeth three times a day after every meal with toothpaste containing *SS*. Bacteria present on the tongue was calculated by portable devices. As a result, toothpaste reduced halitosis, not bacteria on the tongue (Sakagami et al., 2016).

12.6.6 Anti-HIV Activity

The anti-HIV activity of products first, second, and third displayed an effective dose-dependent reduction of the cytopathic effect of HIV infection. However, their anti-HIV activity was notably lower compared to positive controls such as dextran sulfate, curdlan sulfate, azidothymidine and 2′,3′-dideoxycytidine. The order of potency for anti-HIV activity was observed as follows: Product A (Selectivity index = 607) > Product C (Selectivity index = 117) > Product B (Selectivity index = 111) (Ono et al., 2011).

Previous research also indicated a significantly higher antiviral efficacy of the alkaline extract derived from *SS* against HIV in comparison to lignin precursors, tannins, and flavonoids, hinting at its potential application in oral diseases. However, systematic comparative studies with herpes simplex virus (HSV) have been relatively scarce in comparison to those conducted with HIV. In this study, the effect of combining *SS* with other widely recognized antiviral agents was explored to see the possibility to augment their individual activities. *SS* exhibited potent anti-HIV activity, albeit two orders of magnitude less compared to azidothymidine, 2′,3′-dideoxycytidine, dextran sulfate, and curdlan sulfate. Combining *SS* with these antiviral agents resulted in synergistic effects. *SS* and acyclovir demonstrated synergistic anti-HSV activity when combined (Sakagami et al., 2010).

12.6.7 In Lichenoid Dysplasia

Over an extended period, treatment with *SS* led to a gradual reduction in 'white streaks' and salivary levels IL-6 and IL-8. *SS* also notably inhibited the differentiation of macrophages into osteoclasts. For 11 months, a male patient with white lacy streaks in his oral mucosa was given *SS* orally three times a day. An enzyme-linked immunosorbent test was used to diagnose the amounts of IL-6 and IL-8 in saliva, and intraoral photography was used to monitor the development of white streaks. Tarrate-resistant acid phosphatase positive multinuclear cell formation was used to measure osteoclastogenesis in mouse macrophage-like RAW264.7 cells, which was activated by a receptor activator of nuclear factor-κB ligand. Hence, *SS* exhibited promising a therapeutic potential associated with oral diseases (Matsuta et al., 2011a,b).

12.6.8 Immune Regulator

SS extract has been reported to regulate immunity by hindering the expression of JNK and ERK, which are involved in inflammatory signal transduction. In a study conducted by Sato et al. (2021), RAW 264.7 cells were subjected to the treatment of *SS* leaf extract, and LPS alone, or *SS* leaf extract in combination with LPS, and the effect of the treatment was observed; the mRNA and protein expression of TNF-α, IL-1β, MAPKs, and NFκB was also measured. The results revealed that the mRNA expression levels of IL-1β and TNF-α increased in RAW 264.7 cells upon the treatment of *SS* leaf extract. On the contrary, the level of expression of IL-1β and TNF-α dropped when treated with the combination of *SS* leaf extract and LPS in the cells and the protein expression of JNK and ERK expression was also decreased. It was concluded that *SS* leaf extract could be used to regulate immunity by reducing the expression of JNK and ERK protein expressions.

12.6.9 Improving Bone Health

SS appears to exert anabolic effects on bone metabolism, positively influencing bone formation while negatively affecting bone resorption. This dual action suggests that *SS* may hold therapeutic promise for skeletal conditions like osteoporosis, as it can potentially mitigate increased bone resorption while promoting bone formation simultaneously. Research indicated that *p*-coumaric acid stimulates osteoblast proliferation and differentiation, while luteolin inhibits osteoclast differentiation and encourages osteoblast differentiation. The identification and characterization of compounds responsible for the anti-osteo-blastogenesis and/or pro-osteogenesis effects in *SS* could pave the way for developing drug designs aimed at treating osteoporosis (Tomomura et al., 2017).

SS, classified under Group III, is anticipated to pose lower risks in comparison to Kampo Medicines classified under Group II. In Japan, drug stores offer three commercial *SS* products. The first product comprised a 100% pure extract of *SS* infused with Fe(II)-chlorophyllin, wherein Fe(II) replaces Mg(II) at the core through the addition of $FeCl_2$. The second and third products contained Cu(II)-chlorophyllin along with a smaller amount of LCC. Product C is enriched with ginseng extract (1.8 mg/mL) and pine (*Pinus densiflora*) leaf extract (3 mg/mL), incorporating paraben as a preservative (Ono et al., 2011).

12.6.10 UV Protection

In a recent study, UV irradiation triggered 'non-apoptotic' cell death in HSC-2 cells, without activating caspase-3 or causing inter-nucleosomal DNA fragmentation. All the products exhibited a dose-dependent reduction in UV-induced cytotoxicity (Sakagami et al., 2010).

Another study revealed that *SS* exhibited anti-UV activity slightly lower in sodium ascorbate but higher than that of gallic acid, (–) epigallocatechin gallate, chlorophyll a, and chlorophyllin. Combining *SS* with sodium ascorbate resulted in synergistic anti-UV activity. Upon gel filtration fractionation, the majority of the anti-UV activity was recovered which eluted as a single peak in high-performance liquid chromatography (Matsuta et al., 2011a, b).

12.6.11 Radical Scavenging Activity

Studies have revealed that *S. senanensis* extract (SE) has enhanced the radical scavenging activity of vitamin C. According to Sakagami et al. (2008), SE showed synergistic radical scavenging activity with vitamin C. Superoxide was mixed with DPMO and four radical peaks were obtained (DPMO-OOH) in spin adduct. Similarly, when hydroxyl radical was mixed with DPMO, four peaks were obtained (DPMO-OH) in spin adduct. The tests confirmed that SE extract successfully enhanced the superoxide anion and hydroxyl radical scavenging activity synergistically with vitamin C.

12.6.12 CYP3A4-Inhibitory Activity

The activity of SE, grapefruit juice, and chlorophyllin showed a dose-dependent inhibition of the β-hydroxylation of testosterone, typically employed to assess CYP3A4 activity which is generally assessed by 50% inhibitory concentration (IC_{50}). Among them, grapefruit juice demonstrated the highest CYP3A4-inhibitory activity with an IC_{50} of 58 μg/mL, followed by chlorophyllin with an IC_{50} of 124 μg/mL, and then SE with an IC_{50} of 403 μg/mL. Notably, grapefruit juice exhibited a CYP3A4 inhibition level similar to that of Cu(II)-chlorophyllin, while SE inhibitory activity on CYP3A4 was lower compared to the inhibition typically achieved by grapefruit juice and chlorophyllin (Sakagami et al., 2010).

12.6.13 Other Uses

The *SS* leaf, illuminated by the sun, enriches the verdant landscapes of mountains. Throughout ancient times, people have utilized *SS* leaves to wrap food and put into a beverage for consumption, stomatitis, weariness, reduced appetite, halitosis, and body odor. These leaves are also documented in the ancient Chinese classic 'Honzoukoumoku' as herbal medicine. SASA-Health® harnesses an active ingredient from natural *SS* leaves using specialized technology, offering it in the form of medicinal supplements (Daiwabiological Research Institute Co. Ltd, 2012).

12.6.14 Good Health Products

An extract of bamboo grass leaves, Sasa Health has been recognized as a wonder product. It shows multiple actions used for good body odor, drowsiness, loss of taste, and foul breath. *SS* leaf effectively alleviates fatigue and tiredness. It also provides relief from stomach and intestinal disorders, as well as helps mitigate loss of appetite. Stomach and intestine weakness can lead to a decrease in appetite. Moreover, bad breath can stem from various causes, including gum problems and gut issues. *SS* leaf exhibited remarkable deodorizing properties, offering effective odor control for both oral and bodily odors. While bodily odor often goes unnoticed, the phyto-constituents in *SS* leaf contain substances that actively eliminate body odor. Despite their small size, canker sores can be extremely painful. The active ingredient in *SS* leaf combats inflammation and improves the condition of canker sores (Daiwabiological Research Institute Co. Ltd., 2012).

12.7 FUTURE PERSPECTIVES

Given bamboo's extensive historical use and recent scientific validation—especially for the *SS* variety—the future seems hopeful for incorporating *SS*-derived medicines into all-encompassing healthcare frameworks. An increase in the number of studies indicates *SS* health benefits, such as antitumor, antiviral, and antimicrobial effects, etc. Furthermore, there is great potential for improving healthcare results worldwide through the incorporation of medicines developed from *SS* into conventional and modern medicine. Beyond its use in medicine, *SS* can significantly impact sustainable practices in a wide range of industries. *SS* is a leader in environmentally friendly solutions for a variety of industries, like textiles and construction, because of its quick growth, low environmental effect, and adaptability. *SS* is becoming more widely recognized as a beneficial natural antioxidant source, which is causing it to be incorporated into a wide variety of functional foods, drinks, and dietary supplements that are intended to improve health and well-being. Research on *SS* in the creation of natural treatments for oxidative stress–related ailments is being fueled by the growing interest in traditional herbal medicine methods. This is consistent with the worldwide trend towards natural and holistic approaches to healing. As environmental sustainability gains more attention, there is a push to support *SS* cultivation and harvesting through environmentally responsible and sustainable farming practices. This guarantees that this priceless plant species will be preserved for upcoming generations.

Various advantages for present and future generations by incorporating treatments derived from *SS* into holistic healthcare initiatives may lead towards a path of a more sustainable and healthy future. However, there are not enough extensive clinical trials to support the traditional uses of *SS* in treating a range of medical ailments. To prove that treatments based on *SS* are safe and effective, more thorough research is required. More investigation is required to determine the underlying mechanisms that exhibit the therapeutic benefits of *SS*. Comprehending the molecular mechanisms implicated will offer discernment into its pharmacological characteristics. To guarantee constant quality and potency in pharmaceutical goods, it is essential to standardize *SS* extracts and formulations. Creating standardized extraction and preparation procedures will improve the repeatability of research. Method development and validation as per ICH Guidelines confirmed the industrial acceptability of the analytical procedure. Analytical processes for several phyto-constituents and drug moieties has been established in the author's lab (Baldi et al., 2010; Sharma et al., 2021; Singh et al., 2023; Singh and Baldi, 2013; Singh et al., 2019). It is evident from previous studies that agro-ecological zoning (Singh et al., 2021a; Singh et al., 2021b), seasonal and geographical variations (Gautam et al., 2019), good agricultural/collection practices (Singh and Baldi, 2018), and several other factors during cultivation and collection of herbal drugs have significant impact on the concentration of active constituents (Singh et al., 2021d; Singh et al., 2022; Singh et al., 2023). On similar lines, active constituents and quantitation are essential to explore *SS* for further benefits.

The bioavailability of *SS*'s active components has not been well studied. Dosage schedules for medicinal applications will be optimized by research on the components of *SS* absorption, distribution, metabolism, and excretion. Little is known about potential interactions between conventional drugs and items derived from *SS*. To evaluate safety and possible antagonistic or synergistic effects when using bamboo with other medications, more research is needed. There is limited data regarding the long-term safety of extracts or supplements made from *SS*. To track the effects of long-term *SS* consumption and determine its safety profile, longitudinal studies are required.

Research into new therapeutic uses for *SS* that go beyond its conventional ethnomedical use is needed. *SS* potential use in contemporary healthcare will increase with the discovery of new disease indications and treatment targets. Filling in these research voids will improve our knowledge regarding therapeutic benefits of *SS* and make it easier to incorporate *SS* into conventional medical procedures.

12.8 CONCLUSION

This compilation concludes by illuminating the multifaceted potential of *Sasa senanensis*, referred to as *SS* here, a Japanese dwarf bamboo species with a long history of traditional use. There is still a significant scientific study gap that prevents us from properly comprehending and using its medical benefits, despite its long-standing cultural relevance and great therapeutic potential. Pharmacological capabilities of *SS*, which majority include immunological modulation, antioxidant and anticancer effects, antimicrobial effects, and the possibility of treating oral conditions such as lichenoid dysplasia, highlight the plant's importance in modern medicine. Nevertheless, to confirm and clarify the mechanisms behind these reported effects, thorough scientific investigations are necessary. The chapter highlights that to guarantee safety, effectiveness, and uniformity in medicinal applications of *SS*, extensive clinical trials, mechanistic investigations, and standardization of extracts are necessary. Closing information gaps on bioavailability, drug interactions, long-term safety, and dose optimization will also improve our comprehension and application of medicines based on *SS*.

ABBREVIATIONS

ERK	Extracellular signal regulated kinase
Fe	Ferrous
$FeCl_2$	Ferrous chloride
HIV	Human immunodeficiency virus
HSC-2	Hematopoietic stem cell
HSV	Herpes simplex virus
IL-6	Interleukin-6
LCC	Lignin carbohydrate complex
LPS	Lipopolysaccharide
MAPK	Mitogen activated protein kinase
Mg	Magnesium
MM	Meshima Max
NF-kB	Nuclear factor kappa-chain enhancer of activated B-cell; JNK; Jun N-terminal kinase
NK	Natural killer
SS	*Sasa senanensis*
TNF	Tumor necrosis factor

REFERENCES

Auluck, A., Rosin, M. P., Zhang, L., & Sumanth, K. N. 2008. Oral submucous fibrosis, a clinically benign but potentially malignant disease: Report of 3 cases and review of the literature. *Journal of the Canadian Dental Association 74*(8): 735–740.

Baldi, A., & Bansal, P. K. 2020. Traditional and herbal medicines: Understanding & exploration (part II). *Current Traditional Medicine 6*(4): 1–3.

Baldi, A., Patidar, A. K., & Sanadya, J. 2010. Method development and validation for estimation of quetiapine fumarate by RP-HPLC. *Asian Journal of Research in Chemistry 3*(3): 604–607.

Brown, C. G., Beck, S. L., Peterson, D. E., McGuire, D. B., Dudley, W. N., & Mooney, K. H. 2008. Patterns of sore mouth in outpatients with cancer receiving chemotherapy. *Supportive Care in Cancer 17*: 413–428.

Bystriakova, N., Kapos, V., Lysenko, I., & Stapleton, C. M. A. 2003. Distribution and conservation status of forest bamboo biodiversity in the Asia-Pacific region. *Biodiversity and Conservation 12*(9): 1833–1841.

Chuyen, N. Y., Kurata, T., & Kato, H. 1983. Antiseptic activity of *Sasa senanensis* Rehder extract. *Journal of Antibacterial and Antifungal Agents 11*: 69–75.

Daiwabiological Research Institute CO., LTD. 2012. SASA health: Kumazasa leaf extract SASA health. Retrieved from https://www.hmaj.com/pdf/SASA_HEALTH.pdf

Fang, J., Gao, S., Islam, R., Teramoto, Y., & Maeda, H. 2020. Extracts of *Phellinus linteus*, bamboo (*Sasa senanensis*) leaf and chaga mushroom (*Inonotus obliquus*) exhibit antitumor activity through activating innate immunity. *Nutrients 12*(8): 2279.

Gautam, V. K., Datta, M., & Baldi, A. 2019. Effect of geographical and seasonal variations on phenolic contents and antioxidant activity of aerial parts of *Urtica diocia* L. *Current Traditional Medicine 5*(2): 159–167.

Gratani, L., Crescente, F. M., Varone, L., Fabrini, G., & Digiulio, E. 2008. Growth pattern and photosynthetic activity of different bamboo species growing in the *Botanicalgarden of Rome*. *Flora 203*: 77–84.

Hollman, P. 2004. Absorption, bioavailability, and metabolism of flavonoids. *Pharm Biology 42*: 74–83.

Khatun, A., Nakajima, K., Saito, M., Giridharan, V. V., Hara, H., & Konishi, T. 2013. *In vitro* and *in vivo* antioxidant potential of Japanese Kumaisasa bamboo leaf (*Sasa senanensis* Rehd). *Journal of Biochemical and Pharmacological Research 1*(4): 236–240.

Lu, B., Wu, X., Shi, J., Dong, Y., & Zhang, Y. 2006. Toxicology and safety of antioxidant of bamboo leaves. *Food and Chemical Toxicology 44*: 1739–1743.

Matsuta, T., Sakagami, H., Kitajima, M., Oizumi, H., & Oizumi, T. 2011a. Anti-UV activity of alkaline extract of the leaves of *Sasa senanensis* Rehder. *In Vivo 25*(5): 751–755.

Matsuta, T., Sakagami, H., Satoh, K., Kanamoto, T., Terakubo, S., Nakashima, H., Kitajima, M., Oizumi, H., & Oizumi, T. 2011b. Biological activity of luteolin glycosides and tricin from *Sasa senanensis* Rehder. *In Vivo 25*(5): 757–762.

Nagasawa, H., Hattori, A., & Murayama, Y. 2001a. Effects of gonadectomy at different ages and *Sasa* health, bamboo grass leaf extract, on spontaneous motor activity in female and male mice. *In Vivo 15*: 503–511.

Nagasawa, H., Murayama, Y., & Ishigame, H. 2001b. Food restriction and spontaneous motor activity in male mice: Effects of feeding pattern, far-infrared ray and bamboo grass leaf extract. *In Vivo 15*: 309–318.

Nazreen, S., Alam, M. S., Hamid, H., Kaur, G., & Alam, M. M. 2011. Phytochemical investigation of *Bambusa arundinacea* Retz. *International Journal of Natural Product Science 3*: 1–7.

Ono, M., Kantoh, K., Ueki, J., Shimada, A., Wakabayashi, H., Matsuta, T., Sakagami, H., Kumada, H., Hamada, N., Kitajima, M., Oizumi, H., & Oizumi, T. 2011. Quest for anti-inflammatory substances using IL 1β-stimulated gingival fibroblasts. *In Vivo 25*: 763–768.

Pandey, B. P. 1994. A Textbook of Botany: Angiosperms. S. Chand Company Ltd., New Delhi.

Ren, M., Reilly, R. T., & Sacchi, N. 2004. Sasa health exerts a protective effect on Her2/NeuN mammary tumorigenesis. *Anticancer Research 24*: 2879–2884.

Sakagami, H., Amano, S., Kikuchi, H., Nakamura, Y., Kuroshita, R., Watanabe, S., Satoh, K., Hasegawa, H., Nomura, A., Kanamoto, T., Terakubo, S., Nakashima, H., Taniguchi, S., & Ohizumi, T. 2008. Antiviral, antibacterial and vitamin C-synergized radical scavenging activity of *Sasa senanensis* Rehder extract. *In Vivo 22*(4): 471–476.

Sakagami, H., Fukuchi, K., Kanamoto, T., Terakubo, S., Nakashima, H., Natori, T., & Oizumi, T. 2016a. Synergism of alkaline extract of the leaves of *Sasa senanensis* Rehder and antiviral agents. *In Vivo 30*(4): 421–426.

Sakagami, H., Sheng, H., Ono, K., Komine, Y., Miyadai, T., Terada, Y., & Oizumi, T. 2016b. Anti-halitosis effect of toothpaste supplemented with alkaline extract of the leaves of *Sasa senanensis* Rehder. *In Vivo 30*(2): 107–111.

Sakagami, H., Zhou, L., Kawano, M., Thet, M. M., Takana, S., Machino, M., Amano, S., Kuroshita, R., Watanabe, S., Chu, Q., Wang, Q. T., Kanamoto, T., Terakubo, S., Nakashima, H., Sekine, K., Shirataki, Y., Hao, Z. C., Uesawa, Y., Mohri, K., Kitajima, M., Oizumi, H., & Oizumi, T. 2010. Multiple biological complex of alkaline extract of the leaves of *Sasa senanensis* Rehder. *In Vivo 24*: 735–744.

Sasa senanensis PFAF Plant Database. (n.d.). http://pfaf.org/USER/Plant.aspx?Latin Name= Sasa senanensis.

Sasa senanensis- Useful Temperate Plants. (n.d.). https://temperate.theferns.info/plant/Sasa senanensis.

Sato, K., Tatsunami, R., Nakata, A., Komatsu, K. I., Harakawa, S., Nedachi, T., & Wakame, K. 2021. Effects of Kumaizasa (*Sasa senanensis*) leaf extract on innate immune regulation in HEK293 cells and macrophages. *Anticancer Research 41*(8): 4093–4100.

Seki, T., Kida, K., & Maeda, H. 2010. Immuno stimulation-mediated anti-tumor activity of bamboo (*Sasa senanensis*) leaf extracts obtained under 'vigorous' condition. *Evidence-Based Complementary and Alternative Medicine 7*: 447–457.

Seki, T., & Maeda, H. 2010. Cancer preventive effect of Kumaizasa bamboo leaf extracts administered prior to carcinogenesis or cancer inoculation. *Anticancer Research 30*: 111–118.

Sharma, S., Kushwaha, R. S., Naman, S., Patil, U. K., & Baldi, A. 2021. Optimized extraction of oleoresin capsicum and analytical method validation for capsaicin using HPLC. *Indian Journal of Natural Products 35*(1): 38–45.

Singh, M., Kumar, D., Naman, S., Madhavi, N., Singh, P. A., Bajwa, N., & Baldi, A. 2019. Development and validation of HPTLC method for the stimulation estimation of ascorbic acid and gallic acid in amla juice preparation. *Journal Of Drug Delivery and Therapeutics 9*(4): 227–231.

Singh, P. A., Bajwa, N., & Baldi, A. 2021a. Possible role of traditional systems of medicine to manage COVID-19: A review. *Israel Journal of Plant Sciences 68*(1–2): 3–28.

Singh, P. A., Bajwa, N., Chinnam, S., Chandan, A., & Baldi, A. 2022. An overview of some important deliberations to promote medicinal plants cultivation. *Journal of Applied Research on Medicinal and Aromatic Plants 31*: 100400.

Singh, P. A., Bajwa, N., Sharma, L., & Baldi, A. 2023. Farm-level medicinal plants production and economic analysis in the Indian state of Punjab. *Journal of Applied Research on Medicinal and Aromatic Plants 34*: 100486.

Singh, P. A., & Baldi, A. 2018. Good agricultural practices: A prerequisite approach for enhancing the quality of Indian herbal medicines. *Biomedical Journal of Scientific & Technical Research 5*(5): 486–4863.

Singh, P. A., Sood, A., & Baldi, A. 2021b. Determining constraints in medicinal plants adoption: A model geospatial study in the Indian state of Punjab. *Journal of Applied Research on Medicinal and Aromatic Plants 25*: 100342.

Singh, P. A., Sood, A., & Baldi, A. 2021c. Highlighting agro-ecological zones to upscale the production of immunity-booster plants in Punjab. *Journal Research Journal of Agricultural Science 12*: 865–870.

Singh, P. A., Sood, A., & Baldi, A. 2021d. An agro-ecological zoning model highlighting potential growing areas for medicinal plants in Punjab. *Indian Journal of Pharmaceutical Education and Research 55*(2): S492–S500.

Singh, U., & Baldi, A. 2013. Stimulation estimation of quercetin and silymarin development and validation. *International Journal of Pharmaceutical and Biological Archives 4*(3): 527–531.

Tomomura, M., Tomomura, A., Oizumi, T., Yasui, T., & Sakagami, H. 2017. Extract of *Sasa senanensis* Rehder leaf promotes osteoblast differentiation in MC3T3-E1 cells. *International Journal of Dental Medicine 46*: 111–116.

Tsunoda, S., Yamamoto, K., Sakamoto, S., Inoue, H., & Nagasawa, H. 1998. Effects of *Sasa* health, extract of bamboo grass leaves, on spontaneous mammary tumorigenesis in SHN mice. *Anticancer Research 18*: 153–158.

Wong, K. M. 1995. The Bamboos of Peninsular Malaysia. CABI Database, Malaysia.

Yagi, Y., Hara, H., & Hara, A. 2013. Food function of small molecules derived from Sasa. *Food Function 11*: 20–29.

Zhou, L., Hashimoto, K., Satoh, K., Yokote, Y., Kitajima, M., Oizumi, T., Oizumi, H., & Sakagami, H. 2009. Effect of *Sasa senanensis* Rehder extract on NO and PGE_2 production by activated mouse macrophage-like RAW264.7 cells. *In Vivo 23*: 773–778.

13 Nutritional and Medicinal Properties of *Melocanna baccifera*

Konnath Chacko Koshy and Sabulal Baby

13.1 INTRODUCTION

Melocanna baccifera (Roxb.) Kurz is a unique bamboo well known for its 'apple-like' fruits. Its fruits are baccoid and the largest (weighing up to 300 g) in the grass family. This taxon belongs to the tribe Bambuseae (tropical woody bamboos) under the subfamily Bambusoideae (family Poaceae). The genus *Melocanna* Trin. embraces three species: *viz.*, *M. baccifera* (Roxb.) Kurz, *M. arundina* C. E. Parkinson, and *M. clarkei* (Gamble) P. Kumari and P. Singh. *M. baccifera* is distributed from India (Assam), Bangladesh, East Himalaya, and Nepal to Myanmar. It is locally known as *Muli, Saneibi, Moubi, Rieng, Watrai,* or *Wati* in NE India (Banik, 2010). It is also naturalized in countries like Jamaica, Ecuador, Colombia, and southeast Brazil (Vorontsova et al., 2016). Besides, its germplasm has been conserved at least in 14 bambuseta (Koshy, 2010; 2023). This species contains mostly hexaploids (2n = 72) (Alam, 1995; Devi and Sharma, 1993), but octoploids (2n = 96) were also reported (Remya et al., 2015). *M. baccifera* is a bamboo with widely distanced culms originating from sympodial c. 1 m long-necked rhizomes. The term '*Melocanna*' has its origin from the Greek words '*melon*' (meaning, an apple) and '*kanna*' (a reed), means a '*reed-like bamboo with apple-like fruits*' (Banik, 2016; Ruiz-Sanchez and Sosa, 2015). This chapter elucidates the nutritional and medicinal properties of *M. baccifera.*

As other woody bamboos, *M. baccifera* is semelparous and monocarpic. They have a prolonged vegetative phase, at the end of which they flower gregariously, set seeds, and die. The seeds germinate into offsprings. The gregarious flowering and mast seeding in *M. baccifera* occur once in 40 to 50 years of growth (Banik, 2010; Gamble, 1896; Janzen, 1976; Koshy et al., 2022; McClure, 1966; Shibata, 2010). Rats predate the fruits and, owing to their short propagative cycles, proliferate rapidly resulting in rat flooding. As and when the fruit production wanes, the elevated rat population attacks nearby grain crops, causing severe food shortage. Such situations led to famines in 1881, 1912, and 1959, which caused heavy human loss, c. 15,000 (1881), and 10,000–15,000 people (1959) (Hazarika, 1995; Jeeva et al., 2009; Nag, 1999). The famines and high human casualties steered political uprisings too (Adhikari, 2013; Belmain et al., 2010). The northeast Indian state of Mizoram is the geographic center of this natural event (John and Nadgauda, 2002; Shoumatoff, 2008), locally known as *Mautam* or *bamboo death* (Aplin and Lalsiamliana, 2010; Belmain et al., 2010; Govindan et al., 2016; Htwe et al., 2010; Normile, 2010; Ruiz-Sanchez and Sosa, 2015; Wilkins, 2010).

Melocanna baccifera culms are an important source of superior paper pulp. Its higher specific gravity, better fiber length, and higher wall/lumen ratio makes it suitable for papermaking (Alam, 1995; Chaurasia et al., 2016). Its culms are used in house construction and to make woven goods, such as mats, hats, handicrafts, baskets, screens, wall plates, and domestic utensils (Alam, 1995). Young shoots are cooked and eaten fresh or sliced, dried, and preserved for future consumption (Alam, 1995; Lalfakzuala et al., 2007). *M. baccifera* contributes 53.7% to the annual bamboo shoot consumption in the northeastern Himalayan region (Bhatt et al., 2003). Its leaves are used in fermenting liquor (Alam, 1995). A siliceous aggregate, known as Tabashir, found in its culms is used as a tonic in the treatment of respiratory diseases (Alam, 1995). Its glossy stem surface is

 DOI: 10.1201/9781003485933-13

used to heal wounds and the liquid inside its culm is used as a remedy for hiccup, urethritis, fever, and paralysis (Bhardwaj and Gakhar, 2005). *M. baccifera* fruits are fleshy, edible, and are used as food during famines (Alam, 1995). Further, its fruits are used to cure low blood pressure (Bhardwaj and Gakhar, 2005) and as an aphrodisiac (Gentleman, 2008; Shoumatoff, 2008). The native tribes eat the fruits as a vegetable, savor the sweet liquid within the young fruits, and use the white embryo as medicine (Banik, 2010). Apart from human beings, various animals consume *M. baccifera* fruits (Aplin and Lalsiamliana, 2010; Belmain et al., 2010; Htwe et al., 2010; Koshy et al., 2022; Normile, 2010). A latest study on its fruit production dynamics has shown that the fruit predators include mollusks (slug: *Mariaella dussumieri* Gray; snails: *Cryptozona bistrialis* H.H. Beck, *Macrochlamys* sp.), arthropods (borers: *Achroia grisella* Fb., *Blattella germanica* L.), mammals (monkeys: *Macaca radiata* É. Geoffroy; rats: *Rattus* L.; porcupine: *Hystrix indica* Kerr; wild boar: *Sus scrofa* L.; palm civet: *Paradoxurus hermaphroditus* Pallas), and the millipede (*Spinotarsus colosseus* Attems), which consume fruit remnants (Koshy et al., 2022). In northeast India, rats are the major fruit predators, whereas in southern India monkeys are the dominant predators (Koshy et al., 2022). *M. baccifera* shoots, leaves, culms, and fruits are commonly used for nutritional and medicinal purposes.

13.2 SHOOTS

A bamboo shoot is the emerging growth of the rhizome apex into a culm, consisting of young internodes protected by sheaths. The tender portion of the shoot after removing the sheaths is the edible part. Bamboo shoots are nutritious and are eaten fresh, dried, or fermented (Bisht et al., 2012; Mao and Odyuo, 2007; Singhal et al., 2013; Tamang and Tamang, 2009). Fermented bamboo shoots have remarkable health benefits. They have anticancer, antioxidant, antiaging, cardioprotective, weight reducing, and probiotic properties besides their functional food value (Li et al., 2022). They are also used as flavoring agents (Fu et al., 2002; Jeyaram et al., 2009; Mao and Odyuo, 2007; Sarangthem and Singh, 2003a,b; Tamang and Tamang, 2009; Tylewicz et al., 2017), health nutriments, and nutraceuticals (Chandramouli and Vishwanath, 2012; Li et al., 2022; Chongtham et al., 2011; Nirmala et al., 2014). Nowadays, due to their health-promoting and nutrition-enhancing properties, bamboo shoots are also used in food fortification (Chongtham et al., 2024).

The *M. baccifera* shoot is a highly preferred vegetable in the northeastern states of India *viz.* Assam, Manipur, Meghalaya, Mizoram, and Tripura, particularly by the Mizos of Mizoram and the Meiteis of Manipur (Bhatt et al., 2004; Bisht et al., 2015; Singhal et al., 2013). Its shoots are devoid of any strong odor or bitter taste and contain less amounts of cyanogenic glycosides. Young shoots (Figure 13.1a) are collected during the rainy season, their culm sheaths are removed, and the soft inner portion is separated out (Figure 13.1b); these separated shoots are washed or immersed in water for some time. They are eaten fresh, boiled, fried in oil, or mixed with meat and other vegetables. Fresh shoots are used to make pickles or recipes like *Usoi-Ooti* of Manipur, *Mia-gudhog* of Tripura, and *Rawtuai-bai* of Mizoram (Bisht et al., 2015).

Young bamboo shoots remain fresh only for a short period (2–3 days). Their long-term preservation is achieved by drying and fermentation. These processes reduce the content of cyanogenic glycosides in the shoots, and they become more delicious, nutritious, and attain several health benefits. Dried shoots are consumed all through the year as such or are blended with other vegetables and/or meat. Some local recipes (with the name of the tribe and State in parentheses) are *Hithyi* (Apatani tribe, Arunachal Pradesh), *Ikung, Ib* (Adi tribe, Arunachal Pradesh), and *Rawtuai rep* (Mizo tribe, Mizoram) (Bisht et al., 2015). Bamboo shoot fermentation is commonly practiced in the northeastern states of India. The names of a few fermented bamboo products (with the name of the tribe and State in parentheses) are *Pdam* (Khasi, Meghalaya); *Soibum*, *Soidon*, and *Soijin* (Meetei, Manipur); *Hikhu* and *Hiring* (Apatani, Arunachal Pradesh); *Ekung* and *Edung* (Adi, Arunachal Pradesh); *Eup* (Nyishing, Arunachal Pradesh); *Mesu* (Limboo,

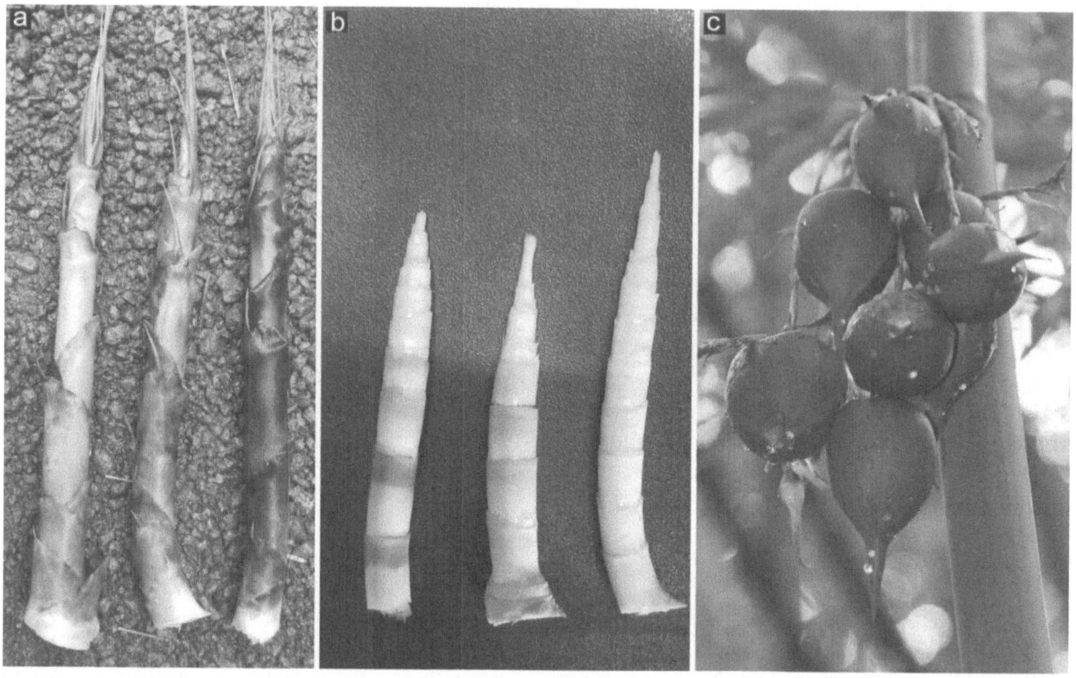

FIGURE 13.1 *Melocanna baccifera:* young shoots and mature fruits. (a) Shoots immediately after extraction with sheaths intact, (b) edible portion of shoots after removing the sheaths, and (c) fruits (photos: B. Gopakumar).

Drajeeling, Sikkim); *Moiya-Koshak* (Debbarma, Tripura); *Midukye* and *Melye-Amiley* (Chakma, Tripura); and *Moiya-Pangsung* (Uchoi, Tripura) (Bisht et al., 2015). The fermentation process reportedly helps to increase the content of essential amino acids, proteins, vitamins, and bioactive compounds (Bisht et al., 2015; Chongtham et al., 2011; Fu et al., 2002; Li et al., 2022; Tylewicz et al., 2017). Examples of traditional recipes using *M. baccifera* shoots are *Rawtuai bawl* (fresh shoots, pork, bean leaves, green chili), Mizoram; *Rawtuai-Kan* (boiled bamboo shoots, oil, spices), Mizo, Mizoram; *Rawtui-bai* (fresh shoots, rice, sodium bicarbonate), Mizo, Mizoram; *Usoi-kangsu* (bamboo slices, potato, fermented fish, dried chilies), Meetei, Manipur; and *Soijin-eromba* (fermented bamboo shoots, potato, fermented fish, dried chilies), Meetei, Manipur.

The major bioactive entities in young bamboo shoots, *viz.*, phytosterols, phenols, and dietary fibers, have health benefits (Singhal et al., 2013). They provide immunity to chronic and degenerative diseases. Phenolic metabolites in bamboo shoots have antioxidant, anti-inflammatory, antiallergic, antiageing, antifatigue, antimicrobial, cardioprotective, and vasodilation properties (Nirmala et al., 2014). Phytosterols and dietary fibers have positive effects on the lipid profile and the regular passage of soft, formed stools (Park and Jhon, 2009). They help reduce total serum cholesterol and low-density lipoprotein cholesterol levels. Dietary fiber content estimated in bamboo shoots is 2.2 to 4.5% (fr. wt.), which is higher than many common vegetables (Chongtham et al., 2011; Nirmala et al., 2014). Cellulose, hemicellulose, and lignin are the major constituents of fiber in bamboo shoots. Dietary fibers are reported to lessen the risk of cancer, cardiovascular diseases, hypertension, obesity, diabetes, and several gastrointestinal problems (Anderson et al., 2009; Brennan et al., 2012; Lattimer and Haub, 2010). Bamboo shoots contain many macro- and micro-nutrients too. They are rich source of potassium, manganese, calcium, chromium, zinc, copper, iron, selenium, phosphorus (Choudhury et al., 2012; Nirmala et al., 2014), and vitamins (A, B_6, E, thiamine, and niacin) (Bisht et al., 2015).

Though chemical screening of shoots of a few bamboo species has been reported, only limited studies are available on *M. baccifera* shoots (Bhatt et al., 2003; Manna and Debnath, 2020; Namasudra et al., 2021, 2024). Bhatt and coworkers (2003) provided energy values of *M. baccifera* shoots as food energy 15.8 MJ/Kg; moisture, 75.5%; protein, 3.62%; fat, 0.57%; and carbohydrates, 6.12% dry mass. In comparison, the average calorie and protein values of cereals and grains are 16.8 MJ/Kg and 148.0 g/Kg; those of leafy and fruit vegetables are 15.8 MJ/Kg and 42.0 g/Kg and tuber and rhizome crops are 15.3 MJ/Kg and 17.0 g/Kg (dry wt. basis), respectively. The values of cereals and grains are considerably higher than the food value of edible bamboo shoots. However, their availability, delicacy, and health benefits make bamboo young shoots crucial supplements to the nutritious diets of tribal communities (Bhatt et al., 2003). Manna and Debnath (2020) studied the secondary metabolites, minerals, heavy metals, proximate composition, and antioxidant activity of *M. baccifera* shoots. Their estimates are (i) secondary metabolites: alkaloids 14.53 ± 0.67 mg/g, phenols 81.33 ± 1.52 mg/g, flavonoids 71.66 ± 0.57 mg/g, tannins 140.02 ± 0.02 mg/g of dry extract; (ii) proximate composition: moisture content 13.02 ± 0.98 g/100 g, ash 9.34 ± 0.56 g/100 g, carbohydrate 43.35 ± 0.57 g/100 g, and total protein 3.9 ± 0.2 g/100 g of dry powder; (iii) minerals and heavy metals: iron 14.47 ± 0.81 mg/100 g, sodium 2.80 ± 0.11 mg/100 g, potassium 47.4 ± 0.69 mg/100 g, calcium 144.74 ± 0.57 mg/100 g, lead 0.023 ± 0.005 mg/100 g, and arsenic below detectable limit. The total fat content of *M. baccifera* shoots was 1.77 ± 0.02 g/100 g of the dry powder. The caloric value of dry powder of a young shoot was 204 kcal/100 g, indicating it as an energy source. In DPPH and H_2O_2 radical scavenging activities, IC_{50} values of 50% shoot ethanolic extract were 69.54 ± 0.005 and 69.19 ± 0.060 µg/mL; IC_{50} values of standard ascorbic acid on similar assays were 57.63 ± 0.57 and 57.93 ± 0.58 µg/mL, respectively. The shoots, rich in secondary metabolites (phenolics, flavonoids, tannins), are potential enough to scavenge DPPH due to their hydrogen donating property (Manna and Debnath, 2020).

M. baccifera shoot methanolic extract revealed the presence of bioactive entities like alkaloids, triterpenoids, flavonoids, glycosides, carbohydrates, proteins, fixed oil, and minerals (sodium, potassium, calcium). Pharmacological studies of the methanolic extract have reportedly shown analgesic, antidiabetic, hepatoprotective, antimicrobial, and central nervous system (CNS) depressant activities (Namasudra et al., 2021). In another study, Namasudra and coworkers (2024) estimated the phytomineral constituents in *M. baccifera* young shoots as carbohydrates (23.69%, w/w), fats (12.50%, w/w), proteins (17.51%, w/w), tannins (23.95 mg/100 g), alkaloids, fixed oil, glycosides, triterpenoids (5.22 mg/100 g), vitamin C (5.90 mg/100 g), flavonoids (8.96 mg/100 g), copper (11.0 mg/100 g), magnesium (35.1 mg/100 g), calcium (90.1 mg/100 g), iron (19.0 mg/100 g), sodium (150.0 mg/100 g), and phosphorous (15.0 mg/100 g) (Namasudra et al., 2024). The variations in the nutritional factors of shoots in these studies can be ascribed to the differences in their collection locations, season, growth stage, and the analysis methodologies adopted. Overall, these reports demonstrated that *M. baccifera* shoots contain considerable amounts of macro-micronutrients and phytochemicals, and hence underpin their potential as a functional food.

13.3 CULMS

A study by Pachuau and coworkers (2014) indicated that *M. baccifera* culms are a green source to make microcrystalline cellulose by hydrolysis. Microcrystalline cellulose is used in industries like pharmaceuticals, cosmetics, plastics, and food. In pharmaceutical drugs, it is used as a diluent and binder for tablets and capsules (Pachuau et al., 2014). Singh and coworkers (2024) extracted bio-oil from *M. baccifera* mature culms through pyrolysis. Bioassays showed that the smoky dark-brown bio-oil (pH 2.46 ± 0.2) was effective in hindering the growth of pathogenic microbial strains. Moreover, the bio-oil, even at 1% concentration, exhibited strong free radical scavenging activity. Bioactive compounds like catechin, catechol, guaiacol, quinol, and 2-methoxy-4-methylphenol make the bio-oil antimicrobial and antioxidant. This bamboo bio-oil is a promising substitute to synthetic chemicals used in antipathogenic products (Singh et al., 2024).

13.4 LEAVES

Govindan and coworkers (2019) studied the chemistry of *M. baccifera* leaves and isolated 19 molecules, *viz.*, 3β-friedelinol, β-sitosterol fatty ester, 3(20)-phytene-1,2-diol, 5(6)-gluten-3α-ol, mixture of α-amyrin and β-amyrin, mixture of dihydrovomifoliol-*O*-β-D-glucopyranoside and icariside B_5, hexacosanal, *E*-phytol, 4-hydroxy benzaldehyde, 7,3′,5′-tri-*O*-methyltricetin, blumenol B, syringic acid, *p*-coumaric acid, tianshic acid, corchorifatty acid F, tricin, 7,3′,5′-tri-*O*-methyltricetin 5-*O*-β-D-glucopyranoside, tricin-5-*O*-β-D-glucopyranoside, and 1-*O*-palmitoyl-3-*O*-(6-sulfo-α-D-quinovopyranosyl)-glycerol (Figure 13.2a,b). Khan and coworkers (2018, 2019) have shown that

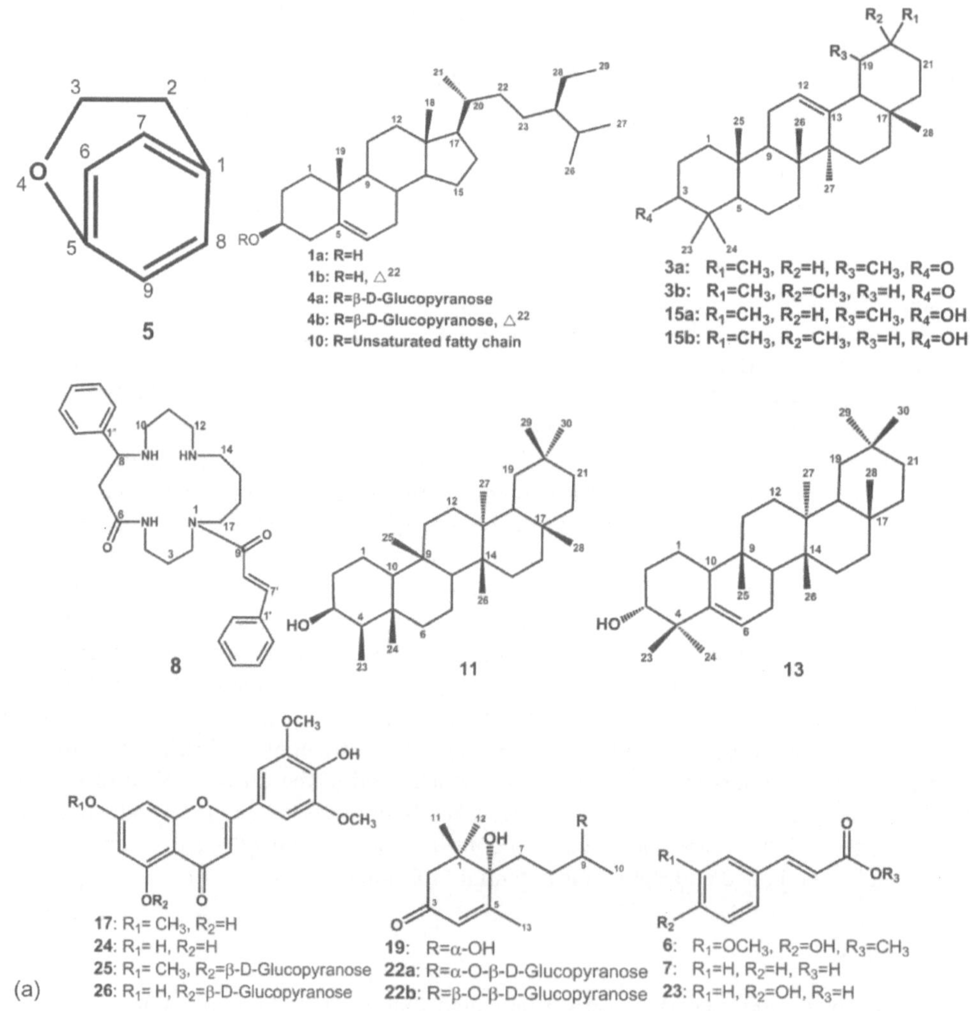

FIGURE 13.2 Secondary metabolites isolated from *M. baccifera* leaves and fruits (a,b). (1a: β-sitosterol, 1b: stigmasterol, 2: hexadecanoic acid, 3a: α-amyrenone, 3b: β-amyrenone, 4a: β-sitosterol-3-*O*-glucoside, 4b: stigmasterol-3-*O*-glucoside, 5: 4-oxabicyclo[3.2.2]nona-1(7),5,8-triene, 6: *E*-ferulic acid methyl ester, 7: cinnamic acid, 8: verbacine, 9: hexacosanal, 10: β-sitosterol fatty ester, 11: 3β-friedelinol, 12: 3(20)-phytene-1,2-diol, 13: 5(6)-gluten-3α-ol, 14: *E*-phytol, 15a: α-amyrin, 15b: β-amyrin, 16: 4-hydroxy benzaldehyde, 17: 7,3′,5′-tri-*O*-methyltricetin, 18: syringic acid, 19: blumenol B, 20: corchorifatty acid F, 21: tianshic acid, 22a: dihydrovomifoliol-*O*-β-D-glucopyranoside, 22b: icariside B_5, 23: *p*-coumaric acid, 24: tricin, 25: 7,3′,5′-tri-*O*-methyltricetin 5-*O*-β-D-glucopyranoside, 26: tricin-5-*O*-β-D-glucopyranoside, 27: 1-*O*-palmitoyl-3-*O*-(6-sulfo-α-D-quinovopyranosyl)-glycerol (Govindan et al., 2019). *(Continued)*

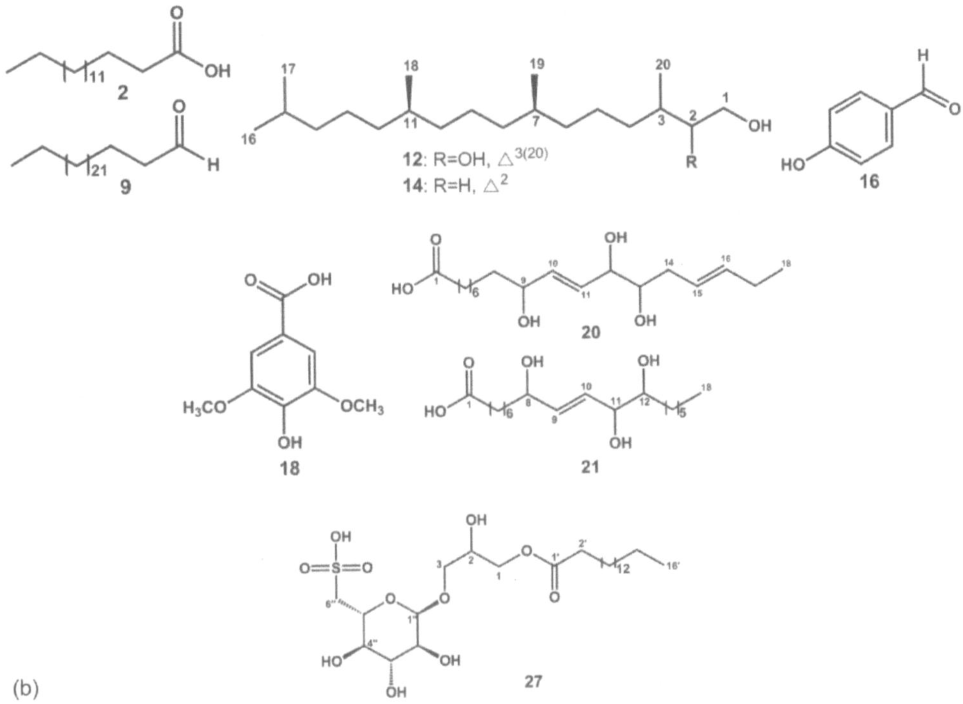

FIGURE 13.2 *(Continued)*

M. baccifera leaf alcoholic extract functions as a nontoxic phytoprophylactic. Further, the leaf alcoholic extract proved to be efficient in preventing bacterial infection in fish fingerlings, endorsing its immunomodulatory potential (Khan et al., 2022).

13.5 FRUITS

The mature fruit of *M. baccifera* is a pear-shaped baccate caryopsis, ovoid to globose with a tapering curled beak, pericarp fleshy and thick, glabrous and smooth, brownish, brownish-green to pure green (Figure 13.3). Mature fruits (at 42 days) measured 9.93 ± 1.51 cm long (n = 30), 5.49 ± 0.62 cm

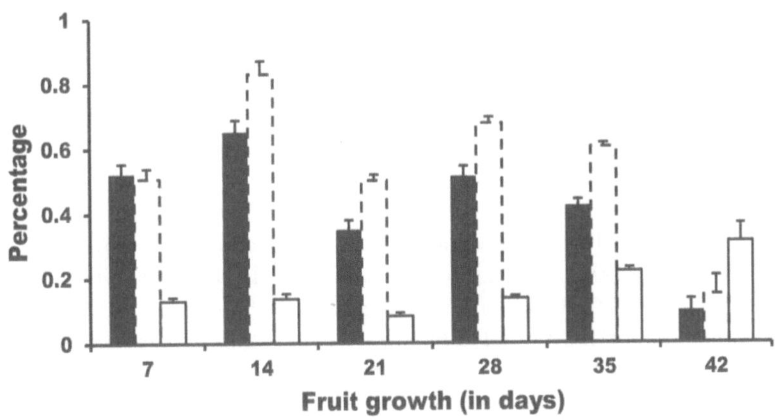

FIGURE 13.3 Sugar profile in *M. baccifera* pericarp. Filled bar: Glu, dotted bar: Fru, unfilled bar: Suc; n = 4 (Govindan et al., 2016).

wide (n = 30), and weighed 104.39 ± 26.96 g (n = 30, initial years of flowering) to 49.45 ± 25.75 g (n = 500, later years of flowering) (Koshy et al., 2022). Alam (1995) reported the fruit dimensions as 3.5–11.0 cm long, 2.2–6.0 cm wide, and weighed 7–151 g (n = 10,000); a few fruits weighed up to 300 g. The quantity of fruit production in individual clumps varied from 37.19 to 456.67 kg (n = 8). On an average, a clump during its lifetime produced 4,796.38 ± 3,420.26 (n = 8) fruits with an estimated weight of 237.18 ± 169.13 kg (n = 8) (Koshy et al., 2022). Kuddus and coworkers (2011) isolated olean-12-en-28-carboxy-3β-acetate, 3-oxo-olean-12-en-28-al, β-sitosterol glucoside, stigmasterol glucoside, isochavicinic acid, and eugenol from *M. baccifera* fruits; the fruit extracts showed notable antioxidant and moderate antimicrobial activities (Kuddus et al., 2013).

Govindan and coworkers (2016) conducted a comprehensive study on the nutritional aspects of *M. baccifera* fruits. Its pericarp showed various amino acids and the major ones were lysine (397.67 μg/g, fr. wt.), glutamic acid (367.00 μg/g, fr. wt.), histidine (192.33 ± 45.62 μg/g, fr. wt.), arginine (152.33 ± 17.79 μg/g, fr. wt.), serine (60.33 ± 15.37 μg/g, fr. wt.), and glycine (60.33 ± 25.79 μg/g, fr. wt.). Fruit liquids at different growth stages also showed free amino acid patterns (Govindan et al., 2016). Lysine, the major amino acid in fruits, is extensively used as an additive in beverages, foodstuffs, animal feeds, and pharmaceuticals. Glutamic acid and its derivatives are known anticancer agents (Ali et al., 2013). Dietary glutamic acid is expected to decrease blood pressure (Stamler et al., 2009), besides acting as a precursor for γ-aminobytyric acid (GABA) synthesis and as a neurotransmitter in mammalian brain (Fagg and Foster, 1983; Rindi and Ferrari, 1959). Arginine is known to improve male and female fertility. It helps the body to make nitric oxide, which promotes blood circulation to the genitalia and ovaries (Wu et al., 2009, 2013). *M. baccifera* fruits (pericarps, seeds, fruit liquids) contain only very low free protein; bound protein in the fruit parts was also low (Govindan et al., 2016). This finding of very low protein disproved the earlier hypothesis that high-protein fruits cause elevated rat predation (Aplin and Lalsiamliana, 2010; Banik, 2010; Das, 2008; Gentleman, 2008; Normile, 2010; Shoumatoff, 2008).

M. baccifera fruits demonstrated sugar (glucose (Glu)-fructose (Fru)-sucrose (Suc)) profiles in its growth stages (Figure 13.3). In 7- and 42-days-old fruit pericarps, the Glu-Fru-Suc contents were 0.50%, 0.51%, and 0.13% and 0.10%, 0.15%, and 0.31%, respectively. The fruit liquid and inner seed in mature fruits also showed Glu-Fru-Suc patterns (Govindan et al., 2016). Sugars are major components that determine the quality and sweetness of fruits. Various predators, including slugs, snails, borers, mammals, and millipedes were observed relishing *M. baccifera* fruits (Banik, 2010; Janzen, 1976; Jeeva et al., 2009; Koshy et al., 2022; Majumdar et al., 2015), and sugars are a crucial factor attracting them. Sugars, mainly Suc, not only impart taste to the fruits but also provide energy to humans and animals who consume them.

Phenolics influence the antioxidant effects of a food source. The phenolic content in *M. baccifera* fruits was quantified as 2.06 ± 0.003 mg ferulic acid equivalent/g, dry wt., n = 3 (Govindan et al., 2016). Ferulic acid is the only phenolic acid detected in *M. baccifera* fruits; its content in bound and esterified fractions was quantified as 8.65 ± 0.36 mg/g, dry wt. and 45.92 ± 6.86 μg/g, dry wt., respectively (Govindan et al., 2016). The bound ferulic acid content in *M. baccifera* fruits is higher than cereals (rice, maize, wheat; 0.8–2.0 mg/g, dry wt.) (Adom and Liu, 2002; Sosulski et al., 1982). Ferulic acid, known for various biological activities, reduces the risk of diseases such as cancer, heart disease, and diabetes, and therefore is a beneficial dietary element for human well-being (Adom and Liu, 2002; Baby et al., 2013). *M. baccifera* fruits on lipid analysis showed mostly saturated fatty acids, with palmitic acid (C16:0) as the major component (0.13 g/100 g, dry wt., n = 3) (Govindan et al., 2016). Palmitic acid is a therapeutic agent with anticancer effects against human leukemic cell lines (Harada et al., 2002). *M. baccifera* fruits showed the presence of vitamin B series (B_3 2.68 mg/100 g, fr. wt.; B_2 2.13 mg/100 g, fr. wt., n = 3); vitamin C or other fat-soluble vitamins (A, D, E, K) were not detected (Govindan et al., 2016). Vitamin B_3 (niacin) is known to reduce serum lipid and cholesterol levels (Capuzzi et al., 2000; Schachter, 2005). Vitamin B_2 (riboflavin) helps the body to release energy by the breakdown of carbohydrates, proteins, and fats. *M. baccifera* fruit is potassium rich (831.5 mg/100 g, dry wt., n = 3). It is an essential element for the functioning of the key cellular systems in the human body. It also protects from stroke-associated mortality, and helps in maintaining electrolyte and fluid balance in the body. *M. baccifera* fruits also contain essential minerals like Mg, Fe, Cu, Zn, and Mn in trace quantities (Govindan et al., 2016, 2019).

Govindan and coworkers (2019) isolated eight secondary metabolites from *M. baccifera* fruits, *viz.*, 4-oxabicyclo[3.2.2]nona-1(7),5,8-triene, verbacine, mixture of β-sitosterol and stigmasterol, mixture of α-amyrenone and β-amyrenone, mixture of β-sitosterol-3-*O*-glucoside and stigmasterol-3-*O*-glucoside, E-ferulic acid methyl ester, cinnamic acid, and hexadecanoic acid (Figure 13.2a,b). Verbacine (17-membered macrocyclic lactam alkaloid), the major isolate of fruits, is rarely reported from natural sources (Drandarov, 1995). Biological activity studies demonstrated its acetylcholinesterase inhibition and cytotoxic activities (Govindan et al., 2019).

13.6 CONCLUSION

The food value of *M. baccifera* bamboo shoots is lower compared to cereals, grains, leafy and fruit vegetables, and tuber crops. But, their delicacy, availability, and associated health benefits make them robust supplements to nutritious diets and bamboo-fortified food products. The major metabolites rendering nutritional and therapeutic potential to *M. baccifera* fruits are sugars (Glu, Fru, Suc), amino acids (lysine, glutamic acid, arginine), fatty acids (palmitic acid), phenolic acids (ferulic acid), vitamins (B_3), and minerals (potassium). The protein content in *M. baccifera* fruits is very low. As evidenced by rat feeding studies, the fruit alone is not a complete food, but along with other protein complements they are valuable food additives. This information is helpful for the better utilization of the fruits during future flowering events. The isolation of 27 secondary molecules *viz.*, terpenoids, steroids, flavonoids, alkaloid, nor-isoprenoids, phenolic acids, fatty acid derivatives, from *M. baccifera* leaves and fruits is very significant. *M. baccifera* leaf alcoholic extract functions as a phytoprophylactic. Pyrolytic oil from its culms possesses antimicrobial and antioxidant properties. Since bamboo shoots and fruits are rich in nutritional factors and bioactive compounds, they are promising ingredients in the preparation of functional foods. Similar phytochemical, biological activity, and nutritional studies on other bamboos could discover new secondary metabolites, drug precursors, and nutritional additives.

ACKNOWLEDGMENT

We acknowledge (i) Science and Engineering Research Board (SERB), Government of India (grant number SR/S0/PS/0015/2011) and (ii) Kerala State Council for Science, Technology and Environment, Government of Kerala, India (plan funded project KSCSTE-JNTBGRI-P007A) for financial assistance to our studies on *Melocanna baccifera*.

REFERENCES

Adhikari, U. 2013. Bamboo flowering, human security and the state: A political ecological study of the impact of cyclical bamboo flowering on human security and the role of state in Mizoram. PhD diss., University of North Bengal, Darjeeling.

Adom, K. K., & Liu, R. H. 2002. Antioxidant activity of grains. *Journal of Agricultural and Food Chemistry* *50*(21): 6182–6187.

Alam, M. K. 1995. *Melocanna baccifera* (Roxb.) Kurz. *Plant Resources of South-East Asia 7*: 126–129.

Ali, I., Wani, W. A., Haque, A., & Saleem, K. 2013. Glutamic acid and its derivatives: Candidates for rational design of anticancer drugs. *Future Medicinal Chemistry 5*(8): 961–978.

Anderson, J. W., Baird, P., Davis, R. H. Jr, Ferreri, S., Knudtson, M., Koraym, A., Waters, V., & Williams, C. L. 2009. Health benefits of dietary fiber. *Nutrition Reviews 67*(4): 188–205.

Aplin, K., & Lalsiamliana, J. 2010. Chronicle and Impacts of the 2005–09 Mautam in Mizoram. Rodent Outbreaks: Ecology and Impacts. International Rice Research Institute, Los Banos, Philippines, 13–42.

Baby, S., Johnson, A. J., Govindan, B., Lukose, S., Gopakumar, B., & Koshy, K. C. 2013. UV induced visual cues in grasses. *Scientific Reports 3*(1): 2738.

Banik, R. L. 2010. Biology and silviculture of Muli bamboo *Melocanna baccifera* (Roxb.) Kurz. Department of Science & Technology. *National Mission on Bamboo Applications 2010*: 1–206.

Banik, R. L. 2016. *Melocanna* Trin.: Silviculture of South Asian Priority Bamboos. Springer, Singapore, 211–233.

Belmain, S. R., Chakma, N., Sarker, N. J., Sarker, S. U., Sarker, S. K., & Kamal, N. Q. 2010. The Chittagong story: Studies on ecology of rat floods and bamboo masting. In: *Rodent Outbreaks: Ecology and Impacts*. International Rice Research Institute, Los Banos, Philippines, 49–64.

Bhardwaj, S., & Gakhar, S. K. 2005. Ethnomedicinal plants used by the tribals of Mizoram to cure cuts & wounds. *Indian Journal of Traditional Knowledge 4*(1): 75–80.

Bhatt, B. P., Singha, L. B., Sachan, M. S., & Singh, K. 2004. Commercial edible bamboo species of the North-Eastern Himalayan region, India. Part I; Young shoot sales. *Journal of Bamboo and Rattan 3*(4): 337–364.

Bhatt, B. P., Singha, L. B., Singh, K., & Sachan, M. S. 2003. Some commercial edible bamboo species of North East India: Production, indigenous uses, cost-benefit and management strategies. *Journal of the American Bamboo Society 17*(1): 4–20.

Bisht, M. S., Nirmala, C., & Meetei, O. S. 2015. Bamboo shoots for food in North-East India: Conventional and contemporary. In: *Proceedings of 10th World Bamboo Congress, Korea*, World Bamboo Organization, Boston, Massachusetts, 393–402.

Bisht, M. S., Nirmala, C., & Vyas, P. 2012. Bamboo Shoot A neglected natural resource: A source of food and prosperity for North-East India. In: *Proceedings of National Seminar: Recent Advances in Natural Product Research*. Mizoram University, Aizawl, 18–22.

Brennan, M. A., Derbyshire, E. J., Brennan, C. S., & Tiwari, B. K. 2012. Impact of dietary fibre-enriched ready-to-eat extruded snacks on the postprandial glycaemic response of non-diabetic patients. *Molecular Nutrition & Food Research 56*: 834–837.

Capuzzi, D. M., Morgan, J. M., Brusco, O. A. Jr, & Intenzo, C. M. 2000. Niacin dosing: Relationship to benefits and adverse effects. *Current Atherosclerosis Reports 2*: 64–71.

Chandramouli, S., & Vishwanath, S. 2012. Bamboo shoots - An emerging new age health food. *Forestry Bulletin 12*: 21–28.

Chaurasia, S. K., Singh, S. P., Naithani, S., & Srivastava, P. 2016. A comprehensive study on proximate chemical composition of *Melocanna baccifera* (Muli Bamboo) and it's suitability for pulp and paper production. *Forest Research 5*: 168.

Chongtham, N., Bisht, M. S., & Haorongbam, S. 2011. Nutritional properties of bamboo shoots: Potential and prospects for utilization as a health food. *Comprehensive Reviews in Food Science and Food Safety 10*: 153–165.

Chongtham, N., Santosh, O., & Bajwa, H. K. 2024. Innovative technologies in food fortification: A bamboo perspective. In 12th World Bamboo Congress, Taiwan, World Bamboo Organization, Boston, Massachusetts.

Choudhury, D., Sahu, J. K., & Sharma, G. D. 2012. Value addition to bamboo shoots: A review. *Journal of Food Science and Technology 49*: 407–414.

Das, B. 2008. Indian state faces famine after plague of rats. https://www.reuters.com/article/economy/indian-state-faces-famine-after-plague-of-rats-idUSDEL120141/. (accessed July 27, 2024).

Devi, T. S., & Sharma, G. J. 1993. Chromosome numbers in some bamboo species of Manipur. *Bamboo Information Centre-India Bulletin 3*: 16–21.

Drandarov, K. 1995. Verbacine and verballocine, novel macrocyclic spermine alkaloids from *Verbascum pseudonobile* Stoj. et Stef. (Scrophulariaceae). *Tetrahedron Letters 36*(4): 617–620.

Fagg, G. E., & Foster, A. C. 1983. Amino acid neurotransmitters and their pathways in the mammalian central nervous system. *Neuroscience 9*: 701–719.

Fu, S. G., Yoon, Y., & Bazemore, R. 2002. Aroma-active components in fermented bamboo shoots. *Journal of Agricultural and Food Chemistry 50*: 549–554.

Gamble, J. S. 1896. The Bambuseae of British India. Annals of the Royal Botanic Garden, Calcutta. *Bengal Secretariat Press Calcutta 7*: 1–132.

Gentleman, A. 2008. One million face famine after rats feast on crops. https://www.theguardian.com/world/2008/mar/23/india. (accessed July 27, 2024).

Govindan, B., Johnson, A. J., Nair, S. N. A., Gopakumar, B., Mallampalli, K. S. L., Venkataraman, R., Koshy, K. C., & Baby, S. 2016. Nutritional properties of the largest bamboo fruit *Melocanna baccifera* and its ecological significance. *Scientific Reports 6*: 26135.

Govindan, B., Johnson, A. J., Viswanathan, G., Ramaswamy, V., Koshy, K. C., & Baby, S. 2019. Secondary metabolites from the unique bamboo, *Melocanna baccifera*. *Natural Product Research 33*: 122–125.

Harada, H., Yamashita, U., Kurihara, H., Fukushi, E., Kawabata, J., & Kamei, Y. 2002. Antitumor activity of palmitic acid found as a selective cytotoxic substance in a marine red alga. *Anticancer Research 22*: 2587–2590.

Hazarika, S. 1995. Strangers of the Mist: Tales of War and Peace from India's Northeast. Penguin Books, New Delhi.

Htwe, N. M., Singleton, G. R., Thwe, A. M., & Lwin, Y. Y. 2010. Rodent population outbreaks associated with bamboo flowering in Chin State, Myanmar. In: Singleton, G. R., Belmain, S. R., Brown P. R., & Hardy, B. (eds) Rodent Outbreaks: Ecology and Impacts. International Rice Research Institute, Los Banos, Philippines, 79–98.

Janzen, D. H. 1976. Why bamboos wait so long to flower. *Annual Review of Ecology and Systematics 7*(1976): 347–391.

Jeeva, S., Kiruba, S., Lalthruatluanga, H., Prasad, M. N. V., & Rao, R. R. 2009. Flowering of *Melocanna baccifera* (Bambusaceae) in northeastern India. *Current Science 96*(9): 1165–1166.

Jeyaram, K., Singh, T. A., Romi, W., Devi, A. R., Singh, W. M., Dayanidhi, H., Singh, N. R., & Tamang, J. P. 2009. Traditional fermented foods of Manipur. *Indian Journal of Traditional Knowledge 8*(1): 115–121.

John, C. K., & Nadgauda, R. S. 2002. Bamboo flowering and famine. *Current Science 82*(3): 261–262.

Khan, M. I. R., Saha, R. K., & Saha, H. 2018. Muli bamboo (*Melocanna baccifera*) leaves ethanolic extract a non-toxic phyto-prophylactic against low pH stress and saprolegniasis in *Labeo rohita* fingerlings. *Fish & Shellfish Immunology 74*(2018): 609–619.

Khan, M. I. R., Saha, R. K., & Saha, H. 2022. Quantification of phytochemicals in Muli bamboo [*Melocanna baccifera* (Roxb.) Kurz] leaf and shoot extracts and immunomodulatory effect of dietary ethanolic leaves extract on *Labeo rohita* (Hamilton). *Indian Journal of Experimental Biology 60*: 101–111.

Khan, M. I. R., Singh, M., & Monsang, S. J. 2019. Prophylactic potential of ethanolic leaves extract of Muli bamboo (*Melocanna baccifera*) against *Aeromonas hydrophila* infection in *Labeo rohita* (Hamilton). *Journal of Entomology and Zoology Studies 7*(2): 288–291.

Koshy, K. C. 2010. Bamboos at TBGRI. Tropical Botanic Garden and Research Institute, Thiruvananthapuram.

Koshy, K. C. 2023. Germplasm resources of bamboos. In: Genetics, Genomics and Breeding of Bamboos. CRC Press, Boca Raton, 26–115.

Koshy, K. C., Gopakumar, B., Sebastian, A., Nair, S. A., Johnson, A. J., Govindan, B., & Baby, S. 2022. Flower-fruit dynamics, visitor-predator patterns and chemical preferences in the tropical bamboo, *Melocanna baccifera. PLoS One 17*: e0277341.

Kuddus, M. R., Rumi, F., Haque, M. R., Hassan, M. A., & Rashid, M. A. 2013. Assessment of antioxidant, antimicrobial and cytotoxic properties of fruits of *Melocanna baccifera* (Roxb.) Kurz. *Turkish Journal of Pharmaceutical Sciences 10*: 185–192.

Kuddus, M. R., Rumi, F., Kaisar, M. A., Rahman, M. S., Hasan, C. M., Hassan, M. A., & Rashid, M. A. 2011. Secondary metabolites from *Melocanna baccifera* (Roxb.). *Asian Journal of Chemistry 23*: 85–88.

Lalfakzuala, R., Lalramnghinglova, H., & Kayang, H. 2007. Ethnobotanical usages of plants in western Mizoram. *Indian Journal of Traditional Knowledge 6*: 486–493.

Lattimer, J. M., & Haub, M. D. 2010. Effects of dietary fibres and its component on metabolic health. *Nutrients 2*: 1266–1289.

Li, S., Tian, Y., Sun, M., Liu, J., Bai, Y., Liu, X., & Guo, Y. 2022. Characterization of key aroma compounds in fermented bamboo shoots using gas chromatography-olfactometry-mass spectrometry, odor activity values, and aroma recombination experiments. *Foods 11*: 2106.

Majumdar, K., Nath, A. J., Gupta, A. K., & Datta, B. K. 2015. Bamboo Invasion: Threat to primate conservation in North East India. *Current Science 108*: 1969–1971.

Manna, K., & Debnath, B. 2020. Phytochemicals, nutritional properties and anti-oxidant activity in young shoots of *Melocanna baccifera*, a traditional vegetable of Northeast India. *Journal of Food, Agriculture & Environment 18*: 59–65.

Mao, A. A., & Odyuo, N. 2007. Traditional fermented foods of the Naga tribes of Northeastern India. *Indian Journal of Traditional Knowledge 6*: 37–41.

McClure, F. A. 1966. The Bamboos: A Fresh Perspective. Harvard University Press, Cambridge, 1–288.

Nag, S. 1999. Bamboo, Rats and Famines: Famine relief and perceptions of British paternalism in the Mizo Hills (India). *Environment and History 5*: 245–252.

Namasudra, U., De, B., & Bhaumik, D. 2024. Quantitative estimation of food value of edible young shoot of *Melocanna baccifera. International Journal of Multidisciplinary Research 6*: 1–5.

Namasudra, U., De, B., Bhaumik, D., & Sarkar, S. 2021. A review on *Melocanna baccifera. International Journal of Current Science Research and Review 4*: 154–157.

Nirmala, C., Bisht, M. S., & Laishram, M. 2014. Bioactive compounds in bamboo shoots: Health benefits and prospects for developing functional foods. *International Journal of Food Science & Technology 49*: 1425–1431.

Normile, D. 2010. Holding back a torrent of rats. *Science 327*: 806–807.

Pachuau, L., Vanlalfakawma, D. C., Tripathi, S. K., & Lalhlenmawia, H. 2014. Muli bamboo (*Melocanna baccifera*) as a new source of microcrystalline cellulose. *Journal of Applied Pharmaceutical Science 4*: 87–94.

Park, E. J., & Jhon, D. Y. 2009. Effects of bamboo shoot consumption on lipid profiles and bowel function in healthy young women. *Nutrition 25*: 723–728.

Remya, K. R. V., Mathew, P. J., & Koshy, K. C. 2015. Polyploidy in Muli bamboo - *Melocanna baccifera. Journal of Cytology and Genetics 16*: 1–5.

Rindi, G., & Ferrari, G. 1959. The γ-aminobutyric acid and glutamic acid content of brains of rats treated with toxopyrimidine. *Nature 183*: 608–609.

Ruiz-Sanchez, E., & Sosa, V. 2015. Origin and evolution of fleshy fruit in woody bamboos. *Molecular Phylogenetics and Evolution 91*: 123–134.

Sarangthem, K., & Singh, T. N. 2003a. Transformation of fermented bamboo (*Dendrocalamus hamiltonii*) shoots into phytosterols by microorganisms. *Journal of Food Science and Technology 40*: 622–625.

Sarangthem, K., & Singh, T. N. 2003b. Biosynthesis of succulent bamboo shoots of *Bambusa balcooa* into phytosterols and biotransformation into ADD. *Acta Botanica Sinica 45*: 114–117.

Schachter, M. 2005. Strategies for modifying high-density lipoprotein cholesterol: A role for nicotinic acid. *Cardiovascular Drugs and Therapy 19*: 415–422.

Shibata, S. 2010. Examination into records and periodicity concerning to *Melocanna baccifera* flowering with 48-year interval. *Japanese Journal of Ecology 60*: 51–62.

Shoumatoff, A. 2008. Waiting for the plague. http://www.vanityfair.com/news/2007/12/famine200712. (accessed July 27, 2024).

Singh, K. S., Nayak, S., Lalnunmawii, E., Devi, M. B., Shagolsem, B. S., & Gouda, S. 2024. Pyrolytic oil from Muli bamboo (*Melocanna baccifera,* Roxb.): Biological potential and possible functional attributes. *Journal of Analytical and Applied Pyrolysis 180*: 106528.

Singhal, P., Bal, L. M., Satya, S., Sudhakar, P., & Naik, S. N. 2013. Bamboo shoots: A novel source of nutrition and medicine. *Critical Reviews in Food Science and Nutrition 53*(5): 517–534.

Sosulski, F., Krygier, K., & Hogge, L. 1982. Free, esterified, and insoluble-bound phenolic acids. 3. Composition of phenolic acids in cereal and potato flours. *Journal of Agricultural and Food Chemistry 30*(2): 337–340.

Stamler, J., Brown, I. J., Daviglus, M. L., Chan, Q., Kesteloot, H., Ueshima, H., Zhao, L., & Elliott, P., & INTERMAP Research Group. 2009. Glutamic acid, the main dietary amino acid, and blood pressure: The INTERMAP study (International collaborative study of macronutrients, micronutrients and blood pressure). *Circulation 120*(3): 221–228.

Tamang, B., & Tamang, J. P. 2009. Traditional knowledge of biopreservation of perishable vegetables and bamboo shoots in Northeast India as food resources. *Indian Journal of Traditional Knowledge 8*(1): 89–95.

Tylewicz, U., Inchingolo, R., & Rodriguez-Estrada, M. T. 2017. Food aroma compounds. In: Nutraceutical and Functional Food Components: Effects of Innovative Processing Techniques. Academic Press, London, 297–334.

Vorontsova, M. S., Clark, L. G., Dransfield, J., Govaerts, R., & Baker, W. B. 2016. World checklist of bamboos and rattans. INBAR Technical Report No. 37, International Network of Bamboo and Rattan, Beijing, 5–452.

Wilkins, A. 2010. Massive plagues of rats swarm across India every fifty years. http://io9.com/5694107/massive-plagues-of-rats-swarm-across-india-every-fifty-years (accessed July 27, 2024).

Wu, G., Bazer, F. W., Davis, T. A., Kim, S. W., Li, P., Rhoads, J. M., Satterfield, M. C., Smith, S. B., Spencer, T. E., & Yin, Y. 2009. Arginine metabolism and nutrition in growth, health and disease. *Amino Acids 37*: 153–168.

Wu, G., Bazer, F. W., Satterfield, M. C., Li, X., Wang, X., Johnson, G. A., Burghardt, R. C., Dai, Z., Wang, J., & Wu, Z. 2013. Impacts of arginine nutrition on embryonic and fetal development in mammals. *Amino Acids 45*: 241–256.

14 Utilization of Bamboos for Pharmaceutical and Nutraceutical Industry

Shivam Pandey, Shefali Pundir, and Himani Sharma

14.1 INTRODUCTION

All species of perennial evergreen plants of the grass family, Pinaceae's subfamily *Bambusoideae* are collectively referred to as bamboo (Goyal et al., 2013). Bamboo exists naturally in terrestrial areas and can reach heights of over 40 meters or 100 feet. The length of the plant varies, with some species growing as small as 30 centimeters (Smith, 1993). It is made up of oval leaves and a hollow culm or stem with nodes or junctions between the stem's segments. Most bamboos thrive in warm, humid climates with plenty of moisture and fertile soil, but some can also withstand temperatures as low as 20°C (Handique et al., 2010). They grow in most types of soils, except for alkaline soils, marshes, and plains, as well as hilly and high-altitude mountainous environments. Even in the winter, the bamboo's culm, branches, and leaves maintain their green color throughout its life (Bystriakova et al., 2003). Bamboo only lives for around 20 years and, depending on the species, it flowers once every 70 to 120 years (Sharma, 1987). Regardless of their geographic location, every specimen of bamboo species produces flowers simultaneously. More than 1,500 species of bamboo are found in various parts of the world; these are classified into 60–70 different taxonomy. Around 50% of these bamboo species are cultivated in Asia, predominantly in the Indo-Burmese region, which harbors 136 species comprising 23 taxa that are considered indigenous to India (Goyal et al., 2010). Africa is home to an extremely small quantity of bamboo species in contrast to the abundance of such species in South America and the Asia-Pacific region. An estimated 136 species, both indigenous and exotic, find their home in India and more than half of the known species are found in the northeastern states (Ghosh and Sinha, 2018; Nimachow et al., 2010).

Bamboo was once known as "poor man's timber," but is now referred to as "green gold of forests," "friend of the people," and "plant with a thousand faces." In addition to being regarded as a quick absorber of atmospheric carbon, bamboo possesses mechanical and physical qualities that allow it to be utilized in the production of goods typically made from replanted or native wood, including building materials, furniture, agricultural tool, cables, panels, and plates (Banik, 2015). Many industries worldwide are involved in the manufacturing of various bamboo-based goods such as flooring, pulp, panels, charcoal, toothbrushes, everyday items, etc. (Kumar et al., 2021). The global bamboo market size was estimated to be at USD$59.30 billion by 2021 but currently, its growth is being driven by growing infrastructure development, rising consumer awareness, and a healthy lifestyle of people. According to a new estimate, the market size of bamboo-related products will increase to almost USD$121.81 billion by the year 2029. The United States, Canada, Germany, France, the United Kingdom, Indonesia, China, Japan, and South Korea are a few countries other than India that are majorly expected to contribute to this growth of the bamboo market. With increasing awareness about the benefits of bamboo, pharmaceutical, nutraceutical, and cosmetic industries will be some of the sectors that will majorly contribute to the utilization of bamboo besides construction, wood and furniture, paper and pulp, textile, medical, agriculture, and food packaging.

DOI: 10.1201/9781003485933-14

FIGURE 14.1 Image of bamboo plant.

It has been increasingly realized that bamboo species are a rich source of nutritional components that make them priority plants for food and medicine. A wide range of bioactive substances have been reported in bamboo leaves and shoots that impart exceptional pharmacological properties. These include flavonoids, polysaccharides, phenolic acids, amino acids, and volatile oils, that are responsible for therapeutic properties of bamboo like antivirus, anti-inflammatory, antioxidation, antiosteoporosis, antihyperglycemic, blood circulation enhancement, blood lipid regulation, and hepatoprotective properties (Choudhury et al., 2012; Pande and Pandey, 2008). Consequently, the entire bamboo plant (Figure 14.1), including culm, rhizome, bark shaving, roots, shoots, seeds, and leaves, has been prominently used in traditional systems of food and medicine.

With recent advances in the industrial sectors, food, pharmaceutical, and cosmetics industries from all over the world have been interested in utilizing bamboo resources owing to its nutritional and medicinal possibilities (Chandra et al., 2004).

14.2 MEDICINAL VALUE OF BAMBOO

Bamboo stems and leaves contribute to the major nutritional value of bamboo as they have been found to contain many phytoactive ingredients. The bamboo shoot is divided mainly into three sections (i.e., the sheath, tender bamboo shoot, and basal bamboo shoot) (Zhang et al., 2024). Only after removing the surface of the bamboo shoot is the soft cylindrical stem of the bamboo visible as it is coated in many epidermal hairs (He et al., 2020). Both bamboo shoots and leaves are rich in minerals, amino acids, and are a good source of dietary fibers (Kimura et al., 2022; Wang et al., 2012). The leaf extracts of bamboo have been found to have anti-inflammatory, antidiabetic, anti-fatigue, antioxidation, antiosteoporosis, antivirus, blood lipid regulation, anti-inflammatory, hepatoprotective, and neuroprotective properties. They also protect the liver (Carey et al., 2009; Ding et al., 2010). In the traditional systems, bamboo seeds, roots, leaves, branches, juice, and shoots have been used to treat an array of ailments, including rhinorrhagia (nosebleed), cooling fever, phlegm, vomiting, stomach and liver diseases, and laryngitis. Thus, it is reasonable to state that every piece of bamboo is a treasure as well as a therapeutic plant (Lalthansangi, 2022).

14.3 PHARMACEUTICAL AND NUTRITIONAL POTENTIAL OF BAMBOO SHOOTS

Bamboo shoots are the young culms that have just emerged from the ground; they have a hollow, cylindrical form and are cream yellow in color. They are also known as the meristematic tissue or aerial bud of bamboo plants. The use of edible bamboo shoots as food dates to ancient China, approximately 3,000 years ago (Figure 14.2). Considering their high dietary fiber content and low caloric content, bamboo shoots are an excellent complement to a balanced diet. In many other Asian countries, it is regarded as one of the healthiest and tastiest dishes (Seki and Maeda, 2010).

Based on their vast expertise, the inhabitants of the northeastern regions consume fresh shoots of bamboos like *Bambusa nutans, B. tulda, Chimonobambusa callosa, Dendroclamus sikkimensis, D. giganteus, Phyllostachys mannii,* and *Melocanna baccifera* (as vegetables or mixed with pork, or as pickle). The Tripura dish "Miagudhog," the Manipuri cuisine "Usoi-Ooti," and the Mizoram cuisine "Rawtui-bai" are a few of the well-known dishes prepared with fresh bamboo shoots (Sood et al., 2013). Dried bamboo shoots are another incredibly common practice for nourishment by various tribes. The dried shoots are used as a spice in a wide range of vegetable and meat recipes. Young bamboo shoots are frequently fermented by methods which are specific to a tribe, region, species, or variety of bamboo. All these methods are prevalent in the traditional knowledge system of different regions (Satya et al., 2012). The small-scale industries in these regions have already started manufacturing and marketing bamboo shoot–based products like candies, canned juice, flavored pickles, vinegar, etc.

Studies have shown antimicrobial, anti-inflammatory, antidiabetic, antioxidant, and cardioprotective potential of bamboo shoots, besides many other properties (Jayarambabu et al., 2023; Singhal et al., 2013). Phytocomponents like flavonoids and phenolic acids exhibit inhibitory effects against pathogens, while also blocking inflammatory mediators and pathways to reduce inflammation. The development of anti-inflammatory, antibacterial, and antifungal medications from bamboo shoots extracts are potential pharmaceutical applications.

Bamboo shoots are also rich in bioactive substances, including several flavonoids, tricin, isovitexin, isoorientin, vitexin, and orientin, as well as catechin, phenolic acids, caffeic acid, ferulic acid, protocatechuic acid, *p*-hydroxybenzoic acid, *p*-coumaric acid, syringic acid, chlorogenic acid, vanillic acid, and gallic acid. These compounds are linked to different medicinal applications. For instance, orientin has been connected to a lower risk of cardiovascular illness and isovitexin has been demonstrated to have a benefit against liver damage. Another component, phytosterol, found

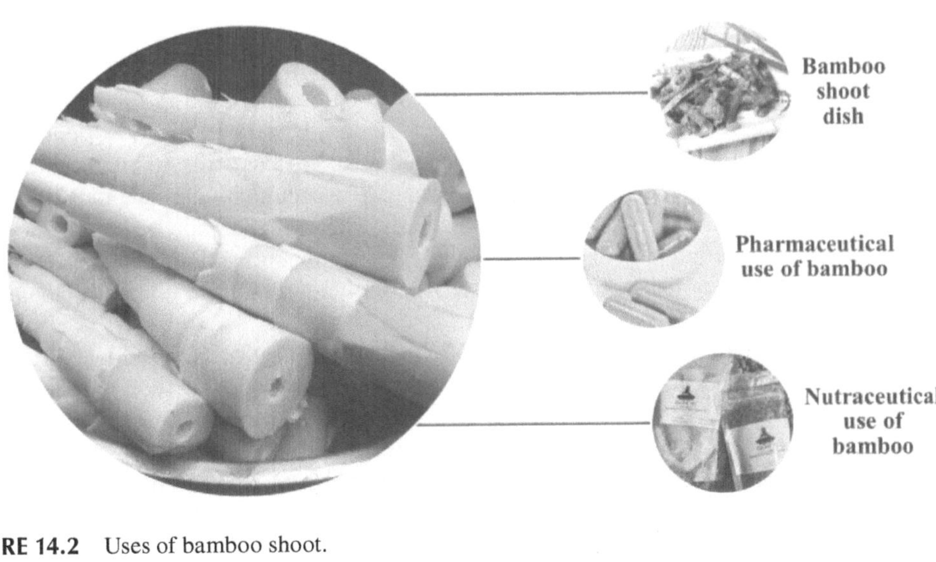

FIGURE 14.2 Uses of bamboo shoot.

in bamboo shoots, is known to reduce cholesterol, promote weight loss, prevent cancer, and improve appetite and digestion, hence making it a raw material of interest in herbal drug development. Studies have confirmed the presence of bioactive components in bamboo shoots improves glucose metabolism and insulin sensitivity research (Goyal et al., 2017). This, along with low sugar content in bamboo, also makes it suitable for product development for those following a sugar-restricted diet. With low calories and high fiber, amino acids, beta-carotene, proteins, and vitamins, these properties of bamboo shoots are another much interesting area in the food and pharma industries for products that can help control blood cholesterol levels and prevent colon cancer. The bamboo shoots also have cardioprotective properties. By lowering blood pressure, it improves lipid profiles and lowers cholesterol. The potential cardiovascular advantages of bamboo shoot extracts could be investigated in various pharmaceutical formulations (Chongtham et al., 2011).

However, it is crucial to remember that not all bamboo shoots are equal in terms of nutrients, and some could contain toxic substances if improperly processed. This calls for a thorough understanding of specific bamboos species, and their nutritive and antinutritive properties before utilizing the shoots for product development in food or pharma industries.

Bamboo shoots are of much interest to the food industry because they have both therapeutic and nutritional benefits and growing consciousness on good health and living. Numerous nutrients, including reducing and non-reducing sugars, proteins, lipids, co-enzymes, minerals, vitamins (A, B_3, B_1, C), enzymes, lactic acid, and citric acid (fermented products) make it an exceptional dietary supplement. More than 10 different mineral elements, including zinc (Zn), manganese (Mn), copper (Cu), magnesium (Mg), cobalt (Co), nickel (Ni), and lysine, have been confirmed in bamboo shoots (Lu et al., 2022). These findings have bolstered the development of different varieties of nutritional food products from bamboo shoots.

14.4 PHARMACEUTICAL AND NUTRITIONAL POTENTIAL OF BAMBOO LEAVES

The leaves of several bamboo species have been utilized to treat a variety of illnesses from ancient times to the present. They also been demonstrated to have uses in the food, pharmaceutical, medical, and agro-based industries. By advancing various processing techniques and many technologies through scientific and engineering expertise, bamboo leaves can serve as a wide range of substitutes for diverse chemical, synthetic, or other materials in commercial or structural applications (Bajwa et al., 2021). Some of the bioactives, like poylphenols, tricine, orientin, caffeic acid, isoorientin, isovitexin, vitexin, rutin, naringin-7-rhamnoglucoside, chlorogenic acid, quercetin, luteolin, and polyhydroxycoumaric acid, are responsible for the therapeutic potential of bamboo leaves, hence making them suitable for medical applications (Das et al., 2021). Many antioxidants found in bamboo leaves, including flavonoids, phenolic compounds, and chlorophyll, aid in the body's defense against free radicals and reduction of oxidative stress. Orientin, homoorientin, iso-orientin, tricin, homovitexin, vitexin, and the phenolic acids chlorogenic acid, caffeic acid, *p*-coumaric acid, and ferulic acid that comprise C-glycoside flavonoids are majorly associated with the antioxidant property of bamboo leaves (Tripathi et al., 2015). Consequently, including bamboo leaf extract in nutraceutical supplements can promote general health by preventing oxidative damage. The antioxidant and antimicrobial potential of leaf extracts has also been shown to be of use in the development of a variety of skincare products (Alias et al., 2023).

Studies have also reported the antiobesity and antidiabetic impact of bamboo leaf extracts by enhancing insulin sensitivity and controlling lipid metabolism, while also showing anti-inflammatory qualities (Indira et al., 2023; Verschoyle et al., 2006; Zubair et al., 2013). The phytosterol and flavonoid content are also responsible for enhancing the lipid profile, improving cardiovascular health and promoting detoxification (Muniappan and Sundararaj, 2003). The utilization of bamboo leaf extracts in nutraceutical formulations has the potential to support metabolic health and weight management.

TABLE 14.1
Pharmacological Activity of Different Bamboos Utilized for Drug Delivery and Other Biomedical Applications

Bamboo Species	Pharmacological Activity
Bambusa arundinacea	Antimicrobial, antifertility, anti-inflammatory, antiulcer, antioxidant, antihyperglycemic, analgesic
Dendrocalamus strictus (Roxb.)	Antimicrobial
Phyllostachys nigra var. *henonis*	Anticarcinogenic
P. aurea	Antimicrobial activity
P. edulis	Antidiabetic properties
P. heterocycla cv. pubescens	Antimicrobial activity against foodborne pathogens
P. nigra var. *henonis*	Anticarcinogen against acrylamide
P. pubescens. gramineae	Antifungal activity

Some interesting research has established the potential of bamboo leaves for utilization in cancer treatment. Tricine, also known by its chemical name 5,7,4′-trihydroxy-3′,5′-dimethoxyflavone, is a basic flavone that is found bamboo leaves and has a glycosidic form. This compound has been realized as a cancer chemotherapy preventative agent (Sangeetha et al., 2015). Bamboo leaves have also been used in the treatment of ulcers. A study confirmed the antiulcer properties of methanolic extracts of *B. arundinacea* leaves against paw edema caused by carrageen (Lourenco et al., 2019). Bamboo leaves also have adaptogenic qualities that ease tension and encourage rest. Bamboo leaf-based nutraceutical compositions may aid in stress reduction and mental health. Bamboo leaves also contain silica, a mineral that aids in the formation of collagen and is also good for health. The extract of these leaves also improves hydration and general health. Dietary fibers and prebiotic substances included in bamboo leaves help to maintain the gut microbiota balance, improve nutrient absorption, and support regularity in the digestive system. Supplementing nutraceutical goods with bamboo leaf extract presents an opportunity to take advantage of its many health-promoting qualities for enhanced general well-being and prevention of diseases (Van Hoyweghen et al., 2010).

Bamboo leaves also exhibit a vast nutraceutical potential. Herbal teas or infusions can be made by drying and steeping bamboo leaves in hot water. Through this procedure, the flavor and nutrients from the leaves are extracted, resulting in a fragrant and revitalizing drink. The leaves can also be used in powdered form after pulverization. This powder can be used in a variety of culinary applications as a flavoring or as a nutritional supplement. It is generally used to include the nutritious value of the leaves and add a delicate bamboo flavor to soups, sauces, smoothies, or baked products. Additionally, bamboo leaves can be treated to separate their active ingredients and utilized to create tinctures or extracts. Because of their possible health benefits, these concentrated forms of bamboo leaf extract can be added to drinks. Table 14.1. shows pharmacological activities of different species of bamboo in drugs and other allied applications.

14.5 CHALLENGES AND FUTURE PROSPECTS

While the utilization of bamboo in the pharmaceutical and nutraceutical sectors exhibits promising prospects for the future, there are both constraints and prospects in enhancing the efficacy of the extraction, separation, and characterization procedures of bioactive compounds derived from bamboo. To enhance the therapeutic efficacy of chemicals, novel methodologies ought to prioritize yield optimization and hazard reduction in accordance with the green chemistry principle. This could only be accomplished through academic and industrial collaboration. Furthermore, this partnership will have positive societal effects.

The variation in the bioactive chemical composition of species and even of individual plants represents a further significant obstacle. The consistency and effectiveness of product formulation and production is another aspect that needs to be resolved before commercial utilization of bamboo extracts. Moreover, the commercial medicines developed from bamboo should also meet the guidelines so that they can be fully utilized according to their potential. A very significant consideration has to be given to a manner that prevents any negative impact on biodiversity and the environment. An overabundance of bamboo can have negative environmental effects.

14.6 CONCLUSION

The pharmaceutical and nutritional industries find bamboo to be a perfect raw material because of its wide range of bioactive compounds and high nutritional content. Multiple compounds found in bamboo have been shown to have medicinal effects. Antioxidants, anti-inflammatory, antimicrobial, and anticancer agents are just a few of the many pharmaceutical products that these bioactive chemicals may find application for. Moreover, as bamboo contains a wide range of nutritional elements, it is a material that has the potential to be widely used in dietary supplements as well as different foods and culinary recipes. The medical sector and wellness therapies are using bamboo and its numerous products more and more because of new technology advancements and varied extraction methods. More research into the phytoactive ingredients of bamboo will open up the scope for a wide range of pharmaceutical and nutraceutical that can be developed.

REFERENCES

Alias, A. H. D., Razak, N. Q. A., Yusoff, M. H. M., Chin, K. H., Kamal, M. L., Uyup, N. H., & Shafie, M. H. 2023. Antioxidant and antimicrobial activities of essential oils extracted from bamboo (*Bambusa vulgaris*) leaves and its application in skincare products: A review. *Biocatalysis and Agricultural Biotechnology 54*: 102930.

Bajwa, H. K., Santosh, O., & Nirmala, C. 2021. Bamboo shoot for food and nutritional security. *Journal of Pharmacognosy and Phytochemistry 10*(4S): 24–30.

Banik, R. L. 2015. Bamboo Silviculture. In: *Bamboo: The Plant and Its Uses*. Cham: Springer International Publishing 10: 113–174.

Bystriakova, N., Kapos, V., Lysenko, I., & Stapleton, C. M. A. 2003. Distribution and conservation status of forest bamboo biodiversity in the Asia-Pacific Region. *Biodiversity & Conservation 12*: 1833–1841.

Carey, W. M., Dasi, J. M. B., Rao, N. V., & Gottumukkala, K. M. 2009. Anti-inflammatory activity of methanolic extract of *Bambusa vulgaris* leaves. *International Journal of Green Pharmacy 3*(3): 234–238.

Chandra, A. K., Ghosh, D., Mukhopadhyay, S., & Tripathy, S. 2004. Effect of bamboo shoot, *Bambusa arundinacea* (Retz.) Willd. on thyroid status under conditions of varying iodine intake in rats. *Indian Journal of Experimental Biology 42*(8): 781–786.

Chongtham, N., Bisht, M. S., & Haorongbam, S. 2011. Nutritional properties of bamboo shoots: Potential and prospects for utilization as a health food. *Comprehensive Reviews in Food Science and Food Safety 10*(3): 153–168.

Choudhury, D., Sahu, J. K., & Sharma, G. D. 2012. Value addition to bamboo shoots: A review. *Journal of Food Science and Technology 49*: 407–414.

Das, B., Singh, P. K., & Munna, J. 2021. Miraculous medicinal properties of bamboo. *Journal of Medicinal Plants Studies 9*(5): 28–30.

Ding, H. X., Gao, Y. Y., Chao, H. J., & Xia, D. H. 2010. Effect of polysaccharide from Moso bamboo leaves on blood lipid of mice with hyperlipemia. *Food Science 9*: 60.

Ghosh, M., & Sinha, B. 2018. Policy analysis for realizing the potential of timber production from trees outside forests (TOF) in India. *International Forestry Review 20*(1): 89–103.

Goyal, A. K., Kar, P., & Sen, A. 2013. Advancement of bamboo taxonomy in the era of molecular biology: A review. In: Biology of Useful Plants and Microbes. Narosa Publication House, New Delhi, 197–208.

Goyal, A. K., Middha, S. K., Usha, T., Chatterjee, S., Bothra, A. K., Nagaveni, M. B., & Sen, A. 2010. Bamboo-infoline: A database for North Bengal bamboo's. *Bioinformation 5*(4): 184.

Goyal, A. K., Middha, S. K., Usha, T., & Sen, A. 2017. Analysis of toxic, antidiabetic and antioxidant potential of *Bambusa balcooa* Roxb. leaf extracts in alloxan-induced diabetic rats. *3 Biotech 7*: 1–11.

Handique, P., Rethy, P., Dutta, B. K., Das, A. K., & Doley, B. 2010. Role of bamboo resources in socio economic development of the tribal people of Arunachal Pradesh with special reference to Nyishi tribe of Papum Pare District. *Journal of Biosciences Research 1*(3): 216–226.

He, K., Li, Q., Li, Y., Li, B., & Liu, S. 2020. Water-insoluble dietary fibers from bamboo shoot used as plant food particles for the stabilization of O/W Pickering emulsion. *Food Chemistry 310*: 125925.

Indira, A., Joshi, B., Oinam, S., Koul, A., & Chongtham, N. 2023. Potential of bamboo in the prevention of diabetes-related disorders: Possible mechanisms for prevention. In: *Bamboo Science and Technology*. Springer Nature Singapore, Singapore, 89–124.

Jayarambabu, N., Velupla, S., Akshaykranth, A., Anitha, N., & Rao, T. V. 2023. *Bambusa arundinacea* leaves extract-derived Ag NPs: Evaluation of the photocatalytic, antioxidant, antibacterial, and anticancer activities. *Applied Physics A 129*(1): 13.

Kimura, I., Kagawa, S., Tsuneki, H., Tanaka, K., & Nagashima, F. 2022. Multitasking bamboo leaf-derived compounds in prevention of infectious, inflammatory, atherosclerotic, metabolic, and neuropsychiatric diseases. *Pharmacology & Therapeutics 235*: 108159.

Kumar, M., Upadhyay, S. K., Negi, R., Sharma, I., & Singh, R. 2021. Phytodiversity, socioeconomic and ethnomedicinal study of bamboos Dendrocalamus Nees in new forest of forest research institute (FRI), Dehradun (Uttarakhand), India. *International Journal of Botany Studies 6*(3): 412–416.

Lalthansangi, R. 2022. Food, memory and identity: Tracing Mizo foodways. *Litinfinite 4*(1): 13–20.

Lourenço, S. C., Moldão-Martins, M., & Alves, V. D. 2019. Antioxidants of natural plant origins: From sources to food industry applications. *Molecules 24*(22): 4132.

Lu, H., Huang, C., Yu, K., & Liu, Z. 2022. Effects of mixed inoculation of *Leuconostoc citreum* and *Lactobacillus plantarum* on suansun (Sour bamboo shoot) fermentation. *Food Bioscience 47*: 101688.

Muniappan, M., & Sundararaj, T. 2003. Antiinflammatory and antiulcer activities of Bambusa arundinacea. *Journal of Ethnopharmacology 88*(2–3): 161–167.

Pande, S. K., & Pandey, S. 2008. Bamboo for the 21st century. *International Forestry Review 10*(2): 134–146.

Sangeetha, R., Diea, Y. K. T., Chaitra, C., Malvi, P. G., & Shinomol, G. K. 2015. The amazing bamboo: A review on its medicinal and pharmacological potential. *Indian Journal of Nutrition 2*(1): 1–7.

Satya, S., Singhal, P., Bal, L. M., & Sudhakar, P. 2012. Bamboo shoot: A potential source of food security. *Mediterranean Journal of Nutrition and Metabolism 5*(1): 1–10.

Seki, T., & Maeda, H. 2010. Cancer preventive effect of Kumaizasa bamboo leaf extracts administered prior to carcinogenesis or cancer inoculation. *Anticancer Research 30*(1): 111–118.

Sharma, Y. M. L. 1987. Inventory and resources of bamboos. *Recent Research on Bamboos 1987*: 14–27.

Singhal, P., Bal, L. M., Satya, S., Sudhakar, P., & Naik, S. N. 2013. Bamboo shoots: A novel source of nutrition and medicine. *Critical Reviews in Food Science and Nutrition 53*(5): 517–534.

Smith, P. M. 1993. The Grass Genera of the World. By L. Watson and MJ Dallwitz. CAB International, Wallingford, Oxford. 1992. 1024 pages. Hardback. ISBN 0 85198 802 4. *Genetics Research 61*(3): 235.

Sood, S., Walia, S., Gupta, M., & Sood, A. 2013. Nutritional characterization of shoots and other edible products of an edible bamboo–*Dendrocalamus hamiltonii. Current Research in Nutrition and Food Science Journal 1*(2): 169–176.

Tripathi, Y. C., Jhumka, Z., & Anjum, N. 2015. Evaluation of total polyphenol and antioxidant activity of leaves of *Bambusa nutans* and *Bambusa vulgaris. Journal of Pharmacy Research 9*(4): 271–277.

Van Hoyweghen, L., Karalic, I., Van Calenbergh, S., Deforce, D., & Heyerick, A. 2010. Antioxidant flavone glycosides from the leaves of *Fargesia robusta. Journal of Natural Products 73*(9): 1573–1577.

Verschoyle, R. D., Greaves, P., Cai, H., Borkhardt, A., Broggini, M., D'Incalci, M., Riccio, E., Doppalapudi, R., Kapetanovic, I. M., Steward, W. P., & Gescher, A. J. 2006. Preliminary safety evaluation of the putative cancer chemopreventive agent tricin, a naturally occurring flavone. *Cancer Chemotherapy and Pharmacology 57*(1): 1–6.

Wang, J., Yue, Y. D., Tang, F., & Sun, J. 2012. TLC screening for antioxidant activity of extracts from fifteen bamboo species and identification of antioxidant flavone glycosides from leaves of *Bambusa textilis* McClure. *Molecules 17*(10): 12297–12311.

Zhang, Y., Wu, L., Li, Y., Yang, J., Yang, H., Zhao, Y., & Chen, G. 2024. Bamboo shoot and its food applications in last decade: An undervalued edible resource from forest to feed future people. *Trends in Food Science & Technology 146*: 104399.

Zubair, M., Bibi, Z., Rizwan, K., Rasool, N., Zahoor, A. F., & Riaz, M. 2013. In vitro antimicrobial and haemolytic studies of *Bambusa arundinaceae* leaves. *Journal of Applied Pharmaceutical Science 3*(4): 111–115.

Index

Note: *Italicized* page references refer to figures and **bold** references refer to tables.

C

D

E

F

G

R

S

T

U

V

W

X

Y

Z